Sustainable Business Change

Vladimir Obradović
Editor

Sustainable Business Change

Project Management Toward Circular Economy

Springer

Editor
Vladimir Obradović
Faculty of Organizational Sciences
University of Belgrade
Belgrade, Serbia

ISBN 978-3-031-23542-9 ISBN 978-3-031-23543-6 (eBook)
https://doi.org/10.1007/978-3-031-23543-6

This Springer imprint is published by the registered company Springer Nature Switzerland AG
The registered company address is: Gewerbestrasse 11, 6330 Cham, Switzerland

Preface

Thinking circular presents a systematic change, not just the nature of a business but the change of the entire value chain for social, environmental and financial benefits. The circular economy concept has been introduced by academics, practitioners and policymakers and recognised international regulatory bodies worldwide as a roadmap towards sustainable development. Many research papers have addressed action areas necessary for business transformation to deliver a sustainable and green future.

Projects present a mechanism to implement an organisational strategy, emphasising the ongoing dialogue between organisational needs in terms of sustainable business change. Applying circular thinking in everyday business means ensuring organisations follow new paths and introduce new solutions during and after their project's activities. Besides the opportunities for existing companies, the circular economy also creates opportunities for start-up projects that can support social development through start-up acceleration and innovations. The motivation for integrating project management and circular business paradigm is not only to learn how project management can contribute to the circular economy but also to understand the impact of emerging business needs on project management.

This book summarises the results derived from many distinguished authors in different fields and different countries (Serbia, UK, Brazil, Netherlands, South Africa, Indonesia, Ukraine, Bosnia and Herzegovina and Russia), who have devoted their knowledge, experience and extraordinary effort to provide an understanding of project management contribution to sustainable business change via circular economy and the effect of circular economy business solutions on project management.

The book is divided into three sections: Part I: People, Project Management Practice and Circular Economy; Part II: CE Projects and Sustainable Business Change; and Part III: Integrating Sustainability into Project Management Methodology. The chapters within these sections are dedicated to very important aspects of circular economy, project management and sustainable business change: business models, human and financial aspects, process perspective, methodology hybridisation, higher education role and project portfolio. Chapter especially emphasises the importance of integrating highlighted fields and points to the development needs.

The first chapter of this book elaborates on factors that stimulate project managers to consider sustainability based on the new roles of project managers. The next chapter explores the role of higher education in transition to a sustainable and circular economy, emphasising a non-linear approach to solving problems and providing solutions in reaching sustainability and the goals of a circular economy. The third chapter reveals the impact of the circular economy business model on developing green human resource management (GHRM) practices in an organisation. It emphasises the significance of human resources in incorporating circular economy in the system to enhance sustainable development. Sustainable business change and circular economy are extremely important for industries whose activities are leading to an enormous depletion of natural resources and the creation of large volumes of waste, affecting environmental sustainability and often also social sustainability. One of the industries that fit this description is the construction industry, and a special chapter integrates sustainable project management into the construction sector to effectively close the loop of the circular economy.

The first chapter of Part II examines the tensions between individual actions and organisational structures in responsible project management and the impact on overall organisational sustainability behaviours. The second chapter of this section relates corporate social responsibility and sustainable development, business, change and implications on project-oriented companies, stressing out companies' environmental and social awareness and the impact on project activities. A circular economy provides changes for existing businesses and the whole economy, providing opportunities for circular start-up projects. The circular economy is a completely new approach in the European Union present in all aspects of society, national, local and enterprise level. Therefore, the next chapter presents indicators of the circular economy and the evaluation system, pointing to indicators for different purposes and needs developed by various scientists, states, government agencies, NGOs and companies. The following chapter aims to analyse whether start-up projects that integrate the principles of the circular economy have a higher probability of launching a crowdfunding campaign and determine the likelihood of chosen parameters to be the ones determining the crowdfunding campaign will have the elements of the circular economy.

Part III is dedicated to the integration of sustainability into project management methodology. Bearing this in mind, the authors of the first chapter put the focus on creating a hybrid methodology for sustainable strategic management of infrastructure programs in the condition of the circular economy. The next chapter presents a recurrent neural network model of circular economic processes based on identifying correlations between phenomena and processes for the impact assessment of project management on circular economic processes. The third chapter identifies and analyses the impact of sustainable project management on the project's success and the development of values for an organisation and society. It presents a conceptual framework for enabling benefits from linking sustainability and project management. Chapter four analyses project portfolio since project portfolio selection is a tool to reach the organisational strategy, intending to determine whether a specific framework will bridge the knowledge gap between circularity conceptualisations and their application in the project portfolio management field. The final chapter of the book presents an overview of opportunity for the integration of project management and circular economy and the benefits that could arise from this integration, focusing on existing forms of circular economy and project management potential to create the road ahead mapping areas as project strategy, processes, tools and competencies, for integration of circularity.

To my beliefs, this book could be promoted among different stakeholders such as governments, financial institutions and academic institutions with the aim to support and assist project management in driving sustainable business change. The book aims to become an important milestone in developing this topic since each of the chapters could be further developed, even become a book itself.

I hope the content will inspire readers from both academia and practice and provide useful insights for further developing competencies in project management, circular economy and sustainable business.

Belgrade, Serbia Vladimir Obradović

Acknowledgements

Capabilities for delivering projects in the context of societal development (CaProSoc) research programme initiated by the Alma Mater Europaea (Maribor, Slovenia) supported this collection. CaProSoc brings together over 70 partners from more than 50 countries working together, envisioning that all actors involved in sustainable societal development are aware of the role that projects have in society and use projects effectively for the benefit of society. It is aspired to impact societal development, help society develop through projects and project management in sectors beyond the economy, make sense of projects and show ways of dealing with the increasing number of societal challenges (e.g. climate change).

Contents

Part I People, PM Practice and Circular Economy

1 **The Role of Higher Education in Transition to a Circular Economy: Journey on the "Yellow Brick Road" to Sustainability** 3
Nataša Petrović, Marko Ćirović, and Flavio Pinheiro Martins

2 **Green Human Resource Management in Circular Economy and Sustainability** 41
Jelena Cvijović

3 **Exploring the Missing Link to Circular Economy in Construction: A Systematic Review of Waste Management Literature** 59
Leeboy Ndhlovu and Luca Sabini

Part II CE Projects and Sustainable Business Change

4 **Responsible Project Management Tensions in a Tier 1 UK Infrastructure Organization** 97
Charles Spooner and Nigel L. Williams

5 **Corporate Social Responsibility, Circular Economy and Sustainable Development: Business Changes and Implications in Project-Oriented Companies** 111
Isidora Milošević, Sanela Arsić, and Anđelka Stojanović

6 **Development and Review of Circular Economy Indicators: Evidence from European Union** 145
Tijana Milanović, Aleksandar Jovović, and Nataša Petrović

7 **Financing Start-Up Projects in Circular Economy: Does Crowdfunding Fit?** 173
Isidora Ljumović and Aida Hanić

Part III Integrating Sustainability into PM Methodology

8 **Methodology Hybridization for Sustainable Strategic Management of Circular Projects and Programs** 197
Sergiy Bushuyev, Natalia Bushuyeva, and Victoria Bushuieva

9 **The System Dynamics Model for the Impact Assessment of Project Management on Circular Economic Processes** 211
Ekaterina Andreevna Khalimon, Irina Stanislavovna Brikoshina, and Mariya Nikolaevna Guseva

10 **A Conceptual Framework for Enabling Benefits from Linking Sustainability and Project Management** 243
Danijela Toljaga-Nikolić

11 **The Impact of the Circular Economy Approach on the Project Portfolio and Selection Process** 269
Ivana Berić

12 **Circular Economy and Project Management: The Road Ahead** 301
Marija Todorović and Vladimir Obradović

Index 315

Part I
People, PM Practice and Circular Economy

Chapter 1
The Role of Higher Education in Transition to a Circular Economy: Journey on the "Yellow Brick Road" to Sustainability

Nataša Petrović, Marko Ćirović, and Flavio Pinheiro Martins

Abstract In the wake of environmental challenges, it becomes questionable whether transfer from linear to a nonlinear or circular economy can be achieved by continuing to employ traditional linear ways of thinking about potential solutions and designing the tools for achieving the desired outcomes that should lead to overall planetary sustainability. Instead, for this important issue, it is necessary to involve a number of stakeholders in the implementation and measurement of circular economy implementation efficiency in practice. In addition to governments and industry, the entire society must participate in this, including higher education institutions. Higher education is necessary because it is a crucial tool for achieving sustainability and sustainable development. The quality of education is very important in economies – in this case, circular economies. Bearing this in mind, the key aspect of this chapter focuses on the scientific research in circular economy education among students of the University of Belgrade, Republic of Serbia. The results obtained can be a useful source of information to other educators, at a time when improving education for sustainability is gaining more and more credibility and scientific ground.

Keywords Circular economy · Sustainability · Higher education · Higher education for circular economy · Curriculum

N. Petrović (✉) · M. Ćirović
Faculty of Organizational Sciences, University of Belgrade, Belgrade, Republic of Serbia
e-mail: natasa.petrovic@fon.bg.ac.rs; marko.cirovic@fon.bg.ac.rs

F. P. Martins
School of Economics, Business Administration and Accounting at Ribeirão Preto – University of São Paulo, Ribeirão Preto, Brazil
e-mail: fpmartins@usp.br

V. Obradović (ed.), *Sustainable Business Change*,
https://doi.org/10.1007/978-3-031-23543-6_1

1.1 Introduction

Follow The Yellow Brick Road
Follow the Yellow Brick Road. Follow the Yellow Brick Road.
Follow, follow, follow, follow,
Follow the Yellow Brick Road.
Follow the Yellow Brick, Follow the Yellow Brick,
Follow the Yellow Brick Road.
We're off to see the Wizard, The Wonderful Wizard of Oz.
You'll find he is a whiz of a Wiz! If ever a Wiz! there was.
If ever oh ever a Wiz! there was The Wizard of Oz is one because,
Because, because, because, because, because.
Because of the wonderful things he does.
We're off to see the Wizard. The Wonderful Wizard of Oz.
Harold Arlen

"The Wizard of Oz"[1] (1939) one of the most famous films of all time represents an American musical fantasy produced by Metro-Goldwyn-Mayer. In 2007 it was included as a "documentary heritage submitted by the United States of America and recommended for inclusion in the Memory of the World Register in 2007" of the United Nations Educational, Scientific and Cultural Organization – UNESCO (UNESCO, 2017). It is the most commercially successful adaptation of the fantasy novel "The Wonderful Wizard of Oz" for children, written by Lyman Frank Baum in 1900 (Fricke, 1989). In this movie, a prominent subject throughout the story, is a path that Dorothy Gale, a ten-year-old girl who is swept away by a tornado (a natural disaster, is it a coincidence?), takes to get to the Land of Oz to return home and ask the Wizard of Oz for help. This particular road is comprised of yellow bricks. It is also mentioned in a poem by Harold Arlen.

The symbolism of the yellow brick road within the scope of academic literature varies from being interpreted as the road to success to the extent of equalising it with happiness (Collins, 2020), or as a pilgrimage road to a Promised Land found in one's deepest wishes (Yourdictionary, 2020), even as the Golden Path – seen as a transit from human egoism to enlightenment (Findanyanswer, 2020). The Oxford English Dictionary (OED – 3rd edition, 2018) defines the American-English phrase yellow brick road as denoting a course of action or series of events viewed as means towards the desirable end (Word Histories, 2019).

Universities, faculties and higher education institutions should provide a "yellow brick road" to sustainable societies and economies in the world plagued with different crises, in which, due to their unforeseeable consequences and threats, those crises that are inflicted on nature stand out. On the other hand, those who study, or the students who accept this road, and who are future decision makers, need to learn as much as possible and apply their knowledge for necessary changes in their business and private lives. Bearing in mind that road to change is regularly achieved through projects (Obradović, 2010), it should be emphasized the needed

[1] It is the most commercially successful adaptation of the fantasy novel "The Wonderful Wizard of Oz" for children, written by Lyman Frank Baum in 1900 (Fricke, 1989).

integration of concepts of sustainability and project management (Toljaga-Nikolić et al., 2020).

The reasons for this lie in the fact "today, as never before, the stability of the Earth's ecology is undermined due to the activities carried out by the human species" (Petrović et al., 2014, 2016), as well as in the increasing consumption of energy and resources with consequent generation of waste and pollutants (Iannuzzi, 2011). Human impacts on the environment have reached the critical point in the violation of the balance of nature and humankind in the twenty-first century and sadly continue to increase (Petrović et al., 2012; Petrović, 2016). Some of the most significant and negative environmental impacts are: decreasing living standards, ozone layer depletion, conservation effects, solid waste increase, radioactive contamination, deforestation, biodiversity loss… (Bonnett, 2007; Mert, 2006; Petrović, 2016; Petrović et al., 2017; Radaković et al., 2017; Robert et al., 1993). The magnitude of the human footprint on the planet currently has a global impact; hence, the majority of the scientists and researchers state that the planet has entered a new era known as the era dominated by humans, also known as the Anthropocene (Franco, 2013).

At the same time, the growth of the world's population has led to an increase in the inefficient consumption of natural resources (Symth, 2006). Thus, it is evident what Bonnett (2007) notes that it is hard to imagine any other set of concerns that influence at this point the wellbeing of human species than those related to the environment. On the other hand, the limits of the possibilities of the planet Earth that it can provide to human civilization are in reciprocity with those that refer to the limits of what the planet can "receive", which primarily refers to environmental devastation and excessive waste generation.

The mentioned above disturbance of natural balance caused by human activities are causing changes in the global ecosystem. This leads to changes that worsen the survival conditions of many populations, including ours – human, and all under one name – environmental crises (United Nations Educational, Scientific and Cultural Organization [UNESCO] – Division of Science, Technical and Environmental Education, 1986). This should be added to other environmental problems such as rise in pollution, global warming, climate change, as a result of growing carbon emissions (Gupta, 2015).

For these reasons, both in theory and in practice, it has been shown that the exciting development of the economy or linear economy that is traditionally based on the "take-make-dispose" model of production and consumption is a total antithesis to the planet's necessary model of sustainability. This is quite understandable given that the linear economy is based on the concept of "use and throw" which has been present since the industrial revolution (Zamarriego, 2017) when products followed the "linear process" of extraction–transformation–distribution–promotion–use–obsolescence (Salguero-Puerta et al., 2019). When it comes to sustainability, it should be emphasized that sustainability is identified globally as an idea, transitional vehicle, and/or as a targeted goal that enables proper response to the rising environmental crisis. Furthermore as a response to social and economic repercussions of that crisis, labelled together as "global change" (Biggs et al., 2011; Borojević et al., 2017a, b; Hugé et al., 2016). Sustainability is directly related to Sustainable Development (SD). Both concepts are similar but also different. According to the United Nations Educational, Scientific and Cultural Organization – UNESCO (2019) – sustainability is regularly thought of as a wishful state that humanity as a whole has to transit to, while sustainable development underlines many ways, manners and the tools to arrive at that wishful state. The most accepted and used sustainable development definition is the definition that has been published in the report of the Brundtland Commission, in which, under sustainable development is considered the development of today's generation according to their needs, without taking a toll on the future generations' equal ability to do so (World Commission on Environment and Development, 1987).

In contrast to the linear economy and its evident negative impacts on the environment and sustainable development – as the only possible development that humanity has on a finite planet – there is the concept of circular economy (CE). This concept is considered as a mechanism for fulfilling local, regional, national and planetary sustainability (Schroeder et al., 2018). Also, the CE can be said to represent an archetype that can achieve the shift of organizational practices to more sustainable solutions (Sillanpää & Ncibi, 2019). Having all this in mind, the application and transition to a CE reduces the depletion of natural resources, reduces the amount of waste generated, reduces the amount of waste in landfills and thus the consequent pollution of water, air and land, global warming, land destruction, as well as improved energy efficiency and rise in the use of renewable energy sources. In this way, the CE represents an environmental, social and economical benefit because it provides the implementation of a management model of sustainable development. According to Hannoura et al. (2006) this implies correlating pursued objectives with actions that are taken in achieving those objectives, while simultaneously keeping the effects of those actions under predefined limitations that would allow achieving ecosystem sustainability. All this is provided by the CE, which includes such product design called environmental design or design for environment, which in the construction of the product affects the reduction of pollution, the application of near-zero waste strategy, recycling, reuse, servicing products and their long-life cycle as well as product observation – not from "cradle-to-grave" but from "cradle-to-cradle". Cradle-to-cradle approach would imply designing the products and/or systems in a manner that should lead to returning the products at the end of their usage and transforming them into new usage of equal/greater value (Ashkin, 2008).

Also, the importance of environmental education for sustainability, incorporating higher education, in particular, should be emphasized, in the age that Sachs (2015) calls the age of sustainable development. Sachs states that sustainable development is an organizational principle for all policies, economics, and ethics. The most important condition for a successful transition to a sustainable development model is the quality of human resources, in which the key role is played by adequate and quality, i.e., good environmental education. Many authors agree that good environmental education is crucial to achieving sustainable development that will provide the necessary way of life for people within the capacity of nature (e.g.,

McCormick et al., 2005; Petrović et al., 2014; UNESCO, 2012). More precisely, to achieve the goals of ecosystem conservation, the behaviour of entire societies towards the biosphere must be transformed so that the long-term task of environmental education is to encourage and strengthen attitudes and behaviours following environmental ethics (IUCN-UNEP-WWF, 1980). It is also necessary that the development of quality higher environmental education for sustainability is based on The Talloires Declaration (Association of University Leaders for Sustainable Future, 1990) which is an official declaration of the necessary commitment of higher education to environmental sustainability, emphasizing the need to introduce literature on sustainability environment both in curricula and syllabi of subjects at faculties, and in lectures, scientific research and the work and activities of faculties and universities (Maletič et al., 2017). All this is because the role of young people, especially the educated ones, must be one of the main approaches, but also a principle in the environmentally sound and sustainable development of all communities and economies. The reason for this is the fact that the participation of young people in solving problems is important not only for their personal development, but also for their adequate involvement in decision-making processes that imply a better quality of functioning of society as a whole, in this case, a sustainable society (Borojević et al., 2015, 2017a, b; Petković et al., 2019).

So, for our research we have selected students of the University of Belgrade who declared that they have introduced circular economy issues in their students' agenda, believing in the hypothesis that both university teachers and students want sustainable economies and responsible environmental opportunities for a growing world population.

The research was performed on the students attending the winter semester of the 2020/2021 academic year. Students took part in an online survey. The survey was conducted at the University of Belgrade, Republic of Serbia. Seventy seven students participated in the survey (62 females and 15 males). The students completed the survey, and the results for each student were calculated using software for the statistical analysis SPSS version 24.

1.2 Literature Review

1.2.1 Paving the Way Through Higher Education

Education plays an essential role in addressing the civilizational challenges of our era. This mission was formally addressed for the first time in the 1972 United Nations Conference on Human Environment, being later reinforced by the 1992 Rio declaration on Environment and Development (Handl, 2012), by the UNESCO's declaration of the 2005–2014 period as the Decade for Sustainable Development Education (Combes, 2005). More recently, the 2030 Agenda and its 17 Sustainable Development Goals (SDGs) framework have brought a more integrated framework that addresses education for sustainable development in a focal point of the target 4.7 on education where by 2030, all learning agents have to adopt the data and tools necessary to promote sustainable development. This includes various knowledge areas, such as education for sustainability, lifestyle and behavioural changes, human rights, gender equality, peaceful and non-violent action and civil movement, global citizenship, and understanding of cultural diversity and how all these contribute to sustainable development (UN, 2015). The agenda inter windedness and global thematic are present in studies on education (Andreoni & Ruiz Vargas, 2020; Cottafava et al., 2019; Franco et al., 2019), artificial intelligence (Vinuesa et al., 2020), energy (Bisaga et al., 2020; Castor et al., 2020; McCollum et al., 2018; Nerini et al., 2018; Santika et al., 2019), sanitation (Alcamo, 2019; Diep et al., 2020; Parikh et al., 2020a), water quality security (Flörke et al., 2019), health (Hall et al., 2020) and specific issues like COVID-19 (Parikh et al., 2020b).

Regarding education, the Agenda connects with many of the 169 targets and indicators of the sustainable development goals: 28 occurrences of the keyword "education" in the agenda. For example, the target 3.7 on health, target 1.2 on poverty and target 8.6 on economic growth are examples of education being called to aid the agenda fulfilment. There are explicit keywords related to education in the "SDG 12. Ensure sustainable consumption and production patterns" indicator 12.8.1, which address national education policies, curricula, teacher and students' education (UN, 2015) for sustainable business, not as usual models. To have leaderships educated for a paradigm change, the focus must be in the educational level where they are trained and acquire their work and life-long competencies. Higher Education is one of the main drivers for economic development and collective well-being through (I) science development in research, (II) knowledge transfer in education and (III) community outreach through service-learning and extension activities (Organisation for Economic Co-operation and Development [OECD] 2010).

In the university campuses, colleges and faculties from a variety of field areas, new leaderships are educated and acquire competencies to enter tomorrow's workforce. In an ideal way, these new professionals would carry with them the technical skills and theoretical repertoire capable of enabling them to solve local socio-economic problems in a contextualized manner and at the same time glimpse the bigger picture of an ever-changing global conjuncture. Also, to accomplish this within the recommended goals for sustainable development, respecting the limits of the planet's physical life support systems (Ćirović, 2019; Rockström et al., 2009). Therefore, sustainable development paradigm change through social collaboration, research and education has gradually become nuclear among the main university activities: the future we want in education for sustainability is only possible through socially, environmentally and economically oriented universities (Beynaghi et al., 2016).

When it comes to a CE, the authors Bugallo-Rodríguez and Vega-Marcote (2020) advocate that higher education has a dual role on it: first by knowledge transmission through teaching and secondly by cascade effect that professionals cause in the society. CE posits itself as one of the best answers for the bigger picture of sustainable development challenges of our era, and it is a fruitful and needed educational scenario for addressing questions that are multidimensional, multi stakeholder dependent, paradoxical, contextual and globally connected. CE is presented as the main philosophy to rewire the business case for sustainable and responsible production and consumption (Goyal et al., 2018; Ünal et al., 2019). The complexity of the matter is compatible with the diversity of knowledge areas available at universities and, specifically, with the holistic view grounded on systems thinking and interdisciplinary approaches. Questions raised from the social-ecological perspective like eco-innovation, sustainable procurement practices, and development of closed-loop supply and value chains, are only feasible to address through non-linear models of problem-solving (Ramísio et al., 2019). Moreover this emphasises the need for the planet as a whole to take an environmentally risk averse position in order not to overshoot its ecological targets by all the different separate issues and corresponding criteria, starting from the traditional energy resources used (Ćirović et al., 2015), safeguarding the environmental mediums most notably soil, air, biota and water (Makajić-Nikolić et al., 2016), while simultaneously trying to achieve some form of international scientific consensus that would result in creating a political will in political actors as seen in European Union (Ćirović, 2018).

1.2.2 Circular Economy Keywords Network

The cluster analyses from VOSviewer are widely used for literature reviews bibliometric and thematic inquiries (Llanos-Herrera & Merigo, 2019; Shah et al., 2019; Yu et al., 2020). In this specific topic, we found more substantial cluster density in the thematic related to the origins of CE constructs, grounded into natural sciences and engineering, like industrial ecology, closed-loops and resource efficiency (United Nations Environment Programme [UNEP], 2006) (Fig. 1.1).

The most vital frequencies besides CE construct are "sustainability" in the same cluster as CE, and "sustainable development" in the same cluster of "higher education", "teaching" and "education"; the way this is intertwined reinforces that CE principles are connected with sustainability, and they can be interfaced by the sustainable development education sector into higher education (Sanchez et al., 2020). Mixed bottom-up and top-down approaches are considered nuclear for changing the university's organizational culture, policies covering all university areas, permanent communication and monitoring of the goals, institutionalization and creation of networks (Ramísio et al., 2019).

Another point of interest in the mapping is the construct "decision making", "innovation" and "life cycle" that appears closely bound to environmental externalities subjects, such as "carbon footprint" and "environmental impact" and other constructs related to industrial and operations lens of sustainability, this transition refers to studies that address CE through the optics of management, operations and on a broader scope, the business education-related subjects. Articles like "Circular value creation architectures: Make, ally, buy, or laissez-faire" (Hansen & Revellio, 2020), "Economic aspirations connected to innovations in carbon capture and utilization

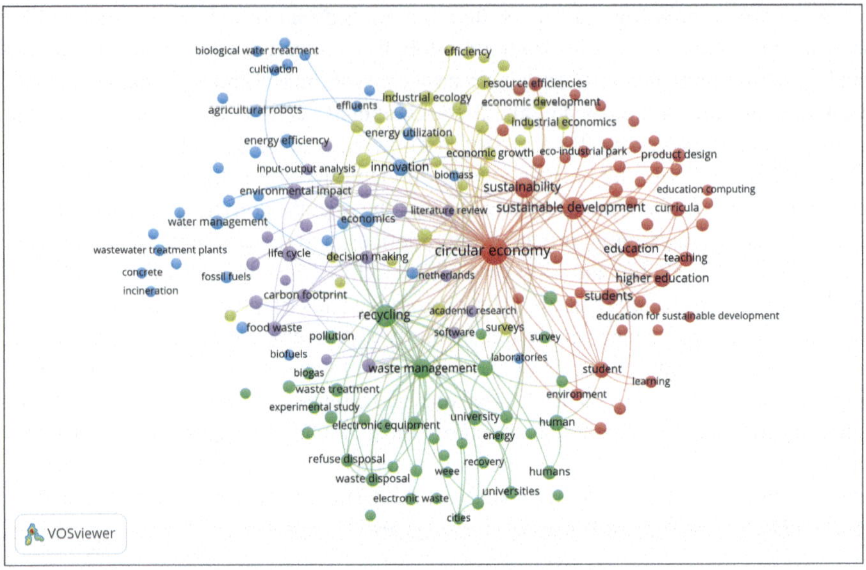

Fig. 1.1 Construct clusters related to Circular Economy. (Source: Elaborated by the authors using VOSviewer)

value chains", and "Activities of circular economy in Japan – towards global multi-value circulation" (Halada, 2020). The major obstacles for the consolidation of the CE into higher education are similar to the ones found in sustainability (Jorge et al., 2015). CE education can be considered both a dimension of Education for Sustainable Development and one of its drivers for the development of competencies. Sanchez et al., (2020) study points out how the CE principles can be mapped and contextualized in different knowledge areas through the sustainability education existing competencies like interdisciplinary learning, systems thinking, competence to deal with uncertainty, cooperative action, conflict resolution, self-motivation and motivating others and long-term, anticipatory, foresight and future thinking.

Global business education networks like Global Compact and the management education platforms like The International Association for Management Development in Dynamic Societies (CEEMAN) and the United Nations Principles for Responsible Management Education (UN-PRME) have a central role in shaping undergraduate competencies towards the integration of sustainability since the business-as-usual paradigm is the primary cause for unsustainability in our world (Von Der Heidt & Lamberton, 2011). Frameworks like the PRME Six Principles (Purpose, Values, Method, Research, Partnership and Dialogue) can foster CE literacy among business students and build upon the already settled system thinking approach (Hester & Adams, 2014) step forward into a circular thinking approach for business (Mendoza et al., 2019).

The "university" construct could be expected to be bound to education theme; nevertheless, it appears interwoven with thematic concepts like "recycling", "waste disposal" and "municipal waste". This highlights that many of the university's sustainability initiatives that somehow use CE principles are likely to be related to the university's management aspects of its environmental footprint. Many of them are related to emerging economies contexts like it is reflected in the research of Adeniran et al. (2017) on waste management at the Lagos University, Nigeria, or in the case of information technology reuse at the Disposal Center and Electronic Waste Reuse (CEDIR) at the University of Sao Paulo, Brazil (Alves & Farina, 2018). The same is evidenced in the research done for developing a model of indicators for composting and biogas generation at the University of Lome, Togo, and the research about waste components assessment realized by the authors Owojori et al. (2020) at the University of Venda, South Africa (Puerta et al., 2020).

The presence of the construct "municipal", "social-economic impacts", and "urban area" points out the presence of the implicit concept of universities as living laboratories for cities, once they can emulate good practices that can be, therefore, replicated in a more superior level at the municipalities (Leal et al., 2019; Verhoef et al., 2020).

This approach indicates that much has been done, at a practical level and case studies in universities, in an approach that considers the educational institutions as organizations and social agents that produce negative social-environmental externalities while pursuing its educational and research goals. Despite education not being the focal point, the living lab approach reinforces the educational paradigm change for integration of sustainability culture (Kılkış, 2017) and CE principles as well in the campuses, bringing legitimacy to the educational approaches since faculty and staff can see in their daily academic routine that their institutions walk the talk on sustainability (Sammalisto et al., 2015), opening opportunities for service-learning initiatives on sustainability (Daub et al., 2020) and fostering awareness in the stakeholders about the emulation possibilities that campuses hold on sustainability innovations (Stephens et al., 2008).

The manifestation of CE at the Universities Living labs seems to be closely bound to project management principles. Despite being transversal and broad, the living labs are grounded in university organizational structure: they can be identified in thematic such as recycling and waste campus management. These familiar places where projects already champion sustainability still face shortcomings when it comes to expanding their grasp on the educational dimension of the campuses. If we split universities' mission into four main areas: research, teaching/learning, community outreach and organizational management, the last one would be where project management towards CE is more easily spotted.

In higher education's teaching and learning dimension, there are still many challenges to address CE adequately; research on how effectively the thematic is being incorporated in the curriculum falls short on precision and represents a yet to be explored gap (Sanchez et al., 2020). Project management can work as an activator to it. Sultanova et al. (2021) describe a contextual solution for bridging the thematic of CE and project management through curricula marginal changes. Authors report that chemistry students could grasp CE goals related to the 2030 Agenda by being exposed to a contextual problem-based learning activity for developing new products. The same kind of bridging constructs through active methodologies is found in engineering courses and specifically related again to green product design (González-Domínguez et al., 2020) and in computer science courses, with the usage of the service-learning approach (Sánchez-Carracedo et al., 2021). The same criticism that arises from rhetoric disconnected from daily practice (walk the talk) also works in the opposite direction: some campuses excel at being sustainable, yet the educational potential of the practices developed is yet to be fully explored, so they "walk", they have solar panels, water reuse, expended recycling frameworks, but still don't properly "talk" about it, don't explore the full potential for training new professionals under these competences. In this sense, the extensive use of active methodologies can foster to shorten this gap and expand the principles of CE to the complete picture of higher education institutions.

1.2.3 Coming of Age Construct: Circular Economy Brick Road

The link between sustainability and CE are undeniable (Murray et al., 2017), and CE is considered a pathway for achieving sustainable development (Geissdoerfer et al., 2017); in the university context integration of CE principles into higher education is considered one step towards sustainable development education (Sanchez et al., 2020). Despite the awareness, relevance and urgency of these questions, and the consensus that universities play an important role in it, there is still a large gap to be bridged in higher education. While research in the subject of the CE is addressed in many worldwide interdisciplinary centres like the Center for Environmental and Sustainability Research (CENSE) and the European School of Sustainability Science and Research (ESSSR), the integration of CE principles into the education dimension is still a very novel approach without a consolidated notion of its status (Sanchez et al., 2020). There is a lack of know-how to implement a CE into higher education, and the main reason for this is because CE in higher education is still having its starting struggles (Mendoza et al., 2019, p. 842). Besides strong similarities with sustainability education or education for sustainable development, the construct "CE education" guards its specificities, since it is a narrow scope for teaching and education in the construct that exists in opposition to the linear economy models and that fosters not only the systems thinking for sustainability but a more detailed version of what we could call circular economy – CE thinking.

We ran a review on two combined strings of keywords. The first one on CE and synonyms and related constructs like "recycling management", "recycling industry" and "closed-loop". The second string had higher education and related constructs like "universities" and "faculties". The keywords inside each string were combined with the Boolean operator "OR" and the strings were combined with the operator "AND". Keywords were searched in articles title, abstract and keywords.

No time scope or file type was delimited. 447 documents were initially found. Data were then loaded into the network visualization software VOSviewer tool to evaluate the constructs through its keyword's incidence and interlinkages (Fig. 1.2).

The majority of the publications are concentrated after 2015, which reinforces the establishment of the 2030 Agenda and the Paris Deal in the research community. The Sustainable Development Goals decade has fostered higher education institutions to address their environmental impacts (Owojori et al., 2020). This is mentioned, for instance, in the studies from Schroeder et al. (2019) referring to a CE as a toolbox for reaching a wide array of Sustainable Development Goals. This paper also highlights the synergies between CE and the SDG 6 (Clean Water and Sanitation), SDG 7 (Affordable and Clean Energy and SDG 12 (Responsible Consumption and Production) that can be bounded respectively, to the scope of subject fields "Engineering", "Energy and Engineering" and "Business, Management and Accounting". Publications like the one from Schroeder et al. (2019) and Stephan et al. (2020) are peripheral for our scope since they scope education from the management lens comprising environmental assessment and projects on campuses, mostly related to waste management, energy and water efficiency (Benltoufa et al., 2017; Gallo et al., 2017; Pardal et al., 2020) (Fig. 1.3).

The aims of the studies are focused on their impact as organizations that have the peculiarity of being educational institutions. On the other hand, works like the papers: "The sustainability matrix: A tool for integrating and assessing sustainability in the bachelor and master theses of engineering degrees" (Sánchez-Carracedo et al., 2020), "*Conceptual and Legal Framework for Promotion of Education for Sustainable Development: Case Study for Ukraine*" (Tetiana & Malolitneva, 2020) and "*Lessons learned in the paths of developing a multidisciplinary certificate program*" (Wang & Van Bueren, 2018). Are nuclear on education, since they address the overlap of CE and education for sustainability and use the Sustainable Development Goals framework as background or the structural axis for the educational programs and projects. CE integration in the curriculum (Kopnina, 2018; Mateus et al., 2020; Qu et al., 2020) and hidden curriculum (Nunes et al., 2018) also appear as a timid yet recent trend.

The keywords-publications linkage spread along the years, clustered by the colours of the VOSviewer analysis (Van Eck & Waltman, 2011), point movement noticeable from constructs related to engineering and industrial sciences, and more consolidated thematic like "Life cycle analysis" and "Ecology" to the main subject of "Circular Economy" in the last two years and the education-related subjects appearing even further and in a much smaller frequency which emphasizes the novelty of the subject.

When we narrowed down the scope in quick prospecting with the keywords "circular economy" and its synonyms, along with "education", on the Scopus database, with the setup "title-abstract-keywords", the output is only 10 articles, after removing the false-positives we have five papers: the construct barely exists (Table 1.1).

Two of these papers address specific educational knowledge fields: the electricity sector (Rokicki et al., 2020) and the construction management sector (Sanchez et al., 2020). Rokicki et al. (2020) found out that a number of doctoral students had

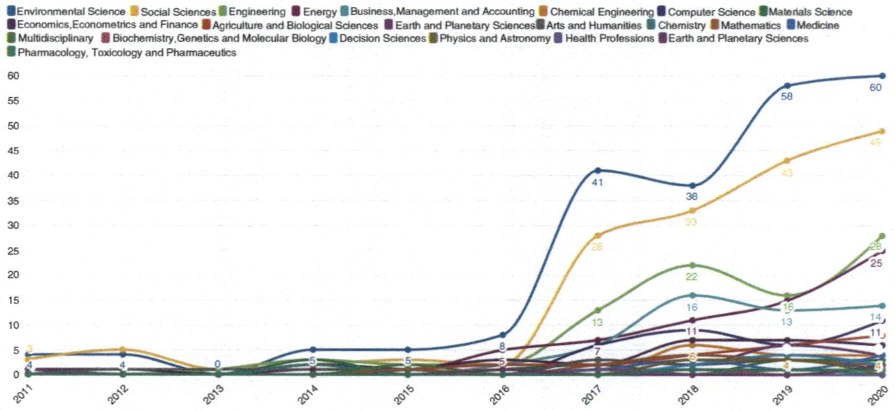

Fig. 1.2 Concentration of publications per subject area* in the years ranging from 2011 to 2020. * Publications refer to two or more subject areas; therefore, the total sum of subject areas are higher than the articles count. (Source: Elaborated by the authors)

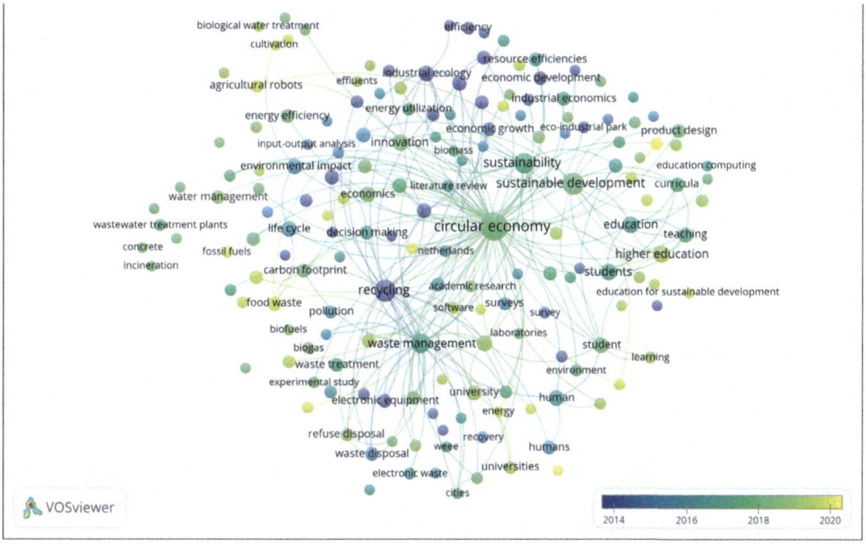

Fig. 1.3 Construct networks and cluster gradient from 2016 to 2019. (Source: Elaborated by the authors using VOSviewer)

a stronger relation with specific CE goals of European Union countries compared to bachelor students in the specific field of electricity. Which reinforces the assumption that the competencies for CE integration are more prevalent in postgraduate than in undergraduate education and also the role of higher education in the nation's sustainability policies (Al-Mansoori & Koç, 2019); a similar perspective was also found in Revinova et al. (2020) that identified universities milieu as a potential platform for countries development models.

Sanchez et al. (2020) also addressed a sector related to engineering, the construction management nevertheless he looked at the five categories of sustainability competencies framework of Olalla and Merino (2019): Learning to Know (LK), to Do (LD), to Live together (LL), to Be (LB), and to Transform oneself and society (LT). Research findings point out that in the specific course, integration of CE principles is needed to foster sustainable development competencies demanded by companies, in a way to future-proof the engineering curriculum, which also makes sense for sustainability in business education (Winfield & Ndlovu, 2019) and in a broader scope for climate change education (Fahey, 2012). Bugallo-Rodríguez and Vega-Marcote (2020) study is focused on the cascading effects of CE teaching and education since it addresses competencies training in CE for preservice teachers. Results show the perception that CE in university should foster the changing individual behaviours: for instance, they point out that some campus campaigns are focused on recycling rather than reducing consumption, which is related to the lack of awareness about the following subjects: bio-waste generation, secondary raw materials, collaborative consumption, and the 9 Rs (rethink, reduce, reuse, repair, refurbish, remanufacture, repurpose, recycle, recover). Even a subject that is technical by

Table 1.1 Main papers addressing circular economy education

Title	Authors	Year	Source
The importance of higher education in the EU countries in achieving the objectives of the circular economy in the energy sector	Rokicki, T., Perkowska, A., Klepacki, B., Szczepaniuk, H., Szczepaniuk, E. K., Bereziński, S., & Ziółkowska, P.	2020	*Energies, 13*(17), 4407. ISSN:1996–1073
Integration of circular economy principles for developing sustainable development competences in higher education: An analysis of bachelor construction management courses	Sanchez, B., Ballinas-Gonzalez, R., Rodriguez-Paz, M. X., & Nolazco-Flores, J. A.	2020	*2020 IEEE global engineering education conference (EDUCON)* (pp. 988–996). IEEE
Circular economy, sustainability and teacher training in a higher education institution	Bugallo-Rodríguez, A., & Vega-Marcote, P.	2020	International Journal of Sustainability in higher education
A methodological framework for the implementation of circular economy thinking in higher education institutions: Towards sustainable campus management	Mendoza, J. M. F., Gallego-Schmid, A., & Azapagic, A.	2019	*Journal of Cleaner Production, 226,* 831–844
Building a business case for implementation of a circular economy in higher education institutions	Mendoza, J. M. F., Gallego-Schmid, A., & Azapagic, A.	2019	*Journal of Cleaner Production, 220,* 553–567

Source: Elaborated by the authors

definition, like eco-design, is approached through the lens of education and awareness: students said that innovation could help with devices to control water and energy consumption, for instance, but this results in using/buying another product (the controlling device). At the same time, the reduction could be achieved through education. The summary of the concerns and points to improve pointed by the students is about understanding the whole material flow process, the recycling awareness stops at the bin, and there is a call for a broader comprehension of the circularity of materials, like the number of recycling times possible and the knowledge about non-renewable resources. This research dialogue with global trends on CE is exemplified by the SDG target 12.8, recommending that "By 2030, ensure that people everywhere have the relevant information and awareness for sustainable development and lifestyles in harmony with nature" one of the indicators of the target is the mainstreaming of education for sustainable development in teacher education (indicator 12.8.1.c) (UN, 2015).

From the summary of the analysis made, it is possible to figure out a roadmap built with the constructs related (Fig. 1.4).

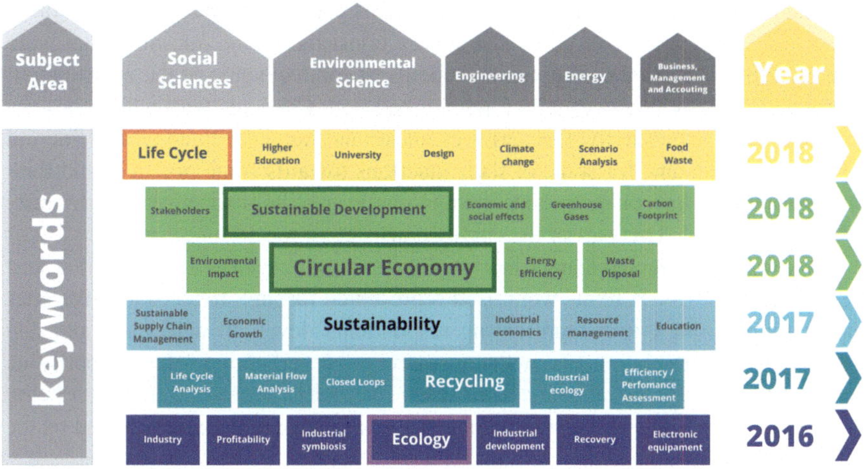

Fig. 1.4 The brick road of circular economy. (Source: Elaborated by authors)

1.3 Research Background and Methodology

During the winter semester of an academic year 2020/2021, at the University of Belgrade, Republic of Serbia, a survey was conducted with an aim to check, survey participants' attitudes towards, CE goals, principles and practices. Additionally, an online survey was organized in order to examine participants' general knowledge about points of convergence and divergence of traditional linear economy and CE with the goals of environmental protection and sustainability. For the purpose of our research, we created a questionnaire for students of the University of Belgrade who, within their studies, had subjects which curricula included topics related to the CE. The online questionnaire was distributed through social networks, and especially through the Facebook public group related to the Circular Economy (https://www.facebook.com/groups/991919851234757). The participants in the survey participated in it freely and willingly. They were assured of anonymity. It has to be noted that this scientific research is the initial stage of a future larger project to innovate, and to improve CE education in higher education.

Survey covered a range of questions, 66 in total, which can be divided in three subject areas. First area covered by the survey consisted of the questions about basic demographics and students' socioeconomic backgrounds. This part of the survey consisted of seven questions. The second area covered by the survey consisted of the questions that should reflect the students' familiarity and knowledge about the general definitions, goals and principles on which CE is based on. Moreover, this part of the survey was used to assess the general environmental practices that participants employ in their daily routines, as a way to examine consistencies between the knowledge students have about the particular topic at hand and their behaviour. More concretely, the idea was to check whether certain discrepancies exist between the knowledge students have and their environmental practices, and if so, to check why such discrepancies come about to exist. This part of the survey consisted of 45 questions. Third area covered by the survey consisted of the questions that were designed to check what can motivate students to conduct some of the environmentally friendly daily practices which are aligned with CE goals with which they are familiar with. Additionally, this series of questions was set with an aim to find out what are the reasons, for some of their inactions towards some practices and their suggestions on what could motivate them more towards the environmentally friendly actions in general. This part of the survey consisted of the 14 questions.

Survey was conducted using Likert scale, which was organised through 58 Likert items. Standard Likert items were used represented through five consecutive integers ranging from 1 to 5 representing respondents' levels of agreement and disagreement with the statements, with equivalent polar verbal opposites (verbal value for the integer 1 being "absolutely disagree", verbal value for the integer 5 being "absolutely agree"), symmetrically centred over a neutral middle represented by the integer value 3.

1.4 Research Results

Regarding the results of the research, firstly descriptive statistic will be used to summarise the responses gained, and then the results of the observed statistically significant correlations gained through cross tabulation will be presented.

1.4.1 On Demographics

After analysing questions that regarded the demographic part of the survey several points have to be made and clarified. Firstly, in regard to gender of the respondents, 80.5% of the respondents declared themselves as female, while 19.5% declared themselves as male. None of the respondents refused to declare their gender. This might seem at a first glance as a disparity in gender representation, but actually regarding that previous research is reflective on this same point, that participants declaring as female are more often to take part in similar surveys than the one's declaring as male (Borojević et al., 2017a, b; Curtin et al., 2000; Moore & Tarnai, 2002; Singer et al., 2000; Smith, 2008), this female/male ratio of respondents was expected.

When the age of the survey participants is at matter, respondents have a very homogenised structure in terms of age so more than 93% of participants are born in 1996 or after.

Regarding the participants' place of residence, the survey included participants coming from 30 cities from Republic of Serbia. Nonetheless majority of them reside in Belgrade, 43 or 55.85% to be exact.

According to Statistical office of Republic of Serbia (2020), reports for first half of the year 2020 note that the average monthly salaries and wages after taxes calculated for 2020 amounted to 485 euros (57 376 RSD). Meaning that the household with two working adults on average accumulates 970 euros. In order to reflect that in the survey conducted, participants were asked to give rough estimation on their household income. Idea behind this was to check whether specific environmental actions, that serve for the purposes of CE, depend or not, on the person's financial wellbeing. In that regard 40.58% of respondents declared household income above the average, and 59.42% declared the household income that is below the average.

1.4.2 On Familiarity and Knowledge of Circular Economy

Within the next part of the survey, questions that were there to examine pre-existing knowledge students had about CE, its goals, principles and practices, were presented next.

First of such questions was about the term itself. When asked whether they were familiar with the term "Circular economy", and whether they know what does the term represent, 93.51% of participants responded positively, while only 6.49% declared that they are not familiar with the term.

At this point of the survey participants that responded positively to the question whether they know what the CE is, were asked to give some form of a definition made in their own words of the term, or precisely to explain what CE represents to them. Among the other answers the word used most in students' definition of the required term, ranked first by the usage were combination of the words "economy" and "waste" with 50 noted usages out of 72 provided definitions, or 69.44% of the times. While the term "resources" was used 20 times to describe what the CE represents, or 27.78% of the times, ranking it as a second most used term at this question. Ranking third the term "environment" was used in the definition students provided, with 13 mentions or 18.06% of times. All other terms are marginally used in comparison to these three such as the "life cycle" with five mentions, or "innovation" with two mentions.

This data, although roughly, gives an insight into what do students think when CE is in question and to which other issues, they connect it to. Primarily, it can be noted that participants of the study relate this subject to better waste management in production and service processes. Which is pretty much in line with the findings of Kirchherr et al. (2017) and their review of definitions of CE in scientific literature, where they examine 114 different definitions used, as well as with the findings of Ghisellini et al. (2016) that provided the review of 155 articles dealing with the subject of CE. Lieder and Rashid (2016) provides a space for a similar conclusion, as well as Blomsma and Brennan (2017), which deal with the issue of CE concept emergence. It can be concluded that students' perceptions about the CE concept does not differ that much from the stated in presented in the observed scientific literature, when the keywords mentioned in existing operative definitions are at question.

When the Likert items that are used in this part of the survey are analysed, several notes can be made.

First two Likert items were there to check whether students recognise the difference between linear and circular economic concepts, and whether they distinguish their environmental impact, harms and benefits. And whether they in general associate economic activities as something that goes in line or against environment.

The statements that were presented to survey participants to express their level of agreement or disagreement for this particular matter, were firstly "Linear economy is in direct conflict with the environmental protection goals", and "Circular economy is in direct conflict with the environmental protection goals". As all other following statements participants were asked to state their level of disagreement or agreement with a statement on a Likert scale from 1 to 5 (1 being strongly disagree, 5 being strongly agree) (Table 1.2).

So, it can be noted regarding the mean of $M = 3.81$ and $SD = 1.06$ for the first statement and comparable mean of $M = 1.99$ and $SD = 1.2$ for the second statement, where overall knowledge and familiarity with the concept of CE is in question, students recognise the difference between the environmental impacts, these two different economical approaches and concepts have. Hence students associated and declared a much higher level of agreement with a statement referring to linear economical approach in comparison to the circular approach, expressing their belief that the first has higher negative environmental impacts than the other.

Additionally, the aim of the survey, mentioned previously, was to check firstly the declared students' knowledge about the concept of CE and their factual knowledge about CE. Therefore, the idea was to examine whether there is statistically significant correlation between the expressed level of the agreement with these two statements, and the statements made by the students at the previous questions "Are you familiar with the concepts of circular economy?"

For the first of the two statements, there is no such correlation. And regarding the overall mean of $M = 3.81$, it can be noted that students are very well informed about the environmental harms of traditional linear economy and its contrasting goals to the goals of environmental protection. This is not coming as a surprise, regarding that these harms nowadays are being accepted and thought at almost all levels of education in the Republic of Serbia.

Table 1.2 Comparison of attitudes toward the environmental effects of linear and circular economy approaches

Statement	Mean (M)	Standard deviation (SD)
Linear economy is in direct conflict with the environmental protection goals	3.81	1.06
Circular economy is in direct conflict with the environmental protection goals	1.99	1.2

On the other hand, the aim was to examine whether the students that responded positively to the question "Are you familiar with the concepts of circular economy?" relate more to the environmental benefits of CE approach. On that note, they do. There was statistically significant correlation observed when the T test has been performed. Students that stated that they are familiar with the concept of the CE gave answers with a mean result of $M = 1.9$ and SD = 1.18, for the second statement, while the students that declared that they are not familiar with the CE, had a mean result of $M = 3.2$ and SD = 0.84, that suggested T test to be done once again, and statistically significant correlation with a p value of $p < 0.05$ was observed.

The following set of questions within this part of the survey was concentrated in determining the practices student employ, in regard of their behaviour towards waste separation and disposal; usage of resources, and other elements that are part of the CE practices when companies are under consideration, in order to check whether these practices can be reflected in the daily behaviour of the survey participants, regarding majority of them stated the familiarity with the issues of CE.

First subset of Likert items in this part of the survey, included six statements that were dealing with properties of the goods and services, which participants consider when they engage in acquiring specific products and services. Idea was to check what determining factors are, for the participants, when they engage in making such choices. Furthermore, the main intention was to determine whether knowing that certain purchasing choices give more incentive for manufacturers to employ environmentally friendly solutions with introducing circular way of thinking and CE practices, is incentive for participants to choses products and services which would encourage manufacturers to go towards that direction. Moreover, knowing the benefits of choosing such products, the idea was to examine to what lengths are students ready to go, in order to get the products, they want, and that fulfil their needs. At the most extreme, the idea was to examine whether considering the knowledge, students have about different production approaches and about environmental harms of traditional linear way of production, can swing their purchasing habits to other purchasing options that are more environmentally friendly.

These six statements mentioned are presented in the Table 1.3 with corresponding mean and standard deviation.

What can be noted, while studying the mean results of the six statements is that the quality of the product or the service, with a mean result of $M = 4.65$ ranked at the top of the list as a main determining factor for making a final choice while deciding between different goods and services. Second dominating factor for making these decisions, considering participants within the survey, is a price with a mean value of $M = 4.12$. What is encouraging is that participants stated as their third most important factor when choosing between the products and services is an environmental impact that the product or service is having, with a mean value of $M = 3.55$, ranking before the brand of the manufacturer, which ranked last with the mean of $M = 3.09$, and before environmental reputation of the brand which ranked fifth, with the mean of $M = 3.35$. Regarding that the fourth ranked statement was "When purchasing food/wardrobe/electronics/other goods/services, purchasing convenience is one of the determining factors", with a mean result of $M = 3.44$, it can be noted that,

Table 1.3 Determining factors influencing purchasing choices

Statement	Mean (M)	Standard deviation (SD)
When purchasing food/wardrobe/electronics/other goods/services, price is one of the determining factors	4.12	0.71
When purchasing food/wardrobe/electronics/other goods/services, quality is one of the determining factors	4.65	0.51
When purchasing food/wardrobe/electronics/other goods/services, environmental impact is one of the determining factors	3.55	1.02
When purchasing food/wardrobe/electronics/other goods/services, purchasing convenience is one of the determining factors	3.44	1.09
When purchasing food/wardrobe/electronics/other goods/services, environmental reputation of the brand is one of the determining factors	3.35	1.1
When purchasing food/wardrobe/electronics/other goods/services, the brand is one of the determining factors	3.09	1.23

participants of the survey are ready to forego that convenience in order to get products and services that are more environmentally friendly, but in a same time that they are less interested in manufacturers brand environmental reputation.

Above mentioned statements were also compared by the answers classified by the gender, and statistically significant correlation was found in how different genders answered the statement "When purchasing food/wardrobe/electronics/other goods/services, environmental impact is one of the determining factors", where participants that declared themselves as of male gender had a mean result of $M = 2.87$ and SD = 1.06, while participants that declared themselves as of female gender had a mean result of $M = 3.71$ and SD = 0.95 with a p value of $p < 0.001$. This leads to the conclusion that for participants of the female gender, environmental impact of their purchasing choices is of much higher importance and stronger determining factor than it is for the participants of the male gender. Same applies for the importance of environmental reputation of the brand where the mean result for the respondents of the female gender was $M = 3.52$ and SD = 1.02, while the mean result for the participants of the male gender was $M = 2.6$ and SD = 1.12, statistically significant correlation was found with a p value of $p < 0.001$. This is in line with the idea Ruether (1975), Griffin (1978) and Merchant (1981) lay the foundation for, in terms of different social roles gender take in different cultural settings, and specific relation between female gender and ecology, environment and nature, as well as argumentation given by the authors Plumwood (2002) and Mellor (2007).

It is interesting to note that household income did not have a statistically significant role in how students graded these statements.

With the following three statements, idea was to examine the importance of manufacturers' practices and how do these practices if at all do, influence purchasing choices of the survey participants (Table 1.4).

It can be noted, by examining the mean result, in answering these three statements that these issues are still not of highest interest for the survey participants,

Table 1.4 Companies' practices and their influence on choices we make

Statement	Mean (M)	Standard deviation (SD)
Sustainability factors influence my purchasing decisions	3.34	1.06
If I find out that the company is environmentally irresponsible, I stop buying its products	3.16	1.2
It's important for me to know, whether the company, whose product I use, is basing their production on circular economy principles	2.94	1.11

when making purchasing choices. Sustainability factors, ranks highest among above mentioned three determining factors with a mean of $M = 3.34$ and SD = 1.06.

Nevertheless, statistically significant correlation was observed, when the answers of participants who strongly agree with the statement "When purchasing food/wardrobe/electronics/other goods/services, environmental impact is one of the determining factors" were compared with the rest of the participants, when expressing their level of agreement with the second statement "It's important for me to know, whether the company, whose products I use, is basing their production on circular economy principles". The participants that strongly agree with the statement "When purchasing food/wardrobe/electronics/other goods/services, environmental impact is one of the determining factors", had a mean result of $M = 4$ and SD = 1.08, when expressing their views for the statement "It's important for me to know, whether the company, which products I use, is basing their production on circular economy principles", while the rest of the participants had a mean result of $M = 2.73$ and SD = 1. Statistically significant correlation was noted when T test was performed and gave back the results with a p value of $p < 0.001$.

On this note, it can be concluded that the participants who care the most about the environmental impacts that are occurring as a result of the manufacturer practices, care the most whether companies, which products they use, employ CE principles or not. This might be due to the fact that, respondents for which determining factor when making purchasing choices are environmental impacts of their choices, understand environmental benefits of CE the most. On the other hand, although sustainability factor has a higher mean value, there is no such correlation found in the answers, probably because sustainability concepts are something that all the participants are very well familiar with.

In the following subset of statements, the idea was to examine the current daily practices and habits that would reflect participants' behaviour that students employ in regard to how they approach the energy resources, waste generation and disposal. The statements are presented in Table 1.5, not in order of appearance, but by the rank of the obtained answers' mean value. Starting from the statements that had strongest level of agreement by the participants, and finishing with the statements that participants had strongest level of disagreement. Interpretation of the results is provided next.

Comparing the gained results, it can be noted that ten statements had a mean result higher than 4, meaning that, participants express a strong agreement, with these practices. Moreover 21 of the statements had a mean result higher than 3.5

Table 1.5 Environmental practices that reflect CE principles employed by the survey participants

Statement	Mean (M)	Standard deviation (SD)
I check whether the lights are turned off when I am leaving the room	4.75	0.65
If the distance is short, I avoid using my car and rather choose to walk or use the bicycle instead	4.66	0.75
When my electronic utilities stop working, firstly I try to fix them before buying new ones	4.62	0.73
When I print my documents, I use two-sided printing option	4.56	0.73
I turn on the washing machine, only when it is full	4.55	0.82
I turn on the dish washer, only when it is full	4.55	0.78
I use energy saving light sources	4.51	0.79
I use reusable food and liquid containers, when I am having my meal outdoors	4.42	0.8
I would try to decrease my ecological footprint, if I would have more knowledge on how to do it	4.38	0.83
I try to donate old wardrobe, before I decide to dispose it	4.19	1.19
My water heater is set up to economical temperature maintainer	3.77	1.35
I don't take flyers that are being distributed on the streets, because that way I encourage their additional printing	3.74	1.39
I try to reuse or fix my old wardrobe, before I decide to buy a new one	3.73	1.24
My refrigerator is set up to economical temperature maintainer	3.71	1.33
I pay the utility bills electronically over the internet	3.71	1.44
I don't buy bottled water, because I rather use my reusable bottle	3.65	1.4
I try to use my car as less that I can	3.64	1.36
During the winter, I turn off the heating in the rooms that I am not using at the moment	3.62	1.53
I try to use public transportation as much as I can	3.61	1.51
I would be ready to pay the higher price for the product that has environmentally friendly properties	3.57	1.09
I try to donate old electronic utilities, before I decide to dispose them	3.56	1.32
I regularly separate household waste coming from food products, preparing it for recycling	3.34	1.27
I regularly separate household waste by categories for recycling	3.31	1.38
I regularly separate old wardrobe, preparing it for recycling	3.26	1.37
I use recycled paper	3.18	1.36
I regularly separate household waste coming from e-products, preparing it for recycling	2.97	1.37
When my electronic utilities stop working, firstly I try to sell them before buying new ones	2.87	1.52
I use recycled printer tonner	2.53	1.42
I try to find suitable sharing passenger transport services, before I use my car, for greater distances	2.13	1.45

(continued)

Table 1.5 (continued)

Statement	Mean (M)	Standard deviation (SD)
I try to sell my old wardrobe, before I decide to buy a new one	2.05	1.36
I tried to use some of the sharing services that rent kitchen appliances	1.45	0.94

which can be considered that on average students expressed agreement and strong agreement with these statements. If statements that had mean value less than 3 are analysed it can be noted that there are just six such statements. Two out of the six such statements, are in regard to some form of sharing services. This is not such a big surprise, regarding that market for such services is not still really developed in the Republic of Serbia, and that service providers such as FlixBus are just entering the local market. Same applies for the two statements that regard the selling of old products such as an old electronic ones and old wardrobe. There are new different options that are becoming popular for both of these, like programs that companies offer when electronic devices are in question like exchange services "old for new", or programs that typically fashion companies offer and market as environmentally friendly "bring your old wardrobe for recycling and get a discount for the next purchase" (Weber et al., 2017).

More importantly, while analysing the results through cross tabulation and performing the T test, the statistically significant correlation was found, when comparing the answers of participants that expressed a strong agreement with the statement that the determining factor when buying a product for them is an environmental impact that product has, and their willingness to pay more for the product that has better environmental properties. Participants that strongly agree with the first statement had a mean result of $M = 4.46$ with SD = 0.88 and that showed statistically significant correlation with a p value of $p < 001$, while the group that didn't express strong agreement with the first statement had a mean result for the second statement of $M = 3.39$ and SD = 1.09. This leads to the conclusion that those who take into consideration the environmental effects of the goods they are buying, are ready to put additional effort if needed when acquiring such goods and services.

Again here, it is interesting to note, as in previous example, that this is being true regardless of the household income that participants stated, regarding the statistically significant correlation was not found when comparing these two answers.

When analysing the results of this part of the survey another statement comes to an attention, and can be used as useful segway into the third part of the survey, as an introduction to the matter. That statement being "I would try to decrease my ecological footprint, if I would have more knowledge on how to do it" regarding the results of a mean $M = 4.38$ and SD = 0.83, it can be noted that this emphasises the need for additional educational content dealing in this matter. That's the additional reason, why the next part of the survey examines the needs for education for CE in tertiary level of education.

1.4.3 On a Future Behavioural Change and Education for Circular Economy

In the third part of the survey, the idea was to examine, why some of the practices mentioned, participants of the survey don't employ, and what could motivate them, to go an extra step, and start conducting them. Most importantly this is where it was tried to examine whether there is a link between the education, and CE principles. Moreover, the idea was to establish whether education for sustainable development and better education in general about CE can improve daily routines, activities and general conduct in accordance to CE principles.

The statements are presented in Table 1.6, as in previous case, not in order of appearance, but by the rank of the obtained answers mean value. Starting from the statements that had strongest level of agreement by the participants, and finishing with the statements that participants had strongest level of disagreement. Discussion of the results will follow.

At this point of the survey, results will be interpreted starting from the statements that had the lowest mean result, in order to start examining the results from the factors that are considered the least influential when the daily practices of the respondents are in question. It can be noted that last four statements in Table 1.6 had a mean resulted $M < 3$, meaning that respondents disagreed with them at least to some extent if not fully.

Several issues can be noted by examining these statements, firstly that the environmental behaviour of their peers is not something that is influencing the decisions, choices and practices of respondents $M = 2.61$.

Secondly, same can be said for their wider surroundings, regarding that the statement "Some of the things mentioned in the second part of the survey, I don't do, because "nobody does it" had a mean result of $M = 2.32$.

Thirdly, it can be noted that the respondents are aware of the positive environmental impact that practices that were stated in the second part of the survey have, regarding that the mean result of the obtained answers for the statement "Some of the things mentioned in the second part of the survey I don't do, because I don't believe that it has a positive environmental impact" was $M = 2.27$. This leads to the conclusion that participants of the survey are aware of the positive impact these practices have, and do not have to be additionally persuaded in order to start or continue conducting them.

The statement with the lowest mean value of $M = 2.1$, tries to examine whether brother cultural context, and surroundings support for the specific practices has an influence or not when making these decisions, and regarding such a low mean score, it can be noted that it doesn't.

Regarding that all four of these statements deal with the issue of the social surrounding of the participants, it can be concluded collectively out of the results for these four statements, that the peer pressure, close and broader surrounding doesn't have an effect on respondents' environmental behaviour, once they believe that these practices do really have a positive environmental impact, which is the case

Table 1.6 Beyond environmental practices – reasons and motivational factors that can influence change

Statement	Mean (M)	Standard deviation (SD)
Implementing principles of circular economy improves the state of the environmental wellbeing	4.84	0.43
For better implementation of circular economy principles, change of perception coming from the general public must be achieved, through education	4.79	0.5
Implementing principles of circular economy contributes to general societal and economic development	4.79	0.5
I believe that environmental education and education for sustainable development must include education for circular economy	4.67	0.57
I believe that education for circular economy has to have more space devoted to it within the curriculum of tertiary education	4.58	0.64
I believe that education for circular economy encourages implementation of circular economy principles	4.45	0.82
Some of the things mentioned in the second part of the survey I don't do, because there is no suitable infrastructure that would allow me to contribute in that manner to environmental protection	3.96	1.22
Implementing principles of circular economy increases the profit for companies	3.92	0.96
Some of the things mentioned in the second part of the survey I don't do regularly, because I didn't know, they have negative environmental impact	3.75	1.29
Some of the things mentioned in the second part of the survey I don't do regularly, because it's not common among my peers	2.61	1.58
Some of the things mentioned in the second part of the survey I don't do, because "nobody does it"	2.32	1.5
Some of the things mentioned in the second part of the survey I don't do, because I don't believe that it has a positive environmental impact	2.27	1.31
Some of the things mentioned in the second part of the survey I don't do regularly, because other people around me don't do, and I wouldn't have a support for such conduct	2.1	1.44

when CE practices are in question judging by the rest of obtained responses. This can be noted by analysing the results of the first nine statements that all had the mean result higher than 3.75, which testifies that issues that the statements deal with, have some or strong influence towards the respondents' behaviour.

First of such statements being "Some of the things mentioned in the second part of the survey I don't do regularly, because I didn't know, they have negative environmental impact" with a mean value of $M = 3.75$ suggests that the reasons for not engaging in some of the practices, is simply a lack of knowledge about their environmental benefits and it ties with CE. This emphasizes the need for two things, firstly the need for associating such practices to CE, and secondly the need for disseminating the knowledge about their environmental benefits and waste management benefits. Hence emphasizing the need for better education on some of the issues. Along the same lines the following statement "Some of the things mentioned in the second part of the survey I don't do, because there is no suitable infrastructure that would allow me to contribute in that manner to the environmental protection" with a mean result of $M = 3.92$ emphasizes the need for better infrastructure that would support the easier implementation of the CE practices and enabling all the groups of society to contribute to CE causes.

Following up this statement was the statement dealing with the issue of profits companies make that employ CE approaches in their production lines. There was an idea to check whether respondents are aware, that companies that engage in CE practices make higher profits in comparison to their competitors that don't implement the CE approach as noted by Lacy and Rutqvist (2016). And it can safely be concluded that respondents are aware of that fact, regarding the mean result of $M = 3.96$. Meaning that information about that is out there, and that educational process is good in communicating that message across to students. Which is encouraging fact, and could lead potentially to generations that are motivated, once starting their entrepreneur endeavours, to incorporate CE approaches.

The previous was confirmed even more firmly by the following statement "Implementing principles of circular economy increases the profit for companies", that had mean result of $M = 4.45$. Such a high result of the mean, emphasizes the need for the increased number of classes that deal with the CE, regarding the respondents' answers, removing any dilemmas on the issue, if any were left. Concluding in essence that if change towards circular thinking and to CE as a whole is something that society in general wants to achieve, such a change must be back pinned by the educational system as a whole, and especially by higher education. And it can be added to that, that respondents feel the same thing not just in binary sense, whether CE education has to be represented or not within the higher education curriculum. They made clear, almost unified answer to that question. It is rather the question of, after achieving the introduction of CE concepts in education, how to enable CE to be covered to the higher extent than now, in terms of space and time reserved for CE within the curriculum, in terms of quantity of courses and classes dealing with the issues and devoted to CE. It is clear that it has to be to the larger extent than in status quo regarding the mean result $M = 4.58$.

Mean result of $M = 4.67$, with the $SD = 0.57$ for the statement "I believe that environmental education and education for sustainable development must include education for circular economy", allows noticing that the participants of the survey not only understand the connection between environmental education and education for sustainable development and education for CE, but also find it inseparable. This again emphasis the need for CE education in this instance as well.

The second highest ranked mean result of $M = 4.79$ was found when analysing results for two of the statements. One being "For better implementation of circular economy principles, change of perception coming from the general public must be achieved, through education" and the other being "Implementing principles of circular economy contributes to general societal and economic development", such a high mean result can be contributed to participants' understanding of the influence education has, regarding that themselves are undergoing the process of acquiring higher education degree, meaning that they know what kind of influence education left on them, and to what personal attitude changes have they undergone because of it. Reflecting on that it seems that participants believe that same influence and exposure to that kind of education wouldn't leave general public immune to same such changes, relaying, off course on the assumption, that general public is susceptible to same attitude changes as someone who is undergoing the educational process in higher education, once the right information are presented to them.

The other statement with a same mean result of $M = 4.79$, "Implementing principles of circular economy contributes to general societal and economic development" allows noticing that the students link better education to social and economic progress. This might be because of their general belief that education leads to such progress, or/and that the social and economic benefits coming out of implementing CE practices lead to the same result.

The highest-ranking statement with the mean value of $M = 4.84$ was "Implementing principles of circular economy improves the state of the environmental wellbeing". Several notions have to be underlined at this point.

Firstly, this is the only statement with such high mean result, where all the participants of the survey, almost unanimously expressed the strong agreement with a statement.

Secondly this means that during their participation within the survey, even the students that stated that they are not familiar with the concept of CE, during the course of the survey, they got familiar with the concept of CE, and got certain level of understatement of its principles, goals and practices. Meaning even the survey had educational component to it, and by itself provided one more evidence of the importance of the education.

Still, even with such a high mean, there was a statistically significant correlation noted in how these two groups answered to this statement. The group that stated at the beginning of the survey that they didn't know what CE was, expressed their agreement with the mentioned statement with a mean result of $M = 4$ and SD = 0.89, while the group that stated at the beginning of the survey that they know what CE is expressed their agreement with the statement with a mean result of 4.90 and SD = 0.3 and with a statistically significant correlation of p value, $p < 0.001$.

Finally, this also provides one more conclusion, the one being, that the students that participated in the survey understand the positive environmental effects of CE and better resource management and the increase in their usage efficiency in production process, as well as the benefits of better waste management.

1.5 Discussion and Conclusion

For the purposes of the discussion all the statistically significant correlations and their values found during the research will be listed firstly. During the research following correlations were noted:

- Students that are familiar with the concept of the CE recognise that CE is less harmful towards the environment than linear economy ($p < 0.05$)
- Participants of the female gender, environmental impact of their purchasing choices is of much higher importance and stronger determining factor than it is for the participants of the male gender ($p < 0.001$)
- Participants of the female gender, environmental reputation of the brand of their purchasing choices is of much higher importance and stronger determining factor than it is for the participants of the male gender ($p < 0.001$)
- Participants who strongly agree that environmental impact is one of the determining factors for them when making choices between the different options care more whether the manufacturer of the goods is basing its production on CE principles ($p < 0.001$).
- Participants who strongly agree that environmental impact is one of the determining factors for them when making choices between the different options are ready to pay more for a product with such properties ($p < 0.001$)
- Students that were familiar with the concept of the CE believe more in the improvements of environmental state and wellbeing by implementing CE principles in business ($p < 0.001$).

In regard of the first correlation noted, it can be concluded that knowledge is a quintessential factor in creating desired awareness levels that are needed in changing the overall firstly attitudes and as a result of that consequentially behaviour. Put more simply, people have to know in order to create a change of attitude and in order to that change of the attitude results in behaving differently.

Additionally, this correlation testifies of both, the efficacy of the higher education for CE and education for sustainable development, and their importance in transitioning from linear to CE and in transitioning from linear to circular thinking.

In regard to second and third statistical correlation observed, discussion on gender roles and their views when environment is at stake can be opened once again, though this was not the main idea of this chapter, it has to be noted that these findings of our research confirmed previous research "that females generally stressed that the protection of nature and the environment is an important aspect of human existence" (Borojević et al., 2017a, b; Eisler et al., 2003). So, it remains for the future research to determine how to bring male gender to the same level of interest in the subject of relation of: our actions – environmental effects, level at which female gender already is.

In regard to fourth statistical correlation observed, it can be noted that participants that make decisions based on the environmental effects of their choices, know the environmental benefits of introducing CE principles in production process, hence care more whether such principles are incorporated into choices they make. Meaning that from knowledge, and consequently education for CE can improve environmental choices people make. Simultaneously, this is beneficial for the manufacturers as well, regarding that this can prove that does who care recognize the extra efforts put by companies to bring more quality and less environmentally harmful products to the consumer. In the same manner this means that this fact should be a motivational factor, for companies to introduce CE principles into their conducts regarding it will be recognized and rewarded by consumers.

Overall, this means that education for CE and sustainability is overarching factor from which the whole three: environment, population and companies benefit.

Fifth correlation can simply be interpreted as additional evidence for what was noted in the fourth one. Regarding those participants that make decisions based on the environmental effects their choices have are ready to pay more for such products and services, means that there is an additional reward waiting for those who recognize the needs of environmentally aware consumers, and again that from the education for CE and sustainability, all stakeholders have benefits.

Finally, the sixth and in a same time last statistical correlation noted testifies the most about the importance of education for CE and education for sustainability. Firstly, as explained in the previous section of the paper, the statement "Implementing principles of CE improves the state of the environmental wellbeing" was the highest-ranking statement with the mean value of $M = 4.84$. Meaning that educating people about specific benefits CE has, leads them to appreciate its benefits more in comparison to people who do not possess such knowledge. This once again flags the education as a quintessential factor for the change that the planet needs.

In conclusion state of the environment that planet is facing now, is a direct result of human activity. It was never so evident that we live in Anthropocene as now. Human dominated era, has resulted in a massive environmental degradation and a decrease in quality of life for all the species inhabiting planet earth. Most of the human environmental impact is a result of economic progress achieved through firstly agricultural development and secondly through industrial development. That progress and development was driven on the back of the linear ways of production that safeguard little if at all the environment. Meaning that it must be concluded that old ways of linear economic development must be abolished. At the same time, the way of thinking that led the humans to this state has to be abolished as well, and changed with more environmentally friendly options. Hence in that transition, this paper proposes taking "a yellow brick road" that should transfer planet's progress to sustainable practices such as CE offers, as well as adopting the circular way of mind.

The survey presented in the paper shows that the education and more specifically higher education has a means to provide necessary circumstances for the planet to undergo such change, by changing the traditional curriculum evolving around linear practices, to the one that shifts the emphasis to the CE and circular practices.

The obtained preliminary results from our research, encouraged us and other researchers to continue exploring this area of wider implementation of CE in higher education as necessary part of transformation of linear economy to CE. Also, as a result of our study one that must be stressed above others is that education for CE and education for the sustainability are not only effective ways to achieve such a transition, but also, they might be the drivers of such transition, because if we view the road to a sustainable society as a necessary project of civilization then we could conclude as authors Cvijović et al. (2021) had proven in their research that if to the contrary neglecting all of the stakeholder's interests, could lead to a complete project downfall and as so, it should never be considered as a possibility when considering successful project management.

References

Adeniran, A. E., Nubi, A. T., & Adelopo, A. O. (2017). Solid waste generation and characterization in the University of Lagos for a sustainable waste management. *Waste Management, 67*, 3–10.

Alcamo, J. (2019). Water quality and its interlinkages with the Sustainable Development Goals. *Current Opinion in Environmental Sustainability, 36*, 126–140. https://doi.org/10.1016/j.cosust.2018.11.005

Al-Mansoori, R. S., & Koç, M. (2019). Sustainability in higher education: The impact of transformational leadership on followers' innovative outcomes a framework proposal. In *Sustainability on university campuses: Learning, skills building and best practices* (pp. 227–243). Springer.

Alves, D. S., & Farina, M. C. (2018). Disposal and reuse of the information technology waste: A case study in a Brazilian university. *European Business Review*. Retrieved from https://www.emerald.com/insight/content/doi/10.1108/EBR-08-2016-0117/full/html

Andreoni, V., & Ruiz Vargas, V. (2020). Tracking the interlinkages across SDGs: The case of Hill Centered Education Network in Bogota, Colombia. *Sustainability, 12*(19), 7924. https://doi.org/10.3390/su12197924

Ashkin, S. (2008). *Cradle To Grave Vs. Cradle To Cradle*. Retrieved from https://www.cleanlink.com/cp/article/Cradle-To-Grave-Vs-Cradle-To-Cradle%2D%2D9277.

Association of University Leaders for Sustainable Future. (1990). *The Talloires Declaration, 10 Point Action Plan*. Retrieved from http://ulsf.org/wp-content/uploads/2015/06/TD.pdf

Beynaghi, A., Trencher, G., Moztarzadeh, F., Mozafari, M., Maknoon, R., & Leal Filho, W. (2016). Future sustainability scenarios for universities: Moving beyond the United Nations Decade of Education for Sustainable Development. *Journal of Cleaner Production, 112*, 3464–3478. https://doi.org/10.1016/j.jclepro.2015.10.117

Benltoufa, A. N. H. S., Jaafar, F., Maraoui, M., Said, L., Zili, M., Hedfi, H., ... & Salah, H. B. (2017, August). From smart campus to smart city: Monastir living lab. In 2017 International Conference on Engineering and Technology (ICET) (pp. 1–6). IEEE.

Biggs, D., Biggs, R. O., Dakos, V., Scholes, R., & Schoon, M. (2011). Are we entering an era of concatenated global crises? *Ecological Society, 16*(2), 27. Retrieved from http://www.ecologyandsociety.org/vol16/iss2/

Bisaga, I., Parikh, P., Tomei, J., & To, L. S. (2020). Mapping synergies and trade-offs between energy and the sustainable development goals: A case study of off-grid solar energy in Rwanda. *Energy Policy*, 112028. https://doi.org/10.1016/j.enpol.2020.112028

Blomsma, F., & Brennan, G. (2017). The emergence of circular economy: A new framing around prolonging resource productivity. *Journal of Industrial Ecology, 21*(3), 603–614. https://doi.org/10.1111/jiec.12603

Bonnett, M. (2007). Environmental education and the issue of nature. *Journal of Curriculum Studies, 39*(6), 707–721. https://doi.org/10.1080/00220270701447149

Borojević, T., Vuk, D., Petrović, N., & Slović, D. (2015). The level of youth activism: Case study of the city of Belgrade. *Management: Journal of Sustainable Business and Management Solutions in Emerging Economies, 20*(76), 23–30.

Borojević, T., Maletič, M., Petrović, N., Radaković, J. A., Senegačnik, M., & Maletič, D. (2017a). Youth Attitudes Towards Goals of a New Sustainable Development Agenda Postawy młodzieży wobec celów Nowej Agendy na rzecz zrównoważonego rozwoju. *Problemy Ekorozwoju - Problems of Sustainable Development, 12*(2), 161–172.

Borojević, T., Maletič, M., Stanojević, K., Đoković, A., Radaković, J. A., & Milenković, N. (2017b, Septembar 25–28). Uloga mladih u zaštiti životne sredine i održivosti lokalne zajednice. U

G. Ćirović (Ed.), *Zbornik radova XLIV Simpozijuma o operacionim istraživanjima – SYM-OP-IS 2017*, (pp. 466-470), Zlatibor, Srbija. Visoka građevinsko-geodetska škola: Beograd. Retrieved from http://symopis.vggs.rs/razno/Zbornik_radova_SYM-OP-IS_2017.pdf. [In Serbian].

Bugallo-Rodríguez, A., & Vega-Marcote, P. (2020). Circular economy, sustainability and teacher training in a higher education institution. *International Journal of Sustainability in Higher Education*. Retrieved from https://www.emerald.com/insight/1467-6370.htm

Castor, J., Bacha, K., & Nerini, F. F. (2020). SDGs in action: A novel framework for assessing energy projects against the sustainable development goals. *Energy Research & Social Science, 68*, 101556. https://doi.org/10.1016/j.erss.2020.101556

Ćirović, M. D. (2018). Risk analysis of the European Union 2030 greenhouse gas emission target compliance. *International Journal of Global Warming, 16*(1), 64–85. https://doi.org/10.1504/IJGW.2018.094311

Ćirović, M. D. (2019). Planetary boundaries: A sustainability framework. In W. Leal Filho, A. Azul, L. Brandli, P. Özuyar, & T. Wall (Eds.), *Climate action. Encyclopaedia of the UN sustainable development goals*. Springer Nature Switzerland AG. Springer. https://doi.org/10.1007/978-3-319-05885-9

Ćirović, M., Makajić-Nikolić, D., Petrović, N., Vujošević, M., & Kuzmanović, M. (2015). European Union oil import dependency risk analysis. *Polish Journal of Environmental Studies, 24*(1), 75–81. https://doi.org/10.15244/pjoes/26105

Collins. (2020). *Definition of 'the yellow brick road'*. Retrieved from https://www.collinsdictionary.com/dictionary/english/the-yellow-brick-road

Combes, B. P. (2005). The United Nations decade of education for sustainable development (2005–2014): Learning to live together sustainably. *Applied Environmental Education and Communication, 4*(3), 215–219. https://doi.org/10.1080/15330150591004571

Cottafava, D., Cavaglià, G., & Corazza, L. (2019). Education of sustainable development goals through students' active engagement. *Sustainability Accounting. Management and Policy Journal, 10*(4), 521. https://doi.org/10.1108/SAMPJ-05-2018-0152

Curtin, R., Presser, S., & Singer, E. (2000). The effects of response rate changes on the index of consumer sentiment. *Public Opinion Quarterly, 64*(4), 413–428.

Cvijović, J., Obradović, V., & Todorović, M. (2021). Stakeholder management and project sustainability - A throw of the dice. *Sustainability, 13*(17), 9513. https://doi.org/10.3390/su13179513

Daub, C.-H., Hasler, M., Verkuil, A. H., & Milow, U. (2020). Universities talk, students walk: Promoting innovative sustainability projects. *International Journal of Sustainability in Higher Education, 21*(1), 97–111. https://doi.org/10.1108/IJSHE-04-2019-0149

Diep, L., Martins, F. P., Campos, L., Hofmann, P., Tomei, J., Lakhanpaul, M., & Parikh, P. (2020). Linkages between sanitation and the sustainable development goals: A case study of Brazil. *Sustainable Development, 29*, 339. https://doi.org/10.1002/sd.2149

Eisler, A. D., Eisler, H., & Yoshida, M. (2003). Perception of human ecology: Cross-cultural and gender comparisons. *Journal of Environmental Psychology, 23*(1), 89–101.

Fahey, S. J. (2012). Curriculum change and climate change: Inside outside pressures in higher education. *Journal of Curriculum Studies, 44*(5), 703–722. https://doi.org/10.1080/00220272.2012.679011

Findanyanswer. (2020). *What does the yellow brick road in The Wizard of Oz represent?* Retrieved from https://findanyanswer.com/what-does-the-yellow-brick-road-in-the-wizard-of-oz-represent#:~:text=Hugh%20Rockoff%20suggested%20in%201990,Dorothy%20inherits%20from%20the%20Wicked

Flörke, M., Bärlund, I., van Vliet, M. T., Bouwman, A. F., & Wada, Y. (2019). Analysing trade-offs between SDGs related to water quality using salinity as a marker. *Current Opinion in Environmental Sustainability, 36*, 96–104. https://doi.org/10.1016/j.cosust.2018.10.005

Franco, A. M. A. (2013, December). The balance of nature and human impact. In K. Rohde (Ed.), *Integrative and Comparative Biology, 53*(6), (pp. 1017–1019). https://doi.org/10.1093/icb/ict099

Franco, I., Saito, O., Vaughter, P., Whereat, J., Kanie, N., & Takemoto, K. (2019). Higher education for sustainable development: Actioning the global goals in policy, curriculum and practice. *Sustainability Science, 14*(6), 1621–1642. https://doi.org/10.1007/s11625-018-0628-4

Fricke, J. (1989). *The wizard of Oz: The official 50th anniversary pictorial history*. Warner Books.

Gallo, M., Marotta, V., Magrassi, F., Taramasso, A. C., & Del Borghi, A. (2017). University campus waste prevention and reduction: A circular-economy approach. *Economics and Policy of Energy and the Environment, 235-252*, 235. https://doi.org/10.3280/EFE2017-001012

Geissdoerfer, M., Savaget, P., Bocken, N. M., & Hultink, E. J. (2017). The circular economy–A new sustainability paradigm? *Journal of Cleaner Production, 143*, 757–768. https://doi.org/10.1016/j.jclepro.2016.12.048

Ghisellini, P., Cialani, C., & Ulgiati, S. (2016). A review on circular economy: The expected transition to a balanced interplay of environmental and economic systems. *Journal of Cleaner Production, 114*, 11–32. https://doi.org/10.1016/j.jclepro.2015.09.007

González-Domínguez, J., Sánchez-Barroso, G., Zamora-Polo, F., & García-Sanz-Calcedo, J. (2020). Application of circular economy techniques for design and development of products through collaborative project-based learning for industrial engineer teaching. *Sustainability, 12*(11), 4368.

Goyal, S., Esposito, M., & Kapoor, A. (2018). Circular economy business models in developing economies: Lessons from India on reduce, recycle, and reuse paradigms. *Thunderbird International Business Review, 60*(5), 729–740. https://doi.org/10.1002/tie.21883

Griffin, S. (1978). *Woman and nature: The roaring inside her*. The Women's Press.

Gupta, S. D. (2015). Dirty and clean technologies. *Journal of Agricultural and Applied Economics, 47*(1), 123–145. https://doi.org/10.1017/aae.2014.1

Halada, K. (2020). Activities of Circular Economy in Japan-towards global multi-value circulation-. *International Journal of Automation Technology, 14*(6), 867–872. https://doi.org/10.20965/ijat.2020.p0867

Hall, N. L., Creamer, S., Anders, W., Slatyer, A., & Hill, P. S. (2020). Water and health interlinkages of the sustainable development goals in remote indigenous Australia. *npj Clean Water, 3*(1), 1–7. https://doi.org/10.1038/s41545-020-0060-z

Handl, G. (2012). Declaration of the United Nations conference on the human environment (Stockholm Declaration), 1972 and the Rio Declaration on Environment and Development, 1992. *United Nations Audiovisual Library of International Law, 11*. Retrieved from https://legal.un.org/avl/pdf/ha/dunche/dunche_e.pdf

Hannoura, A. P., Cothren, G. M., & Khairy, W. M. (2006). The development of a sustainable development model framework. *Energy, 31*(13), 2269–2275. https://doi.org/10.1016/j.energy.2006.01.022

Hansen, E. G., & Revellio, F. (2020). Circular value creation architectures: Make, ally, buy, or laissez-faire. *Journal of Industrial Ecology*, 1250–1273. https://doi.org/10.1111/jiec.13016

Hester, P. T., & Adams, K. M. (2014). *Systemic thinking: Fundamentals for understanding problems and messes*. Springer.

Hugé, J., Block, T., Waas, T., Wright, T., & Dahdouh-Gueba, F. (2016). How to walk the talk? Developing actions for sustainability in academic research. *Journal of Cleaner Production, 137*, 83–92. https://doi.org/10.1016/j.jclepro.2016.07.010

Iannuzzi, A. (2011). *Greener products – The making and marketing of sustainable brands*. CRC Press.

IUCN-UNEP-WWF. (1980). *World Conservation Strategy*. Retrieved from https://portals.iucn.org/library/efiles/edocs/WCS-004.pdf

Jorge, M. L., Madueño, J. H., Cejas, M. Y. C., & Peña, F. J. A. (2015). An approach to the implementation of sustainability practices in Spanish universities. *Journal of Cleaner Production, 106*, 34–44. https://doi.org/10.1016/j.jclepro.2014.07.035

Kılkış, Ş. (2017). Comparative analyses of sustainable campuses as living laboratories for managing environmental quality. *Management of Environmental Quality: An International Journal*.

Retrieved from https://www.emerald.com/insight/content/doi/10.1108/MEQ-06-2015-0107/full/html

Kirchherr, J., Reike, D., & Hekkert, M. (2017). Conceptualizing the circular economy: An analysis of 114 definitions. *Resources, Conservation and Recycling, 127*, 221–232. https://doi.org/10.1016/j.resconrec.2017.09.005

Kopnina, H. (2018). Teaching sustainable development goals in The Netherlands: A critical approach. *Environmental Education Research, 24*(9), 1268–1283. https://doi.org/10.1080/13504622.2017.1303819

Lacy, P., & Rutqvist, J. (2016). *Waste to wealth: The circular economy advantage*. Springer.

Leal Filho, W., Salvia, A. L., Pretorius, R. W., Brandli, L. L., Manolas, E., Alves, F., et al. (Eds.). (2019). *Universities as living labs for sustainable development: Supporting the implementation of the sustainable development goals*. Springer.

Lieder, M., & Rashid, A. (2016). Towards circular economy implementation: A comprehensive review in context of manufacturing industry. *Journal of Cleaner Production, 115*, 36–51. https://doi.org/10.1016/j.jclepro.2015.12.042

Llanos-Herrera, G. R., & Merigo, J. M. (2019). Overview of brand personality research with bibliometric indicators. *Kybernetes*. Retrieved from https://www.emerald.com/insight/content/doi/10.1108/K-02-2018-0051/full/html

Makajić-Nikolić, D., Petrović, N., Ćirović, M., Vujošević, M., & Presburger-Ulnikovic, V. (2016). The model of risk assessment of greywater discharges from the Danube River ships. *Journal of Risk Research, 19*(4), 496–514, ISSN 1366-9877, IF (2016) - 1.340. https://doi.org/10.1080/13669877.2014.988286

Maletič, D., Borojević, T., Petrović, N., Maletič, M., & Senegačnik, M. (2017, Septembar 25–28). Kvalitet viso koškolskog ekološkog obrazovanja za održivost. In U G. Ćirović (Ed.), *Zbornik radova XLIV Simpozijuma o operacionim istraživanjima - SYM-OP-IS 2017* (pp. 460–465), Zlatibor, Srbija. Visoka građevinsko-geodetska škola: Beograd. Retrieved from http://symopis.vggs.rs/razno/Zbornik_radova_SYM-OP-IS_2017.pdf. [In Serbian].

Mateus, D. M., Pinho, H. J., Nogueira, I. M., Rosa, M. A., Cartaxo, M. A., & Nunes, V. M. (2020). Participation of students in the project Valorbio: A case study to accelerate the implementation of sustainability principles in the curriculum. *International Journal of Sustainability in Higher Education, 21*(2), 244–263. https://doi.org/10.1108/IJSHE-09-2019-0254

McCollum, D. L., Echeverri, L. G., Busch, S., Pachauri, S., Parkinson, S., Rogelj, J., et al. (2018). Connecting the sustainable development goals by their energy inter-linkages. *Environmental Research Letters, 13*(3), 033006. https://doi.org/10.1088/1748-9326/aaafe3

McCormick, K., Muhlhauser, E., Norden, B., Hansson, L., Foung, C., Arnfalk, P., Karlsson, M., & Pigretti, D. (2005). Education for sustainable development and the Young Masters Program. *Journal of Cleaner Production, 13*(10–11), 1107–1112. https://doi.org/10.1016/j.jclepro.2004.12.007

Mellor, M. (2007). Ecofeminism: Linking gender and ecology. In *The SAGE handbook of environment and society*, 66–77. https://doi.org/10.4135/9781848607873.n4

Mendoza, J. M. F., Gallego-Schmid, A., & Azapagic, A. (2019). A methodological framework for the implementation of circular economy thinking in higher education institutions: Towards sustainable campus management. *Journal of Cleaner Production, 226*, 831–844. https://doi.org/10.1016/j.jclepro.2019.04.060

Merchant, C. (1981). Earthcare: Women and the environment. *Environment: Science and Policy for Sustainable Development, 23*(5), 6–40.

Mert, M. (2006). *Determination of consciousness level of high school students on the environmental training and solid wastes topics*. Hacettepe University.

Moore, D. L., & Tarnai, J. (2002). Evaluating nonresponse error in mail surveys. In R. M. Groves, D. A. Dillman, J. L. Eltinge, & R. J. A. Little (Eds.), *Survey nonresponse* (pp. 197–211). Wiley.

Murray, A., Skene, K., & Haynes, K. (2017). The circular economy: An interdisciplinary exploration of the concept and application in a global context. *Journal of Business Ethics, 140*(3), 369–380. https://doi.org/10.1007/s10551-015-2693-2

Nerini, F. F., Tomei, J., To, L. S., Bisaga, I., Parikh, P., Black, M., et al. (2018). Mapping synergies and trade-offs between energy and the Sustainable Development Goals. *Nature Energy, 3*(1), 10–15. https://doi.org/10.1038/s41560-017-0036-5

Nunes, B. T., Pollard, S. J., Burgess, P. J., Ellis, G., De los Rios, I. C., & Charnley, F. (2018). University contributions to the circular economy: Professing the hidden curriculum. *Sustainability, 10*(8), 2719. https://doi.org/10.3390/su10082719

Obradović, V. (2010). *Model upravljanja promenama primenom metodologije projektnog menadžmenta*, Ph.D. Thesis. Fakultet organizacionih nauka. [In Serbian].

Olalla, C. B., & Merino, A. (2019). Competences for sustainability in undergraduate business studies: A content analysis of value-based course syllabi in Spanish universities. *The International Journal of Management Education, 17*(2), 239–253. https://doi.org/10.1016/j.ijme.2019.02.006

Owojori, O., Edokpayi, J. N., Mulaudzi, R., & Odiyo, J. O. (2020). Characterisation, recovery and recycling potential of solid waste in a university of a developing economy. *Sustainability, 12*(12), 5111. https://doi.org/10.3390/su12125111

Pardal, A., Romeira, T., & Durão, A. (2020). Eco green campus: Challenges and opportunities. The study case of Polytechnic Institute of Beja. In *E3S Web of Conferences* (Vol. 171, p. 01010). EDP Sciences.

Parikh, P., Diep, L., Gupte, J., & Lakhanpaul, M. (2020a). COVID-19 challenges and WASH in informal settlements: Integrated action supported by the sustainable development goals. *Cities, 107*, 102871. https://doi.org/10.1016/j.cities.2020.102871

Parikh, P., Diep, L., Hofmann, P., Tomei, J., Campos, L. C., Teh, T. H., et al. (2020b). Synergies and trade-offs between sanitation and the sustainable development goals. *SSRN Electronic Journal*. https://doi.org/10.2139/ssrn.3552872

Petković, J., Petrović, N., Dragović, I., Stanojević, K., Radaković, J. A., Borojević, T., & Kljajić Borštnar, M. (2019). Youth and forecasting of sustainable development pillars: An adaptive neuro-fuzzy inference system approach. *PloS One, 14*(6), e0218855. eISSN 1932-6203. https://doi.org/10.1371/journal.pone.0218855

Petrović, N. (2016). Ekološki menadžment, treće izmenjeno i dopunjeno izdanje. Fakultet organizacionih nauka. [In Serbian].

Petrović, N., Slović, D., & Ćirović, M. (2012). Environmental performance indicators as guidelines towards sustainability. *Management - Journal for Theory and Practice Management, 17*(64), 5–14. https://doi.org/10.7595/management.fon.2012.0026

Petrović, N., Jeremić, V., Petrović, D., & Ćirović, M. (2014). Modeling the use of Facebook in environmental higher education. In G. Mallia (Ed.), *The social classroom: Integrating social network use in education* (pp. 100–119). IGI Global. https://doi.org/10.4018/978-1-4666-4904-0.ch006

Petrović, N., Savić, G., Andrijašević, D., Stanojević, M., Ćirović, M., Slović, D., & Radaković, J. A. (2016). Evaluating eco-efficiency of beverage packaging materials: A data envelopment analysis approach. *Fresenius Environmental Bulletin, 25*(8), 2958–2963.

Petrović, N., Bošnjak, I., & Nedeljković, S. (2017). Disaster risk reduction for sustainable development goals. *European Project Management Journal, 7*(2), 27–35.

Plumwood, V. (2002). *Feminism and the mastery of nature*. Routledge.

Puerta, L. S., Díaz, J. C. L., Moreno, V. M., & Montes, F. J. L. (2020). Implementation of circular economy for the evaluation of sustainability in waste management on the University Campus of Lome (Togo). In XXX Jornadas Luso-Espanholas de Gestão Científica: cooperação transfronteiriça. Desenvolvimento e coesão territorial. Livro de resumos (p. 224). Instituto Politécnico de Bragança.

Qu, D., Shevchenko, T., & Yan, X. (2020). University curriculum education activities towards circular economy implementation. *International Journal of Scientific and Technology Research, 9*(5), 200–206.

Radaković, J. A., Petrović, N., Milenković, N., Stanojević, K., & Đoković, A. (2017). The improvement of students' higher environmental and climate change knowledge: A case study. *Polish Journal of Environmental Studies, 26*(6), 2711–2719. ISSN 1230-1485. https://doi.org/10.15244/pjoes/69645

Ramísio, P. J., Pinto, L. M. C., Gouveia, N., Costa, H., & Arezes, D. (2019). Sustainability Strategy in Higher Education Institutions: Lessons learned from a nine-year case study. *Journal of Cleaner Production, 222*, 300–309. https://doi.org/10.1016/j.jclepro.2019.02.257

Revinova, S., Ratner, S., Lazanyuk, I., & Gomonov, K. (2020). Sharing economy in Russia: Current status, barriers, prospects and role of universities. *Sustainability, 12*(12), 4855. https://doi.org/10.3390/su12124855

Robert, L., Flood, E., & Carson, R. (1993). *Dealing with complexity*. Plenum Publishing.

Rockström, J., Steffen, W., Noone, K., Persson, Å., Chapin, F. S., Lambin, E. F., et al. (2009). A safe operating space for humanity. *Nature, 461*(7263), 472–475. https://doi.org/10.1038/461472a

Rokicki, T., Perkowska, A., Klepacki, B., Szczepaniuk, H., Szczepaniuk, E. K., Bereziński, S., & Ziółkowska, P. (2020). The importance of higher education in the EU countries in achieving the objectives of the circular economy in the energy sector. *Energies, 13*(17), 4407. https://doi.org/10.3390/en13174407

Ruether, R. R. (1975). *New woman, new earth: Sexist ideologies and human liberation* (p. 11336). Seabury Press.

Sachs, J. D. (2015). *The age of sustainable development*. Columbia University Press.

Salguero-Puerta, L., Leyva-Díaz, J. C., Cortés-García, F. J., & Molina-Moreno, V. (2019). Sustainability indicators concerning waste management for implementation of the circular economy model on the University of Lome (Togo) Campus. *International Journal of Environmental Research and Public Health, 16*(12), 2234. https://doi.org/10.3390/ijerph16122234

Sammalisto, K., Sundström, A., & Holm, T. (2015). Implementation of sustainability in universities as perceived by faculty and staff–a model from a Swedish university. *Journal of Cleaner Production, 106*, 45–54. https://doi.org/10.1016/j.jclepro.2014.10.015

Sanchez, B., Ballinas-Gonzalez, R., Rodriguez-Paz, M. X., & Nolazco-Flores, J. A. (2020, April). Integration of circular economy principles for developing sustainable development competences in higher education: An analysis of bachelor construction management courses. In *2020 IEEE Global Engineering Education Conference (EDUCON)* (pp. 988–996). IEEE.

Sánchez-Carracedo, F., López, D., Martín, C., Vidal, E., Cabré, J., & Climent, J. (2020). The sustainability matrix: A tool for integrating and assessing sustainability in the bachelor and master theses of engineering degrees. *Sustainability, 12*(14), 5755. https://doi.org/10.3390/su12145755

Sánchez-Carracedo, F., Segalas, J., Bueno, G., Busquets, P., Climent, J., Galofré, V. G., ... & Vidal, E. (2021). Tools for embedding and assessing sustainable development goals in engineering education. *Sustainability, 13*(21), 12154. https://doi.org/10.3390/su132112154

Santika, W. G., Anisuzzaman, M., Bahri, P. A., Shafiullah, G. M., Rupf, G. V., & Urmee, T. (2019). From goals to joules: A quantitative approach of interlinkages between energy and the Sustainable Development Goals. *Energy Research & Social Science, 50*, 201–214. https://doi.org/10.1016/j.erss.2018.11.016

Schroeder, P., Dewick, P., Kusi-Sarpong, S., & Hofstetter, J. S. (2018). Circular economy and power relations in global value chains: tensions and trade-offs for lower income countries. *Resources, Conservation and Recycling, 136*, 77–78. https://doi.org/10.1016/j.resconrec.2018.04.003

Schroeder, P., Anggraeni, K., & Weber, U. (2019). The relevance of circular economy practices to the sustainable development goals. *Journal of Industrial Ecology, 23*(1), 77–95. https://doi.org/10.1111/jiec.12732

Shah, S. H. H., Lei, S., Ali, M., Doronin, D., & Hussain, S. T. (2019). Prosumption: Bibliometric analysis using HistCite and VOSviewer. *Kybernetes*. Retrieved from https://www.emerald.com/insight/content/doi/10.1108/K-12-2018-0696/full/html

Sillanpää, M., & Ncibi, C. (2019). *The circular economy: Case studies about the transition from the linear economy*. Academic Press.

Singer, E., Van Hoewyk, J., & Maher, M. P. (2000). Experiments with incentives in telephone surveys. *Public Opinion Quarterly, 64*(2), 171–188.

Smith, G. (2008). *Does gender influence online survey participation?: A record-linkage analysis of university faculty online survey response behaviour*. ERIC Document Reproduction Service, No. ED 501717.

Statistical office of the Republic of Serbia. (2020). *Average salaries and wages per employee, August 2020*. Retrieved from https://www.stat.gov.rs/en-us/vesti/statisticalrelease/?p=5906&a=24&s=2403?s=2403.

Stephan, A., Muñoz, S., Healey, G., & Alcorn, J. (2020). Analysing material and embodied environmental flows of an Australian university - Towards a more circular economy. *Resources, Conservation and Recycling, 155*, 104632. https://doi.org/10.1016/j.resconrec.2019.104632

Stephens, J. C., Hernandez, M. E., Román, M., Graham, A. C., & Scholz, R. W. (2008). Higher education as a change agent for sustainability in different cultures and contexts. *International Journal of Sustainability in Higher Education, 9*(3), 317–338. https://doi.org/10.1108/14676370810885916

Sultanova, D., Sanger, P. A., & Maliashova, A. (2021). Introducing real-world projects into a chemical technology curricula. In *International Conference on Interactive Collaborative Learning* (pp. 362–370). Springer, Cham.

Symth, C. J. (2006). Environment and education: A view of a changing scene. *Environmental Education Research, 12*(3–4), 247–264. https://doi.org/10.1080/13504620600942642

Tetiana, H., & Malolitneva, V. (2020). Conceptual and legal framework for promotion of education for sustainable development: Case study for Ukraine. *European Journal of Sustainable Development, 9*(2), 42. https://doi.org/10.14207/ejsd.2020.v9n2p42

Toljaga-Nikolić, D., Todorović, M., Dobrota, M., Obradović, T., & Obradović, V. (2020). Project management and sustainability: Playing trick or treat with the planet. *Sustainability, 12*(20), 8619. https://doi.org/10.3390/su12208619

Ünal, E., Urbinati, A., & Chiaroni, D. (2019). Managerial practices for designing circular economy business models: The case of an Italian SME in the office supply industry. *Journal of Manufacturing Technology Management*. Retrieved from https://www.emerald.com/insight/content/doi/10.1108/JMTM-02-2018-0061/full/html

United Nations - UN. (2015). *Transforming our world: The 2030 agenda for sustainable development. General Assembly 70 session*. Retrieved from https://sustainabledevelopment.un.org/post2015/transformingourworld/

United Nations Educational, Scientific and Cultural Organization - UNESCO. (2012). *ESD Sourcebook. Learning & Training Tools*, No. 4. UNESCO. Retrieved from http://unesdoc.unesco.org/images/0021/002163/216383e.pdf

United Nations Educational, Scientific and Cultural Organization - UNESCO. (2017). *Memory of the World - The Wizard of Oz (Victor Fleming 1939), produced by Metro-Goldwyn-Mayer.* Retrieved from http://www.unesco.org/new/en/communication-and-information/memory-of-the-world/register/full-list-of-registered-heritage/registered-heritage-page-8/the-wizard-of-oz-victor-fleming-1939-produced-by-metro-goldwyn-mayer/#c188264

United Nations Educational, Scientific and Cultural Organization – UNESCO. (2019). *Sustainable Development*. Retrieved from https://en.unesco.org/themes/education-sustainable-development/what-is-esd/sd#:~:text=Sustainability%20is%20often%20thought%20of,research%20and%20technology%20transfer%2C%20education

United Nations Educational, Scientific and Cultural Organization - UNESCO - Division of Science, Technical and Environmental Education. (1986). *The balance of life kind: An introduction to the notion of human environment*. Retrieved from http://www.unesco.org/education/pdf/333_3.pdf

United Nations Environment Programme - UNEP. (2006). *Circular economy: An alternative for economic development*. UNEP DTIE.

Von Der Heidt, T., & Lamberton, G. (2011). Sustainability in the undergraduate and postgraduate business curriculum of a regional university: A critical perspective. *Journal of Management & Organization, 17*(5), 670–690.

Van Eck, N. J., & Waltman, L. (2011). Text mining and visualization using VOSviewer. *arXiv preprint arXiv:1109.2058*. Retrieved from https://arxiv.org/abs/1109.2058

Verhoef, L. A., Bossert, M., Newman, J., Ferraz, F., Robinson, Z. P., Agarwala, Y., …, & Hellinga, C. (2020). Towards a learning system for university campuses as living labs for sustainability. In *Universities as living labs for sustainable development* (pp. 135–149). Springer.

Vinuesa, R., Azizpour, H., Leite, I., Balaam, M., Dignum, V., Domisch, S., et al. (2020). The role of artificial intelligence in achieving the sustainable development goals. *Nature Communications, 11*(1), 1–10. https://doi.org/10.1038/s41467-019-14108-y

Wang, H. W., & Van Bueren, J. A. (2018, June). Lessons learned in the paths of developing a multidisciplinary certificate program. In *ASEE Annual Conference and Exposition*, Conference Proceedings (Vol. 2018). Salt Lake City. https://doi.org/10.18260/1-2-90766

Weber, S., Lynes, J., & Young, S. B. (2017). Fashion interest as a driver for consumer textile waste management: Reuse, recycle or disposal. *International Journal of Consumer Studies, 41*(2), 207–215. https://doi.org/10.1111/ijcs.12328

Winfield, F., & Ndlovu, T. (2019). "Future-proof your Degree" Embedding sustainability and employability at Nottingham Business School (NBS). *International Journal of Sustainability in Higher Education*. Retrieved from https://www.emerald.com/insight/content/doi/10.1108/IJSHE-10-2018-0196/full/pdf?title=future-proof-your-degree-embedding-sustainability-and-employability-at-nottingham-business-school-nbs

Word Histories. (2019). *Meaning and Origin of The Phrase 'Yellow Brick Road'*. Retrieved from https://wordhistories.net/2019/08/03/yellow-brick-road/

World Commission on Environment and Development – WCED. (1987). *Our common future*. Oxford University Press.

Yourdictionary. (2020). *Yellow-brick-road definitions*. Retrieved from https://www.yourdictionary.com/yellow-brick-road

Yu, Y., Li, Y., Zhang, Z., Gu, Z., Zhong, H., Zha, Q., et al. (2020). A bibliometric analysis using VOSviewer of publications on COVID-19. *Annals of Translational Medicine, 8*(13), 816. https://doi.org/10.21037/atm-20-4235

Zamarriego, L. (2017). Viaje al centro del ecodiseño. *Circle*, 2. Retrieved from https://www.revistacircle.com/2017/03/08/ecodiseno-michael-braungart/

Chapter 2
Green Human Resource Management in Circular Economy and Sustainability

Jelena Cvijović

Abstract It is already widely known that the world is facing a serious and difficult struggle against climate change and environmental degradation. As a result of continuous efforts of scientists and researchers to find a solution to these problems, two fields developed in the last couple of years: circular economy and green human resource management. Although these two concepts received substantial amount of attention lately, it is clear that circular economy and green human resource management evolved in opposite directions. Referring to the gap existing in literature, the aim of this chapter is to explore the human aspect of circular economy, specifically green human resource management. With the intention of contributing to better understand the human aspect of circular economy, this chapter aims to establish the relationship between circular economy, green human resource management and sustainability.

Keywords Green human resource management · Circular economy · Sustainability

2.1 Introduction

Climate change and environment degradation are burning issues of our modern world. Scientists agree that without a doubt, the main contributors of these problems are human activities (Intergovernmental Panel on Climate Change, 2014). In order to overcome these challenges, European Commission (2019) adopted The European Green Deal – A new growth strategy with the aim of making European Union's economy sustainable, by transforming it "into a modern, resource-efficient and

J. Cvijović (✉)
Faculty of Organizational Sciences, University of Belgrade, Belgrade, Republic of Serbia
e-mail: jc20195012@student.fon.bg.ac.rs

V. Obradović (ed.), *Sustainable Business Change*,
https://doi.org/10.1007/978-3-031-23543-6_2

competitive economy". In 2020, European Commission adopted a new Circular Economy Action Plan which is the main block of the above-mentioned strategy. Successful transition to green economy will require qualified workforce, and in that sense, people will need to acquire special skills, which will allow them to find jobs in the new economy (European Commission, 2020). Although this Action Plan highlights the importance of human resources in this transition, it falls short in defining the actual human resource practices related to transition towards circular economy. The concept of circular economy has become a popular topic in recent years among scientists as well as politicians, governments, policymakers, corporations, etc. For instance, back in 1996, Germany was the first country to legislate "Closed substance cycle and waste management Act" (Matten, 1996), after which Japan also incorporated circular concept into "Basic law for establishing a recycling-based society" in 2000 (Hongo, 2016). The circular economy concept in China is defined by "Circular economy promotion Law on the People's Republic of China" in 2009 (Lieder & Rashid, 2016).

With linear model of economy characterized by over-exploitation of natural resources, vast accumulation of waste and driven only by economic benefits, the world faces serious problems threating to cause irreversible damage to our ecosystem. The concept of circular economy is invented as alternative to current take-make-waste linear model with the aim to minimize the accumulation of waste through design of products and materials that will stay in use longer (The Ellen MacArthur Foundation, 2015). These goals can be interpreted as three main actions in circular economy, better known as 3Rs – reduction, reuse and recycle (Ghisellini et al., 2016). Although the concept of circular economy is in many cases associated closely with recycling, according to Stahel (2014) this action ought to be the last solution, since recycling begins at the end of product life, while the circular philosophy refers to the very beginning of product life. 4R could be found in European Union Waste Framework Directive, with additional R that stands for recover (European Commission, 2008). Some authors extended this 4R framework, adding one or more activities. For example, Sihvonen and Ritola (2015) proposed 6Rs, while Potting et al. (2017) have gone even further with 9R framework. This ambitious idea of transforming world's economy into circular one is supported by numbers. Estimated by Ellen MacArthur Foundation (2015), shifting towards growth within model could decrease European resource spending by 32% or €600 billion, and generate €1.8 trillion in other economic benefits by 2030.

As mentioned earlier, reviewing the available literature in this particular field, it is clear that the aspect of human resource management in circular economy has been largely neglected (Jabbour et al., 2019b). So far, researches have mainly focused their interest in topic such as: production sector (Koh et al., 2017; Winkler, 2011) including eco-design (Mendoza et al., 2017), green public procurement (Liu et al., 2019a, b; Witjes & Lozano, 2016; Zhu et al., 2013), waste management (Luttenberger, 2020; Salmenperä et al., 2021; Tsai et al., 2020), circular economy indicators (Howard et al., 2018; Yadav et al., 2020), limitations of circular economy concept (Korhonen et al., 2018), eco-industrial parks (Martín Gómez et al., 2018; Wang et al., 2020; Wenbo, 2011) and business incubators (Millette et al., 2020).

This chapter will focus on the human aspect of circular economy, specifically new and emerging concept in human resource management – green human resource management. Greening of human resource management can be understood as process that aims at making contribution to protect and preserve the natural resources and environment. Given the growing concern for the global environment, companies around the world started to adopt green strategies with the goal of becoming "green and competitive" (Charbel José Chiappetta Jabbour et al., 2012). With increasing interest for green human resource management, authors argued that incorporating green practices can play an important role in achieving desired organization sustainability (Jackson et al., 2011; Renwick et al., 2015). Yong et al. (2019) also recognized green human resource management as an option for companies to lower their impact on the environment and contribute to sustainable development.

The aim of this chapter is to fill the gap existing in literature concerning human resource aspect of circular economy. Integrating largely separate literatures of green human resource management and circular economy, this chapter could lay the ground for future research. Further, the author will try to answer the following questions: (1) What is the relationship between green human resource management and circular economy business model? (2) How could green human resource practices in circular economy business model contribute to organization sustainability?

This chapter is divided into four sections. The first section discusses the emerging issue of circular economy followed by innovative circular economy business model by reviewing prominent literature in this field. The second section of this chapter is devoted to green human resource management practices. Starting with definitions of green human resource practices proposed by eminent authors in this field, this short review presents also some positive outcomes these practices can have on organizations' performance. The next section briefly analyses the concept of sustainability and sustainable development as it is relevant for exploring further relationship between green human resource practices and circular economy. The fourth section of this chapter explains the relation between green human resources and circular economy, effects of adopting green human resource management practices on organization sustainability performance, as well as effects of adopting green human resource management practices on development of circular economy business model.

2.2 Circular Economy

Circular economy and innovative business model are main concepts from which the idea of circular economy business model originated (Geissdoerfer et al., 2020). Sharing the concept of closed loop system (Homrich et al., 2018), different school of thoughts can be found in the literature, regarding the basis for circular economy. Some of them are cradle to cradle (McDonough & Braungart, 2002), industrial ecology (Graedel & Allenby, 1995), biomimicry (Benyus, 2002), laws on ecology

(Commoner, 1971), blue economy (Pauli, 2010), regenerative design (Lyle, 1996) and permaculture (Mollison & Holmgren 1978). Geissdoerfer et al. (2017) are of the opinion that the most comprehensive definition of circular economy is given by the Ellen MacArthur Foundation (2013), where circular economy is defined as a regenerative system, dominated by the use of renewable energy sources, while the design of products and materials is aimed at eliminating the accumulation of waste. However, there is no universal definition of circular economy in scientific literature while there are many possibilities for defining it (Lieder & Rashid, 2016; Yuan et al., 2006). For example, Kirchherr et al. (2017) identified and analyzed 114 definitions of circular economy.

The Ellen MacArthur Foundation (2015), leading organization in developing and promoting the idea of circular economy, identified three main principles on which the circular economy concept rests: (1) preserving and enhancing natural capital – the core of circular economy concept is preserving natural capital from further detriment, by delivering utility virtually. When this is not achievable, circular concept will select technologies based on renewable or better-performing resources. (2) Optimizing resource yields – the circular economy system allows products, components and materials to circulate within technical and biological circles. Technical circle is enabled through design that supports remanufacturing, recycling and refurbishing, while biological circle is conceived to enable the recycling of biological materials and returning them to natural. (3) Fostering system effectiveness – using circular economy as a tool for discovering and decreasing negative impacts of production and consumption on human utility and environment.

2.2.1 Circular Economy Business Model

While the term business model has been in use for over sixty years (Bellman et al., 1957), only with development of information technology and creation of electronic businesses this concept started gaining popularity in scientific literature (Wirtz et al., 2016). Good business model is crucial for every successful organization (Magretta, 2002) given that it defines the way a company does business and represents a significant initiator for innovation (Chesbrough, 2010; Teece, 2010). There are two approaches to business model innovation, designing a completely new business model or recomposing the present business model (Zott et al., 2011). According to Chesbrough (2010), dissimilar business models introducing the same idea or technology will produce different financial result; therefore, it is clear that it is of great importance that organizations create possibilities for innovation. Further, as Teece (2010) observes, business models should be designed to "capture value from innovation", because developing new products and technologies without commercialization strategy can lead to economic and profit failure.

Traditionally, the concept of value management relies heavily on creating and capturing value for organizations and customers (Osterwalder & Pigneur, 2010). One of the most mentioned business model frameworks in literature is developed by

Richardson (2008), and consists of value proposition, value creation and delivery system and value capture. Authors studying the field of sustainable business models stress the need to go beyond traditional meaning of value and include all stakeholders' interests, especially the society and environment (Bocken et al., 2014; Yang et al., 2017). Depending on the authors and literature, circular business model can be categorized under sustainable business model (Bocken et al., 2014). Although circular business model is considered to be one archetype or sub-category of sustainable business model, this model has additional characteristics that cannot be classified as sustainable (Geissdoerfer et al., 2018).

According to Geissdoerfer et al. (2020), the term circular business model was introduced in an article by Schwager and Moser (2006) that investigated the development and implementation of specific business model that would support sustainable industrial development. Although the concept of circular business model appeared earlier, it was only in 2015 that it began to gain popularity when exponential growth in published works began to show (Geissdoerfer et al., 2020). Reviewing the available literature, it can be understood that circular business model concept evolved from research fields of closed-loop supply chain (Atasu et al., 2008; Kumar & Malegeant, 2006), sustainable product-service systems (Tukker, 2015) and industrial ecology (Lifset & Graedel, 2002).

ReSOLVE framework developed by Ellen MacArthur Foundation (2015) provides set of circular strategies for companies wanting to move towards new business model. In order to make shift from linear to circular economy, this framework includes the following six actions for successful transition:

1. *REgenerate* – Uses renewable energy and recovers natural resources
2. *Share* – Extends product lifespan through repair, refurbish or remanufacture; maximizes the use of a product by selling it to another consumer; maximizes the utilization of a product through sharing
3. *Optimize* – Minimizes the use of natural resources and raw materials in production process; eliminates the waste throughout the lifespan of a product; maximizes the efficiency of the product
4. *Loop* – Keeps products and materials in closed-loops longer; responsible product use has an advantage over product lifecycle extension, while recycle ought to be the last option
5. *Virtualize* – Delivers the same product or service in a dematerialized state
6. *Exchange* – Uses new technologies and innovative materials and designs

Relying on classification provided by Rosa et al. (2019), previous research work done in the field of circular business models could be divided based on the starting point various authors used in their research. In that respect, Charbel Jose Chiappetta Jabbour et al. (2019a) investigated the relationship between circular business model, based on ReSolve framework, and large-scale data. Similarly, Manninen et al. (2018) developed environmental value proposition table based on ReSolve framework, which organizations can use to confirm the environmental benefits of their business model. Some authors focused on Business Model Canvas proposed by Osterwalder and Pigneur (2010) and used this model to design a business model for circular economy (Lewandowski, 2016). Bocken et al. (2014) developed sustainable business model archetypes to cluster mechanisms and solutions for better circular model design, starting from Richardson's (2008) business model. Reviewing the literature in this particular field it is clear that the human resource aspect of circular economy and circular economy business model has been neglected which is in line with previous research (Jabbour et al., 2019b).

2.3 Green Human Resource Management

Wehrmeyer (1996) was one of the first to address the importance of human resource management in environmental management. He puts human resources in a role of supporter of organization's environmental activities and classifies human resource management functions in three categories: supply competent staff, management of staff and promote organizational dynamics. The term green human resources management first appeared in a study by Renwick et al. (2008) which paved the way for future research in this field. In the following years, researchers began in greater amount to include environmental aspect in human resources in their studies.

2.3.1 Green Human Resource Management Practices

Green human resource management relies on basic human resource practices such as recruitment and selection, training and development, performance management and appraisal, pay and rewards system and organizational culture (Charbel José Chiappetta Jabbour & de Sousa Jabbour, 2016; Charbel José Chiappetta Jabbour & Santos, 2008b; Renwick et al., 2015). These green human resource management practices are defined below.

Attracting and hiring competent staff represent a starting point in human resources management. Green recruitment and selection of employees who are dedicated and sensitive to environment issues increases the possibility that future employees' performance and behaviour will follow organization's green strategy (Tang et al., 2018). Some studies show that applicants are more interested in working for pro-environmental organization (Bauer & Aiman-Smith, 1996) with greater odds for accepting job offer based on company's ecological rating (Aiman-Smith et al., 2001) and concern for environment (Greening & Turban, 2000). Although it is common to find in literature recruitment and selection described in the same context, selection process can be viewed separately. Pham and Paillé (2019) identified selection as a method of determining and assessing applicants' environmental awareness, ecological principles, as well as susceptibility to environmental problems.

Training and education were the topics of studies conducted in the 1990s where researches concluded that environmental training and education are essential to organization management approach and commercial success of business (Marshall

& Mayer, 1992; Hale, 1995; Venselaar, 1995). For employees to be able to accept progressive environmental practices, Sarkis et al. (2010) are of the opinion that environmental training plays an important role in that process. Tang et al. (2018) indicated three aspects of green training: awareness enhancement, knowledge management and climate building. Paillé et al. (2020) explored the effect green human resource management practices have on achieving environmental goals and concluded that training is the best green practice that will motivate employees to make effort towards green objectives.

Tang et al. (2018) are of the opinion that organizations should introduce green performance management standard and therefore set green targets for employees, form green performance index for measurement of employees' green results and use dis-benefits. Negative reinforcements are intended for employees who fail in following organization's green practice. However, warning or suspension may not be the best approach, as Renwick et al. (2013) argue, utilization of negative reinforcements may not motivate or educate employees enough to change their environmental behaviour.

Compensation and reward system is an important tool which aligns corporate objectives and individual interests of employees through rewarding for their good performance (Ahmad & Nisar, 2015). The main reason for providing compensation is to "attract, retain and motivate employees" (Mondy & Noe, 2005). Common practices for rewarding employees for their good environmental performance could be monetary incentives in form of bonuses or salary increase and non-monetary rewards like praise and recognition as well as special benefits like work from home or flexible work hours, while special rewards should be reserved for employees who show advanced environmental initiatives (Hosain & Rahman, 2016).

Besides these most common green human resource practices found in the literature, it is important to mention a couple of others that any pro-environmental organization can adopt. Although individually employees can contribute to organizations' environmental performance, better effect is achieved through teamwork (Daily & Huang, 2001). In that sense, special team are being formed in eco-oriented organizations known as green teams (Charbel José Chiappetta Jabbour et al., 2013). These green teams could be formed on voluntary or obligatory basis (Al Kerdawy, 2018) and could be functional or cross-functional, depending on whether the team members are working in the same or different organizational unit (Charbel José Chiappetta Jabbour et al., 2013). Survey of 94 Brazilian companies with ISO14001 certification revealed that cross-functional teams are more represented than functional team, and confirmed that green teams are important practices in environmental management (Charbel José Chiappetta Jabbour et al., 2013). Providing safe and healthy workplace for employees is the obligation of every organization. Further, employees are more eager to work for a company that provides safe and healthy workplace, and therefore more satisfied with their jobs and have a sense of security (Amrutha & Geetha, 2020). The main role of green health and safety management, according to Hosain and Rahman (2016), is to supply green workplace for everybody, the place that is eco-conscious, socially responsible and resource-efficient (The Society for Human Resource Management, 2009).

2.4 Sustainability and Sustainable Development

Although the idea of sustainability according to some sources (Geissdoerfer et al., 2017) dates back to early eighteenth-century "Sylvicultura oeconomica" (von Carlowitz, 1713), not until the World Commission on Environment and Development (also known as Brundtland Commission) released a publication with their results in 1987 (known as the Brundtland Report), sustainability concept started gaining popularity and interest worldwide. It was actually this report that provided the most commonly used definition of sustainability, emphasising the need to support the development of society to a point that does not endanger natural resources, and ensuring human society and its environment's a long-term sustainability (World Commission on Environment and Development, 1987). This report also encompasses the goals and strategies for sustainable development. In the following 20 years after the Brundtland Report, Johnston et al. (2007) estimated that about three hundred definitions of sustainability and sustainable development emerged, mostly in environmental management and associated disciplines research.

Ubiquitous concept of sustainability in scientific and other literature can be found in form of three pillars supporting sustainable development – economic, environmental and social, informally known as profit, planet and people. The three pillars are designed to complement each other, be equally represented and not mutually exclusive. Origins of this concept cannot be attributed to someone specific but continuous academic research of economy from environmental and social perspective together with strong efforts of United Nations to decuple economic growth from environment degradation and social injustice (Purvis et al., 2018) led to its gradual development. Traditionally observing, rapid economic growth leads to large production, distribution and consumption, creating negative impact on environment and society. Massive exploitation of finite natural resources on one side and inevitable increase in demand for natural resources on the other, created a need for economic sustainability. Designing a system of production that can keep up with current consumption needs, without jeopardizing future demands presents the core of economic sustainability (Lobo et al., 2015). Economic pillar encompasses group of practices underpinning economic growth with consideration for environmental and social aspect of sustainability (Zhai & Chang, 2019). According to Morelli (2011), environmental sustainability could be defined as responsible utilization of natural resources to a point that satisfies the human needs but at the same time not exceeding its capacities. Negative effects of climate change could be already seen in ecosystem, water resources, food and human health, rising of sea levels and oceans becoming more acidic (Du & Kang, 2016). All of this dangerous change makes a compelling argument for environmental sustainability. Third pillar of sustainability received substantially less attention compared to economic or environmental sustainability (Amrutha & Geetha, 2020). The basis for social sustainability in organizations lies in the first six of the United Nations Global Principles framed in two words: human rights and labour. There is obvious disbalance between three pillars of sustainability, more precisely, the social pillar is the least represented and poorly treated compared to the other two (Bubicz et al., 2019).

2.5 Green Human Resource Management, Circular Business Model and Sustainable Performance

Human resources management practices are considered essential for successful implementation of environmental management, aligning organizations' green goals with human resources strategy and practices, and thus helping organization in achieving desired environmental performance (Charbel José Chiappetta Jabbour & Santos, 2008a). Renwick et al. (2013) suggested that implementation of green human resource systems as a whole could have greater impact on environmental performance rather than individual green human resource practices, as it was done in the previous researches. Apart from advantages such as employee retention and organizations' good reputation, that green culture brings along, it is noted that green human resource practices can reduce cost and increase sales (Mehta & Chugan, 2015). Green practices, such as recruitment and selection, can attract competent and qualitative employees who are interested working for environmentally aware organization (Linnenluecke & Griffiths, 2010), and as a result enhance organizations' financial performance. Although there are a certain number of studies concerning link between green human resource practices and organizations' sustainability performance, more precisely the environmental and economic dimension of sustainability, social aspect of it is under-researched and represents the weakest pillar of sustainability (Saeed et al., 2019). Amrutha and Geetha (2020) proposed a model connecting green human resource management and social aspect of sustainability, with mediating role of "employee green behaviour at workplace". These two authors are of the opinion that green human resource practices (such as: recruitment, training, appraisal, rewards and employee involvement) positively affect organizations' social performance, with green behaviour as a mediator.

Based on reviewed literature, specific area of interest for researchers is green supply chain management. This concept is becoming more and more popular for companies who want to boost their environmental performance (Testa & Iraldo, 2010). Authors investigated links between human resources management and green supply chain management (Charbel José Chiappetta Jabbour & de Sousa Jabbour, 2016; Nejati et al., 2017) and their studies show that human resources play important part in making supply chains more sustainable. Nejati et al. (2017) study confirmed that biggest effect on supply chain management has green empowerment and green training and development. Longoni et al. (2016) confirmed the hypotheses mentioned earlier by Renwick et al. (2013), that green human management practices together with green supply chain management will have positive effect on environmental and financial performance, and greater impact is achievable when both systems are implemented together.

Jabbour et al. (2019b) developed an integrative framework capturing the main relationship between green human resource practices and circular economy, arguing that the adoption of green practices in workplace, such as recruitment, selection, training and rewards, may have positive impact on the implementation of circular economy organizational strategy. According to Jabbour et al. (2019b), the basis for integration of green human resource management and circular economy business model lies in two organizational theories – stakeholder theory and resource-based view. On one hand, stakeholder theory emphasizes the linked interactions that exist between an organization and its employees, buyers, suppliers, shareholders and all other interested parties that have a stake at the company (Freeman, 1984). On the other hand, the resource-based view of the firm recognized human resources as an important and irreplaceable organization's resource for achieving strategic competitive advantage (Wright et al., 1994; Kamoche, 1996), while Hart (1995) extended this theory into natural resource-based view, adding natural environment into resource-based view of the firm, and thus argued that sustainability strategies can positively influence organization performance.

Although the concept of circular economy is widely recognized as economic model with primary goal of achieving economic growth without harming the environment, there are some evidences that can be supportive of circular economy contributing to sustainable development. Bocken et al. (2014) referred to closed loop business model as one of the archetypes of sustainable business model. Given that circular business models through innovative solutions like new technology, new product and process designs (The Ellen McArthur Foundation, 2015) can influence environmental organizational performance, it is possible to argue that circular economy can benefit overall sustainable development. Similarly, Zhu et al. (2010) based on the study involving Chinese manufacturing companies concluded that both circular economy practices and circular economy targeted performance can be related to economic and environmental sustainability.

2.6 Discussion and Conclusion

The literature concerning green human resource management and circular economy developed noticeably in the past couple of years. Accelerated development of these two fields can be contributed to burning issues facing humanity such as climate change and environment degradation. However, observing the great amount of literature concerning green human resource management and circular economy, it becomes obvious that these two topics developed in separate directions. Researches focused their attention mostly to technical aspects of circular economy and the human resource aspect remained largely unexplored. There is a visible gap in scientific research related to the role of human resources in development of circular economy confirmed also by Jabbour et al. (2019b).

This chapter represents an attempt towards bridging the gap existing in current scientific as well as other relevant literature mentioned earlier, by connecting two fields – green human resource management and circular economy. More specifically, this chapter explores the relationship between green human resource management and circular economy and identifies the possible effects of implementing green human resource practices in circular economy business model on organization's sustainability.

In order to discover this possible joint relationship of green human resource practices and circular economy and their effects on organization sustainable performance, sustainable supply chains have an important role. Organizations interested in boosting their sustainable performance are considering "greening" the supply chain. Examining the literature concerning the topic of supply chains, it is revealed that human resources play an important part in making supply chains sustainable. Empirical study conducted by Nejati et al. (2017) confirmed this positive effect of green human resources and marked off green empowerment and green training and development as green practices with the greatest effect on sustainability of supply chains. Based on the knowledge that the concept of circular economy is closely associated with sustainable supply chains (Koh et al., 2017), it could be argued that green human resource management can have a positive effect on circular economy business model.

Referring to previous research regarding interactions between green human resources and sustainability, it has been established that green human resource practices have substantial impact on organization's sustainability. Green human resource practices such as recruitment, selection, training, development, appraisal and rewards positively affect organizational sustainable performance. It is also worth mentioning, that greater effect could be achieved if these green practices are implemented as a green human resource system rather than individually.

This chapter has several limitations. First, it is based on secondary data source. Finding and accessing relevant articles was done using academic databases and search engines with key words "circular economy", "green human resource management" and "sustainability". Although it was an intention to encompass wider range of academic articles, some of them were not accessible. Also, snowballing technique was applied which itself has certain shortcomings. Second, only the most common green human resource practices found in articles are selected and further explained. Third, this research is limited to effect of green human resource practices on circular economy development, mostly due to lack of literature for exploration of a reverse relationship.

References

Ahmad, S., & Nisar, T. (2015). Green Human Resource Management: Policies and practices. *Cogent Business & Management, 2*(1). https://doi.org/10.1080/23311975.2015.1030817

Aiman-Smith, L., Bauer, T. N., & Cable, D. M. (2001). Are you attracted? Do you intend to pursue? A recruiting policy-capturing study. *Journal of Business and Psychology, 16*(2), 219–237. https://doi.org/10.1023/A:1011157116322

Al Kerdawy M. M. A. (2018). The role of corporate support for employee volunteering in strengthening the impact of green human resource management practices on corporate social responsibility in the Egyptian firms. *European Management Review, 16*, 1079–1095. https://doi.org/10.1111/emre.12310

Amrutha, V. N., & Geetha, S. N. (2020). A systematic review on green human resource management: Implications for social sustainability. *Journal of Cleaner Production, 247*, 119131. https://doi.org/10.1016/j.jclepro.2019.119131

Atasu, A., Guide, V. D. R., & Van Wassenhove, L. N. (2008). Product reuse economics in closed-loop supply chain research. *Production and Operations Management, 17*(5), 483–496. https://doi.org/10.3401/poms.1080.0051

Bauer, T. N., & Aiman-Smith, L. (1996). Green career choices: The influence of ecological stance on recruiting. *Journal of Business and Psychology, 10*(4), 445–458. https://doi.org/10.1007/BF02251780

Bellman, R., Clark, C. E., Malcolm, D. G., Craft, C. J., & Ricciardi, F. M. (1957). On the construction of a multi-stage, multi-person business game. *Operations Research, 5*(4), 469–503.

Benyus, J. (2002). Biomimicry: Innovation inspired by design.

Bocken, N. M. P., Short, S. W., Rana, P., & Evans, S. (2014). A literature and practice review to develop sustainable business model archetypes. *Journal of Cleaner Production, 65*, 42–56. https://doi.org/10.1016/j.jclepro.2013.11.039

Bubicz, M. E., Barbosa-Póvoa, A. P. F. D., & Carvalho, A. (2019). Incorporating social aspects in sustainable supply chains: Trends and future directions. *Journal of Cleaner Production, 237*, 117500. https://doi.org/10.1016/j.jclepro.2019.06.331

Chesbrough, H. (2010). Business model innovation: Opportunities and barriers. *Long Range Planning, 43*(2–3), 354–363. https://doi.org/10.1016/j.lrp.2009.07.010

Commoner, B. (1971). *The closing circle: Nature, man, and technology* (pp. 11–44). Knopf Press.

Communication from the Commission to the European Parliament, The European Council, The Council, The European Economic and Social Committee and The Committee of the Regions. The European Green Deal COM/2019/640 final.

Communication from the Commission to the European Parliament, The Council, The European Economic and Social Committee and The Committee of the Regions. A new Circular Economy Action Plan For a cleaner and more competitive Europe COM/2020/98 final.

Daily, B. F. & Huang, S. (2001). Achieving sustainability through attention to human resource factors in environmental management. *International Journal of Operations & Production Management, 21*, 1539–1552. http://dx.doi.org/10.1108/01443570110410892

Du, Q., & Kang, J. T. (2016). Tentative ideas on the reform of exercising state ownership of natural resources: Preliminary thoughts on establishing a state-owned natural resources supervision and administration commission. *Jiangxi Social Science, 6*, 160.

Ellen MacArthur Foundation (EMF). (2013). *Towards the circular economy* (Vol. 1). Isle of Wight.

Ellen MacArthur Foundation (EMF). (2015). *Growth within: A circular economy vision for a competitive Europe*. Ellen MacArthur Foundation.

European Commission. (2008). *Directive 2008/98/EC of the European Parliament and of the Council of 19 November 2008 on Waste and Repealing Certain Directives*. Available at: https://eur-lex.europa.eu/legal-content/EN/TXT/?uri=CELEX:32008L0098

Freeman, E. (1984). *Strategic management: A stakeholder approach*. Pitman.

Geissdoerfer, M., Savaget, P., Bocken, N. M. P., & Hultink, E. J. (2017). The Circular Economy – A new sustainability paradigm? *Journal of Cleaner Production, 143*, 757–768. https://doi.org/10.1016/j.jclepro.2016.12.048

Geissdoerfer, M., Vladimirova, D., & Evans, S. (2018). Sustainable business model innovation: A review. *Journal of Cleaner Production, 198*, 401–416. https://doi.org/10.1016/j.jclepro.2018.06.240

Geissdoerfer, M., Pieroni, M. P. P., Pigosso, D. C. A., & Soufani, K. (2020). Circular business models: A review. *Journal of Cleaner Production, 277*, 123741. https://doi.org/10.1016/j.jclepro.2020.123741

Ghisellini, P., Cialani, C., & Ulgiati, S. (2016). A review on circular economy: The expected transition to a balanced interplay of environmental and economic systems. *Journal of Cleaner Production, 114*, 11–32. https://doi.org/10.1016/j.jclepro.2015.09.007

Graedel, T. E., & Allenby, B. R. (1995). *Industrial ecology*. Prentice Hall.

Greening, D. W., & Turban, D. B. (2000). Corporate social performance as a competitive advantage in attracting a quality workforce. *Business and Society, 39*, 254–280.

Hale, M. (1995). Training for environmental technologies and environmental management. *Journal of Cleaner Production, 3*(1–2), 19–23.

Hart, S. L. (1995). A natural-resource-based view of the firm. *The Academy of Management Review, 20*(4), 986–1014.

Homrich, A. S., Galvão, G., Abadia, L. G., & Carvalho, M. M. (2018). The circular economy umbrella: Trends and gaps on integrating pathways. *Journal of Cleaner Production, 175*, 525–543. https://doi.org/10.1016/j.jclepro.2017.11.064

Hongo, T. (2016). Circular Economy Potential and Public–Private Partnership Models in Japan. In V. Anbumozhi & J. Kim (Eds.), *Towards a circular economy: Corporate management and policy pathways* (ERIA Research Project Report 2014-44) (pp. 17–29). ERIA.

Hosain, S., & Rahman, S. (2016). Green human resource management: A theoretical overview. *IOSR Journal of Business and Management, 18*(6), 54–59.

Howard, M., Hopkinson, P., & Miemczyk, J. (2018). The regenerative supply chain: A framework for developing circular economy indicators. *International Journal of Production Research, 57*(23), 7300–7318. https://doi.org/10.1080/00207543.2018.1524166

IPCC. (2014). *Climate Change 2014: Synthesis Report. Contribution of Working Groups I, II and III to the Fifth Assessment Report of the Intergovernmental Panel on Climate Change* [Core Writing Team, R.K. Pachauri and L.A. Meyer (Eds.)]. IPCC, 151 pp.

Jabbour, C. J. C., & de Sousa Jabbour, A. B. L. (2016). Green human resource management and green supply chain management: Linking two emerging agendas. *Journal of Cleaner Production, 112*, 1824–1833. https://doi.org/10.1016/j.jclepro.2015.01.052

Jabbour, C. J. C., & Santos, F. C. A. (2008a). The central role of human resource management in the search for sustainable organizations. *The International Journal of Human Resource Management, 19*(12), 2133–2154. https://doi.org/10.1080/09585190802479389

Jabbour, C. J. C., & Santos, F. C. A. (2008b). Relationships between human resource dimensions and environmental management in companies: Proposal of a model. *Journal of Cleaner Production, 16*(1), 51–58. https://doi.org/10.1016/j.jclepro.2006.07.025

Jabbour, C. J. C., Jabbour, A. B. L. D. S., Teixeira, A. A., & de Souza Freitas, W. R. S. (2012). Environmental development in Brazilian companies: The role of human resource management. *Environmental Development, 3*, 137–147. https://doi.org/10.1016/j.envdev.2012.05.004

Jabbour, C. J. C., Santos, F. C. A., Fonseca, S. A., & Nagano, M. S. (2013). Green teams: Understanding their roles in the environmental management of companies located in Brazil. *Journal of Cleaner Production, 46*, 58–66. https://doi.org/10.1016/j.jclepro.2012.09.018

Jabbour, C. J. C., de Sousa Jabbour, A. B. L., Sarkis, J., & Filho, M. G. (2019a). Unlocking the circular economy through new business models based on large-scale data: An integrative framework and research agenda. *Technological Forecasting and Social Change, 144*, 546–552. https://doi.org/10.1016/j.techfore.2017.09.010

Jabbour, C. J., Sarkis, J., Jabbour, A. B., Renwick, D., Singh, S., Grebinevych, O., Kruglianskas, I., & Filho, M. G. (2019b). Who is in charge? A review and a research agenda on the 'human side' of the circular economy. *Journal of Cleaner Production, 222*, 793–801.

Jackson, S. E., Renwick, D. W. S., Jabbour, C. J. C., & Muller-Camen, M. (2011). State-of-the-art and future directions for green human resource management: Introduction to the special issue. *German Journal of Human Resource Management, 25*(2), 99–116. https://doi.org/10.1688/1862-0000

Johnston, P., Everard, M., Santillo, D., & Robert, K. H. (2007). Reclaiming the definition of sustainability. *Environmental Science and Pollution Research International, 14*(1), 60–66. https://doi.org/10.1065/espr2007.01.375

Kamoche, K. (1996). Strategic human resource management within a resource-capability view of the firm. *Journal of Management Studies, 33*(2), 213–233.

Kirchherr, J., Reike, D., & Hekkert, M. (2017). Conceptualizing the circular economy: An analysis of 114 definitions. *Resources, Conservation and Recycling, 127*, 221–232. https://doi.org/10.1016/j.resconrec.2017.09.005

Koh, S. C. L., Gunasekaran, A., Morris, J., Obayi, R., & Ebrahimi, S. M. (2017). Conceptualizing a circular framework of supply chain resource sustainability. *International Journal of Operations & Production Management, 37*(10), 1520–1540. https://doi.org/10.1108/ijopm-02-2016-0078

Korhonen, J., Honkasalo, A., & Seppälä, J. (2018). Circular economy: The concept and its limitations. *Ecological Economics, 143*, 37–46. https://doi.org/10.1016/j.ecolecon.2017.06.041

Kumar, S., & Malegeant, P. (2006). Strategic alliance in a closed-loop supply chain, a case of manufacturer and eco-non-profit organization. *Technovation, 26*(10), 1127–1135. https://doi.org/10.1016/j.technovation.2005.08.002

Lewandowski, M. (2016). Designing the business models for circular economy—Towards the conceptual framework. *Sustainability, 8*(1), 43. https://doi.org/10.3390/su8010043

Lieder, M., & Rashid, A. (2016). Towards circular economy implementation: A comprehensive review in context of manufacturing industry. *Journal of Cleaner Production, 115*, 36–51. https://doi.org/10.1016/j.jclepro.2015.12.042

Lifset, R., & Graedel, T. E. (2002). Industrial ecology: Goals and definitions. In R. U. Ayres & L. Ayres (Eds.), *A handbook of industrial ecology*. Edward Elgar Publishing; ISBN 978-1-84376-547-9.

Linnenluecke, M. K., & Griffiths, A. (2010). Corporate sustainability and organizational culture. *Journal of World Business, 45*(4), 357–366. https://doi.org/10.1016/j.jwb.2009.08.006

Liu, J., Shi, B., Xue, J., & Wang, Q. (2019a). Improving the green public procurement performance of Chinese local governments: From the perspective of officials' knowledge. *Journal of Purchasing and Supply Management, 25*(3), 100501. https://doi.org/10.1016/j.pursup.2018.05.002

Liu, J., Xue, J., Yang, L., & Shi, B. (2019b). Enhancing green public procurement practices in local governments: Chinese evidence based on a new research framework. *Journal of Cleaner Production, 211*, 842–854. https://doi.org/10.1016/j.jclepro.2018.11.151

Lobo, M.-J., Pietriga, E., & Appert, C. (2015). An evaluation of interactive map comparison techniques. In *Proceedings of the 33rd Annual ACM Conference on Human Factors in Computing Systems - CHI '15* (pp. 3573–3582). ACM Press. https://doi.org/10.1145/2702123.2702130

Longoni, A., Luzzini, D., & Guerci, M. (2016). Deploying environmental management across functions: The relationship between green human resource management and green supply chain management. *Journal of Business Ethics, 151*(4), 1081–1095. https://doi.org/10.1007/s10551-016-3228-1

Luttenberger, L. R. (2020). Waste management challenges in transition to circular economy – Case of Croatia. *Journal of Cleaner Production, 256*, 120495. https://doi.org/10.1016/j.jclepro.2020.120495

Lyle, J. T. (1996). *Regenerative design for sustainable development*. John Wiley & Sons.

Magretta, J. (2002). Why business models matter. *Harvard Business Review, 80*(5), 86–92.

Manninen, K., Koskela, S., Antikainen, R., Bocken, N., Dahlbo, H., & Aminoff, A. (2018). Do circular economy business models capture intended environmental value propositions? *Journal of Cleaner Production, 171*, 413–422. https://doi.org/10.1016/j.jclepro.2017.10.003

Marshall, M. E., & Mayer, D. W. (1992). Environmental training: it's good business. *Business Horizons, 35*(2), 54–57.

Martín Gómez, A. M., Aguayo González, F., & Marcos Bárcena, M. (2018). Smart eco-industrial parks: A circular economy implementation based on industrial metabolism. *Resources, Conservation and Recycling, 135*, 58–69. https://doi.org/10.1016/j.resconrec.2017.08.007

Matten, D. (1996). Enforcing sustainable development by legislation: Entrepreneurial consequences of the new German Waste Management Act. *Sustainable Development, 4*(3), 130–137. https://doi.org/10.1002/(SICI)1099-1719(199612)4:33.0.CO;2-#

McDonough, W., & Braungart, M. (2002). *Cradle to cradle: Remaking the way we make things* (1st ed.). North Point Press.

Mehta, K., & Chugan, P. K. (2015). Green HRM in pursuit of environmentally sustainable business. *Universal Journal of Industrial and Business Management, 3*(3), 74–81. https://doi.org/10.13189/ujibm.2015.030302

Mendoza, J. M. F., Sharmina, M., Gallego-Schmid, A., Heyes, G., & Azapagic, A. (2017). Integrating backcasting and eco-design for the circular economy: The BECE framework. *Journal of Industrial Ecology, 21*(3), 526–544. https://doi.org/10.1111/jiec.12590

Millette, S., Eiríkur Hull, C., & Williams, E. (2020). Business incubators as effective tools for driving circular economy. *Journal of Cleaner Production, 266*, 121999. https://doi.org/10.1016/j.jclepro.2020.121999

Mollison, B., & Holmgren, D. (1978). *Permaculture one. A perennial agriculture for human settlements*. Corgi Books, London, UK & Melbourne, Australia. Transworld Publishers. ISBN:0-552-98075-7

Mondy, R. W., & Noe, R. M. (2005). *Human resource management* (9th ed.). Prentice Hall.

Morelli, J. (2011). Environmental sustainability: A definition for environmental professionals. *Journal of Environmental Sustainability, 1*(1), Article 2. https://doi.org/10.14448/jes.01.0002

Nejati, M., Rabiei, S., & Chiappetta Jabbour, C. J. (2017). Envisioning the invisible: Understanding the synergy between green human resource management and green supply chain management in manufacturing firms in Iran in light of the moderating effect of employees' resistance to change. *Journal of Cleaner Production, 168*, 163–172. https://doi.org/10.1016/j.jclepro.2017.08.213

Osterwalder, A., & Pigneur, Y. (2010). *Business model generation: A handbook for visionaries, game changers, and challengers* (Vol. 5). John Wiley & Sons.

Paillé, P., Valéau, P., & Renwick, D. W. (2020). Leveraging green human resource practices to achieve environmental sustainability. *Journal of Cleaner Production, 260*, 121137. https://doi.org/10.1016/j.jclepro.2020.121137

Pauli, G. A. (2010). *The blue economy: 10 years, 100 innovations, 100 million jobs*. Paradigm Publications.

Pham, D. D. T., & Paillé, P. (2019). Green recruitment and selection: An insight into green patterns. *International Journal of Manpower, 41*(3), 258–272. https://doi.org/10.1108/ijm-05-2018-0155

Potting, J., et al. (2017). *Circular economy: Measuring innovation in the product chain*. Available at: https://www.pbl.nl/sites/default/files/downloads/pbl-2016-circular-economy-measuring-innovation-in-product-chains-2544.pdf

Purvis, B., Mao, Y., & Robinson, D. (2018). Three pillars of sustainability: In search of conceptual origins. *Sustainability Science, 14*(3), 681–695. https://doi.org/10.1007/s11625-018-0627-5

Renwick, D., Redman, T., & Maguire, S. (2008). *Green HRM: A review, process model, and research agenda*. University of Sheffield Management School Discussion Paper No 2008.01

Renwick, D. W. S., Redman, T., & Maguire, S. (2013). Green human resource management: A review and research agenda. *International Journal of Management Reviews, 15*, 1–14. https://doi.org/10.1111/j.1468-2370.2011.00328.x

Renwick, D. W. S., Jabbour, C. J. C., Muller-Camen, M., Redman, T., & Wilkinson, A. (2015). Contemporary developments in Green (environmental) HRM scholarship. *The International Journal of Human Resource Management, 27*(2), 114–128. https://doi.org/10.1080/09585192.2015.1105844

Richardson, J. (2008). The business model: An integrative framework for strategy execution. *Strategic Change, 17*(5–6), 133–144. https://doi.org/10.1002/jsc.821

Rosa, P., Sassanelli, C., & Terzi, S. (2019). Towards Circular Business Models: A systematic literature review on classification frameworks and archetypes. *Journal of Cleaner Production, 236*, 117696. https://doi.org/10.1016/j.jclepro.2019.117696

Saeed, B. B., Afsar, B., Hafeez, S., Khan, I., Tahir, M., & Afridi, M. A. (2019). Promoting employee's proenvironmental behavior through green human resource management practices.

Corporate Social Responsibility and Environmental Management, 26(2), 424–438. https://doi.org/10.1002/csr.1694

Salmenperä, H., Pitkänen, K., Kautto, P., & Saikku, L. (2021). Critical factors for enhancing the circular economy in waste management. *Journal of Cleaner Production, 280*, 124339. https://doi.org/10.1016/j.jclepro.2020.124339

Sarkis, J., Gonzalez-Torre, P., & Adenso-Diaz, B. (2010). Stakeholder pressure and the adoption of environmental practices: The mediating effect of training. *Journal of Operations Management, 28*(2), 163–176. https://doi.org/10.1016/j.jom.2009.10.001

Schwager, P., & Moser, F. (2006). The application of chemical leasing business models in Mexico. *Environmental Science and Pollution Research, 13*(2), 131–137. https://doi.org/10.1065/espr2006.02.294

Sihvonen, S., & Ritola, T. (2015). Conceptualizing ReX for aggregating end-of-life strategies in product development. *Procedia CIRP, 29*, 639–644. https://doi.org/10.1016/j.procir.2015.01.026

Society for Human Resource Management (2009). Understanding the factors that make work gratifying

Stahel, W. R. (2014). *Reuse is the key to the circular economy*. Available at: https://ec.europa.eu/environment/ecoap/about-eco-innovation/experts-interviews/reuse-is-the-key-to-the-circular-economy_en

Tang, G., Chen, Y., Jiang, Y., Paillé, P., & Jia, J. (2018). Green human resource management practices: Scale development and validity. *Asia Pacific Journal of Human Resources, 56*(1), 31–55. https://doi.org/10.1111/1744-7941.12147

Teece, D. J. (2010). Business models, business strategy and innovation. *Long Range Planning, 43*(2–3), 172–194. https://doi.org/10.1016/j.lrp.2009.07.003

Testa, F., & Iraldo, F. (2010). Shadows and lights of GSCM (Green Supply Chain Management): Determinants and effects of these practices based on a multi-national study. *Journal of Cleaner Production, 18*(10–11), 953–962. https://doi.org/10.1016/j.jclepro.2010.03.005

Tsai, F. M., Bui, T.-D., Tseng, M.-L., Lim, M. K., & Hu, J. (2020). Municipal solid waste management in a circular economy: A data-driven bibliometric analysis. *Journal of Cleaner Production, 275*, 124132. https://doi.org/10.1016/j.jclepro.2020.124132

Tukker, A. (2015). Product services for a resource-efficient and circular economy – A review. *Journal of Cleaner Production, 97*, 76–91. https://doi.org/10.1016/j.jclepro.2013.11.049

Venselaar, J. (1995). Environmental training: Industrial needs. *Journal of Cleaner Production, 3*(1–2), 9.

von Carlowitz, H. C. (1713). *Sylvicultura Oeconomica: Hausswirthliche Nachricht und Naturmäßige Anweisung zur Wilden Baum-Zucht*. Johann Friedrich Braun.

Wang, N., Guo, J., Zhang, X., Li, Z., Meng, F., Zhang, B., & Ren, X. (2020). The circular economy transformation in industrial parks: Theoretical reframing of the resource and environment matrix. *Resources, Conservation and Recycling*, 105251. https://doi.org/10.1016/j.resconrec.2020.105251

Wehrmeyer, W. (1996). *Greening people – Human resources and environmental management* (1st ed.). Greenleaf Publishing.

Wenbo, L. (2011). Comprehensive evaluation research on circular economic performance of eco-industrial parks. *Energy Procedia, 5*, 1682–1688. https://doi.org/10.1016/j.egypro.2011.03.287

Winkler, H. (2011). Closed-loop production systems—A sustainable supply chain approach. *CIRP Journal of Manufacturing Science and Technology, 4*(3), 243–246. https://doi.org/10.1016/j.cirpj.2011.05.001

Wirtz, B. W., Pistoia, A., Ullrich, S., & Göttel, V. (2016). Business models: Origin, development and future research perspectives. *Long Range Planning, 49*(1), 36–54. https://doi.org/10.1016/j.lrp.2015.04.001

Witjes, S., & Lozano, R. (2016). Towards a more circular economy: Proposing a framework linking sustainable public procurement and sustainable business models. *Resources, Conservation and Recycling, 112*, 37–44. https://doi.org/10.1016/j.resconrec.2016.04.015

World Commission on Environment and Development. (1987). *Our common future*. Oxford University Press. Available at: https://sustainabledevelopment.un.org/content/documents/5987our-common-future.pdf

Wright, P. M., McMahan, G. C., & McWilliams, A. (1994). Human resources and sustained competitive advantage: A resource-based perspective. *International Journal of Human Resource Management, 5*(2), 301–326.

Yadav, G., Mangla, S. K., Bhattacharya, A., & Luthra, S. (2020). Exploring indicators of circular economy adoption framework through a hybrid decision support approach. *Journal of Cleaner Production, 277*, 124186. https://doi.org/10.1016/j.jclepro.2020.124186

Yang, M., Evans, S., Vladimirova, D., & Rana, P. (2017). Value uncaptured perspective for sustainable business model innovation. *Journal of Cleaner Production, 140*, 1794–1804. https://doi.org/10.1016/j.jclepro.2016.07.102

Yong, J. Y., Yusliza, M. Y., & Fawehinmi, O. O. (2019). Green human resource management. *Benchmarking: An International Journal, 27*(7), 2005–2027. https://doi.org/10.1108/bij-12-2018-0438

Yuan, Z., Bi, J., & Moriguichi, Y. (2006). The circular economy: A new development strategy in China. *Journal of Industrial Ecology, 10*(1–2), 4–8. https://doi.org/10.1162/108819806775545321

Zhai, T. T., & Chang, Y. C. (2019). Standing of environmental public-interest litigants in China: Evolution, obstacles and solutions. *Journal of Environmental Law, 30*, 369–397. https://doi.org/10.1093/jel/eqy011

Zhu, Q., Geng, Y., & Lai, K. H. (2010). Circular economy practices among Chinese manufacturers varying in environmental-oriented supply chain cooperation and the performance implications. *Journal Environment Management, 91*(6), 1324–1331. https://doi.org/10.1016/j.jenvman.2010.02.013

Zhu, Q., Geng, Y., & Sarkis, J. (2013). Motivating green public procurement in China: An individual level perspective. *Journal of Environmental Management, 126*, 85–95. https://doi.org/10.1016/j.jenvman.2013.04.009

Zott, C., Amit, R., & Massa, L. (2011). The business model: Recent developments and future research. *Journal of Management, 37*(4), 1019–1042. https://doi.org/10.1177/0149206311406265

Chapter 3
Exploring the Missing Link to Circular Economy in Construction: A Systematic Review of Waste Management Literature

Leeboy Ndhlovu and Luca Sabini

Abstract The consideration of sustainable issues in construction waste management has gained attention in the sector over the last decade. Nevertheless, this consideration has failed yet to create momentum and produce sensible changes in the industry. The few developed ideas have not been fully integrated with the primary goals of the industry. Aiming at addressing the antecedents towards the implementation of sustainable best practices, we performed a systematic literature review of 234 publications in the last 20 years. The literature review intends to shed light on the construction waste management by exploring its current practices, its triggers and its barriers.

Keywords Waste management · Circular economy · Project management · Construction waste

3.1 Introduction

The construction industry, whether considered as civil engineering or large infrastructure projects, plays a critical role in enhancing the economic growth and livelihoods of local communities and society at large (Moraes et al., 2020). Ghisellini et al. (2018) compliments the construction sector for a critical role in propping up the economy and improving the employment level. For example, in Brazil, the construction sector is regarded as the pillar of the economy as it contributes approximately 14% of the gross domestic product (GDP) and the largest generator of employment (Paz & Lafayette, 2016). Furthermore, the growth of construction

L. Ndhlovu (✉)
School of Engineering and Built Environment, Anglia Ruskin University, Chelmsford, UK
e-mail: ldn108@student.aru.ac.uk

L. Sabini
University of Leeds, Leeds, UK
e-mail: l.sabini@leeds.ac.uk

V. Obradović (ed.), *Sustainable Business Change*,
https://doi.org/10.1007/978-3-031-23543-6_3

activities is attributed to housing needs and infrastructure development projects to sustain the urban population increase and large-scale urbanisation and urban renewal programs (Begum et al., 2007b; Yuan, 2013; Lam et al., 2019; Kolaventi et al., 2019; Wu et al., 2019; Nunes & Mahler, 2020).

Construction activities have accelerated rapidly in both developing and developed countries in the last 20 years. The rapid growing developments in the construction sector (especially in developing countries) are significant contributors to Construction Waste (CW) being landfilled (Jain et al., 2020). Globally, the industry generates over 10 billion tons of CW and contributes about 35% of waste to landfills (Wang et al., 2019; Ghaffar et al., 2020). In the EU, the construction industry is responsible for approximately 30% of the total solid waste generated from all economic activities (European Commission, as cited in Ghisellini et al., 2018). The industry is considered one of the largest waste generators hence the prioritisation of CWM. (Del Río Merino et al., 2010).

In all, although the sector provides many benefits to societies, the massive waste generated by the construction and demolition activities has adverse social and environmental impacts if not properly managed (Ding et al., 2016; Chen et al., 2019). A large amount of construction waste has extensive social, economic and environmental repercussions (Wu et al., 2016). For example, the Chinese construction sector is facing problems of CW due to the increase in the construction activities and the ineffective CWM systems (Yuan, 2013). The increasing construction demands have led to an enormous depletion of natural resources and the production of vast amounts of CW, which mainly impact the environment (Faleschini et al., 2016; Bakchan et al., 2019; Zhang et al., 2019; Bakchan & Faust, 2019).

The environmental effects are due to the sheer size of waste generated, poor land practices poor planning and management, greenhouse gas emissions, low awareness of waste reduction and pollutants (Del Río Merino et al., 2010; Faleschini et al., 2016; Wu et al., 2019; Ghaffar et al., 2020; Oliveira et al., 2019; Guerra et al., 2019; Jain et al., 2020; Vilventhan et al., 2019). The CW disposed on the environment may lead to detriments in human health (especially residents and waste workers) such as respiratory impacts, reduction of lung function in children and premature deaths (Wang et al., 2015; Mahpour, 2018; Oliveira et al., 2019).

As a result, the construction sector is under scrutiny to enhance Sustainable construction waste management (CWM) practices. Several studies suggest that albeit the well-developed strategies, advanced technologies to combat CW generation, policies and awareness of CWM, there is a lack of holistic implementation in the sector (Yuan & Shen, 2011; Ajayi et al., 2015; Wu et al., 2016; Ghisellini et al., 2018). In consequence, this study aims to assess the antecedents and current CWM practices to establish practical measures to enhance CWM best practices. Rapid calls for action to protect the environment are encouraged to avoid landfill failures such as the CW landslide in Shenzhen (Ding et al., 2016) and other environmental impacts of dumping untreated CW (Guerra et al., 2019). Therefore, the research question guiding the study is as follows: What are the current practices and antecedents of construction waste management that help the achievement of circular economy?

3.2 Theoretical Background

3.2.1 Construction Waste Management

CWM is essential in protecting the environment and conserving the rapidly depleting non-renewable natural resources for the generations to come. Sustainability in the construction sector is vital to support the rapid growth with minimum harm to the environment. Sustainability is "development that meets the needs of the present without compromising the ability of future generations to meet their own needs" (WCED, 1987; as cited in Sabini, 2016, p. 1). Therefore, the growing pressure to push towards the development of effective CWM practices in the construction industry to promote sustainable construction. According to Shen et al. (2010; as cited in Ghisellini et al., 2018, p. 427), sustainable construction is "implementing construction projects that involve less harm to the environment (minimising CW generation), increased reuse of CW in the production of construction materials (waste management), beneficial to the society, and profitable to the company". Manowong (2012) states that the concept of sustainable development requests for a balanced drive to achieve environmental, economic and social objectives. The sustainable development for CWM is required to reduce the use of natural resources and close the waste management loop in construction. Seadon (2010; as cited in Ghaffar et al., 2020, p. 2) states, "a sustainable waste management system requires vigorous feedback loops and is concentrated on processes to divert wastes from disposal and convert them to secondary raw materials".

Lam et al. (2019) defines CW as "anything generated as a result of construction (or renovation and demolition) and then abandoned, regardless of whether it has been processed or stockpiled". The CW is generated from building and infrastructure material such as concrete, metals, plastics, timber, asphalt, soil, bricks, tiles and glass (Lu et al., 2016a; Blaisi, 2019). CW is generated during all the stages of the construction lifecycle. CW affects the social, economic and ecological aspects of society, and their appreciation is vital in the successful implementation of CWM practices (Blaisi, 2019; Kolaventi et al., 2019). The CW generated can be reused, processed to recycled materials and the remaining unrecyclable waste is hauled to landfill sites. Construction and demolition (C and D) waste is also known as CW (Lu et al., 2015). The terms are used interchangeably in this study.

The construction waste arises from activities such as land excavation/formation, roadworks, site clearance and civil works (Rodríguez et al., 2007; da Rocha & Sattler, 2009; Yuan & Shen, 2011; Jin et al., 2019; Bakchan et al., 2019). Furthermore, CW generation may be due to extraction of raw materials, manufacture of materials, material cutting to meet construction requirements, damages of material during transportation and construction, and demolition due to construction errors and changes (Del Río Merino et al., 2010; Lu et al., 2016a; Wang et al., 2019).

3.2.2 *Circular Economy Model in Construction Waste Management*

The circular economy (CE) is defined as "an industrial system that is restorative or regenerative by intention and design. It replaces the end of life concept with restoration, shifts toward the use of renewable energy, eliminates the use of toxic chemicals impairing reuse, and aims at eliminating waste through the superior design of materials, products, systems, and business models" (Ellen MacArthur Foundation 2016; as cited in Mahpour, 2018, p. 216). CE provides a platform to develop economic patterns aimed at an increased efficiency of production and consumption of resources. The concept seeks to substitute the conventional production and consumption of materials based on continuous growth and increased processing of CW to new resources (Jin et al., 2019). CE is an emerging end of pipe practice towards minimising the CW adverse impacts to the environment such as degradation and exhaustion of resources (Jin et al., 2019; Mahpour, 2018). Bakshan et al. (2015) identify the diminishing natural resources and challenges of siting new landfills (particular in land-limited and continuously developing urban areas) as the primary drivers towards circular economy in construction. The closed-loop practices of CE ensure CW reuse or recycle rather than disposal in landfills (Zhang et al., 2019). CE minimises CW by integrated recycling of CW into secondary or new resources for the construction industry or other industries (Mahpour, 2018; Huang et al., 2018; Ghaffar et al., 2020; Jain et al., 2020). The implementation of CE in construction is a challenge due to limited research (Jin et al., 2019).

3.3 Methodology

We decided to adopt a systematic literature review (SLR) which Kitchenham and Charters (2007, p. 1) defines as "a means of identifying, evaluating and interpreting all available research relevant to a particular research question, or topic area, or phenomenon of interest". The literature review identifies as a critical methodology for assessing the trend developments of a particular research discipline (Yuan & Shen, 2011). According to Kitchenham and Charters (2007), systematic reviews aim to synthesise and compare evidence across studies to answer specific research questions that consolidate or confirm practices, policies or theoretical relationships. A repeatable unbiased search strategy is applied to capture relevant literature around a particular topic rigorously. Therefore, the systematic review has been arguably the most efficient and high-quality method for identifying, evaluating and interpreting literature (Kitchenham & Charters, 2007).

To find the relevant journals for the systematic literature review, keyword searches were performed within the areas of abstract, title and keywords in the Scopus database. The following keywords are used to gather the papers related to the research topic: "construction waste management", "construction and demolition

Table 3.1 Keyword search results

Keywords	No. of papers
"construction waste management"	136
"construction and demolition waste"	87
"C and DW management"	20
Total	**243**

waste" and "C and DW management". These produced 243 papers (as of March 2020), as shown in Table 3.1.

For the sample selection, the three keyword searches were combined using the Boolean Operator "OR". After connecting the keyword searches obtained, duplicate papers were removed, bringing the number of documents (n = 234). Figure 3.1 shows the number of journals published each year, from the year 1994 to 2020, and it shows an exponential increase with notable spikes in some years. This increase in the number of publications reflects the relevance of this literature review.

The top 12 journal sources with the most publications (over three publications) on the keyword search are shown in Tables 3.2 and 3.3. The research will focus on the four most prominent sources, that is, references with the most publications from the keyword search results: (1) Resources Conservation and Recycling, (2) Journal of Cleaner Production (JCLP), (3) Waste Management and (4) Waste Management and Research. Therefore, the final number of papers used in this research study is 78 from the four journals with the highest number of publications.

The final sample used in this research consists of 78 journals from four different sources, as shown in Fig. 3.2.

The journals are from the year 2004 to March 2020 and shown off according to the number of journals published per year by each source in Fig. 3.3. There has been a constant increase in the occurrence of the keywords over the past 16 years, which is, evidence that authors have been researching more around the subject of "construction waste management". Consequently, the final four journal sources have at least one publication in the past 3 years.

The authors with the leading number of journals are shown in Fig. 3.4. The sample points out that the predominant authors are Lu, W and Li, J. The figure shows authors with a minimum of two journals within the literature review sample.

CWM practises may differ from one country to another; this literature review may be biased towards the dominant nation of publication. Dahlbo et al. (2015, p. 335) state, "the volumes, composition and quality of CW vary between sites, regions and countries, and no general composition can be presented." Consequently, The CWM knowledge developed in one region is not easily adapted and applied to the other areas without considering their contextual differences (Saez et al., 2013). Thus, the need to highlight the sample of journals used by country/territory is shown in Fig. 3.5.

The performance of the analysis of the journals on the research design used, and the results are as shown in Fig. 3.6. The results identified case studies (n = 22),

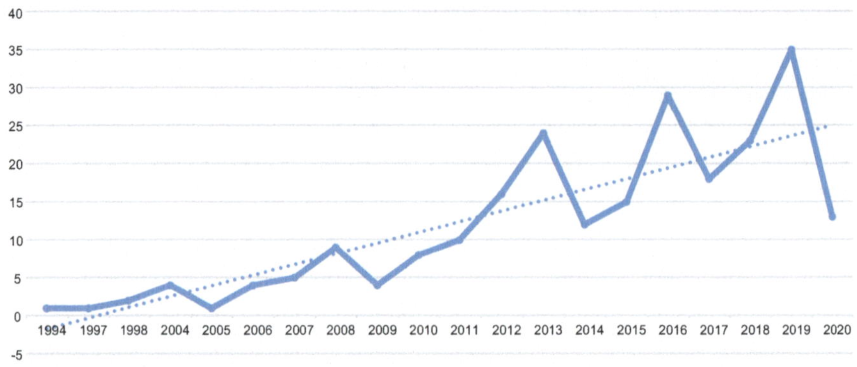

Fig. 3.1 Number of publications per year

Table 3.2 Number of journals containing keywords on a source with over three publications

Source title (with papers greater than 3)	Count
Resources Conservation and Recycling	24
Journal of Cleaner Production	20
Waste Management	18
Waste Management and Research	16
Sustainability Switzerland	9
Wit Transactions on Ecology and the Environment	7
Electronic Journal of Geotechnical Engineering	5
Open Construction and Building Technology Journal	5
Renewable and Sustainable Energy Reviews	5
Construction Innovation	4
Engineering Construction and Architectural Management	4
WSEAS Transactions on Environment and Development	4
Total	**121**

which further divides into single case studies (n = 15) and multiple case studies (n = 7), as the most adopted approach in CWM related research, followed by empirical analysis (n = 20).

The results indicate that all the journals (n = 78) are within the environmental science area of study (see Fig. 3.7). The results are favourable as the research mainly seeks to emphasise on a CWM system with less negative impacts on the environment.

As the research question sought to explore the practices, drivers and barriers of CWM, the SLR approach is adopted to gather a set of relevant papers. For the

Table 3.3 Overview of the discussion section

Category	Number of papers	Main message
What are the current CWM practices?	63	Explains the current CWM practices
What are the factors that contribute to the implementation of sustainable CWM best practices?	36	The factors which promote sustainable CWM practices are explained and ways of enhancing them
What are the factors that impede the adoption of sustainable CWM best practices?	32	The major hindrances to sustainable CWM practices are identified together with their causes and mitigations

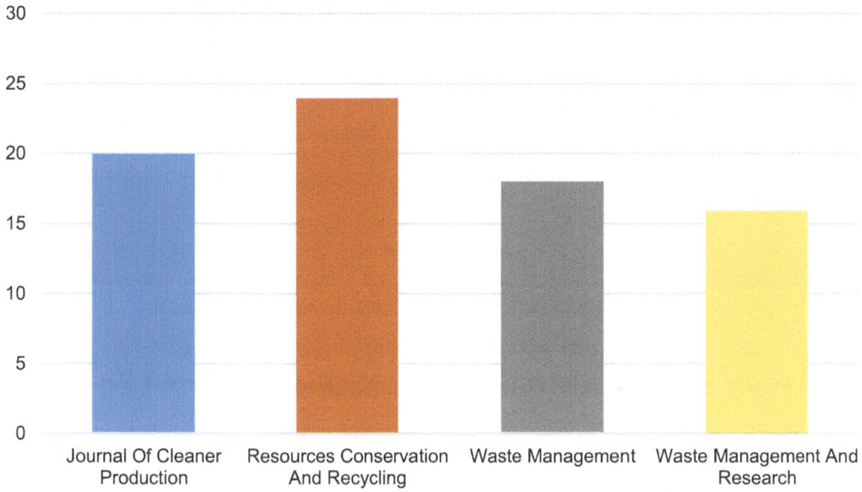

Fig. 3.2 An overview of journals by source

content analysis of the documents, NVivo is used to create three critical nodes extracted from the research question. These parent nodes are practices, triggers and barriers to CWM, further broken down to specific subtopics (child nodes) which are sorting, recycling, landfill, green building, disposal, CWR, CE, bill of quantities, education, legislation, technical, market economy, behaviours and attitudes. Then the papers were analysed for concepts and content relating to the identified nodes.

The relevant content in the data sample was identified and coded to their respective nodes. NVivo word frequency query was performed on the nodes for 100 most frequently occurring words (with stemmed words) with a minimum length of five letters. Therefore, Fig. 3.8 shows the word cloud generated from the search results. The word cloud gives an overview of dominating concepts as well as the visual representation of what the discussion section seeks to address.

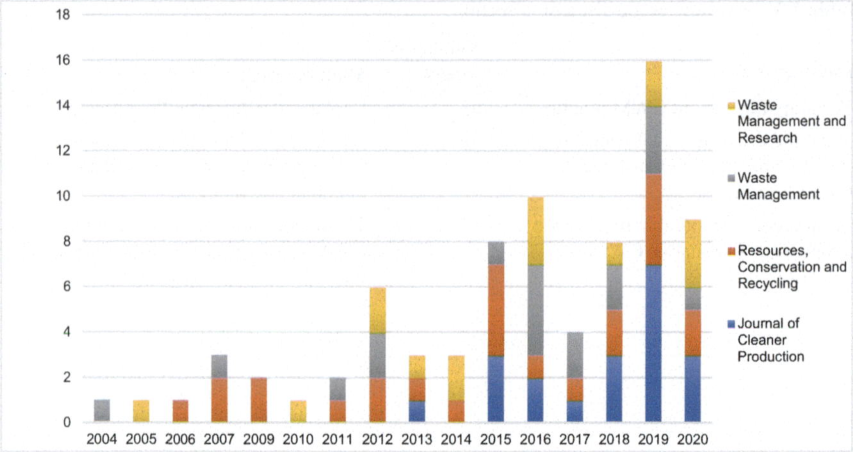

Fig. 3.3 Overview of papers per year by source

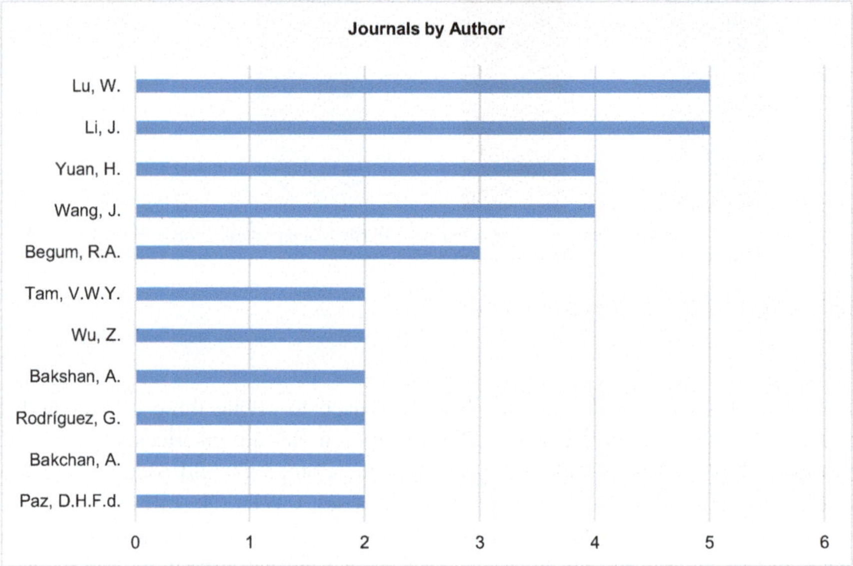

Fig. 3.4 Authors with more than two publications in the sample

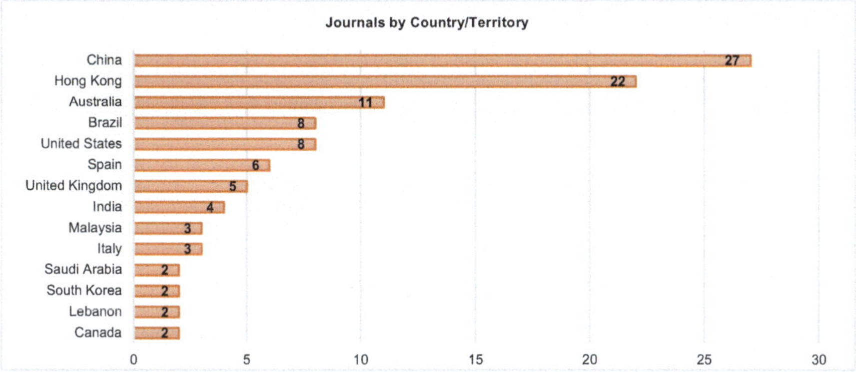

Fig. 3.5 Journals by Country of affiliation of the authors

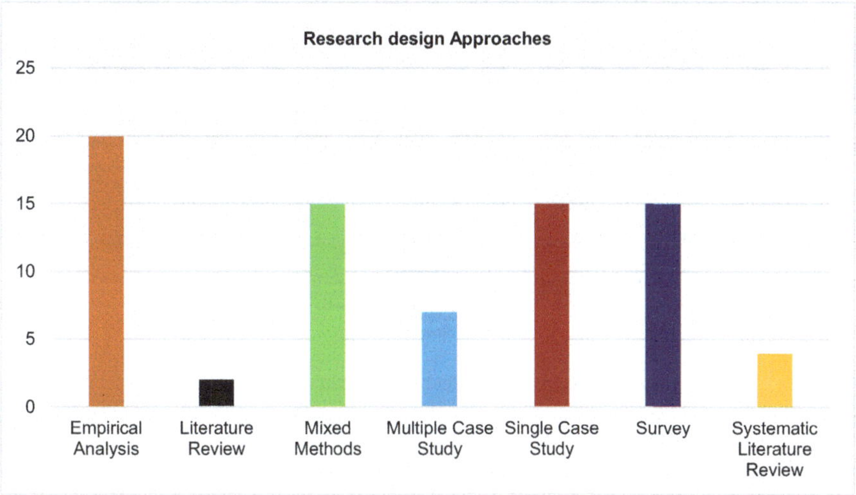

Fig. 3.6 Journals by research design approaches

3.4 Discussion

The findings and discussions section reviews the selected papers based on the data analysis across three major categories. (1) What are the current CWM practices? (2) What are the factors that contribute to the implementation of sustainable CWM best practices? (3) What are the factors that impede the adoption of Sustainable CWM best practices? Table 3.3 below shows the number of papers used to produce each category and the main message each category attempts to address.

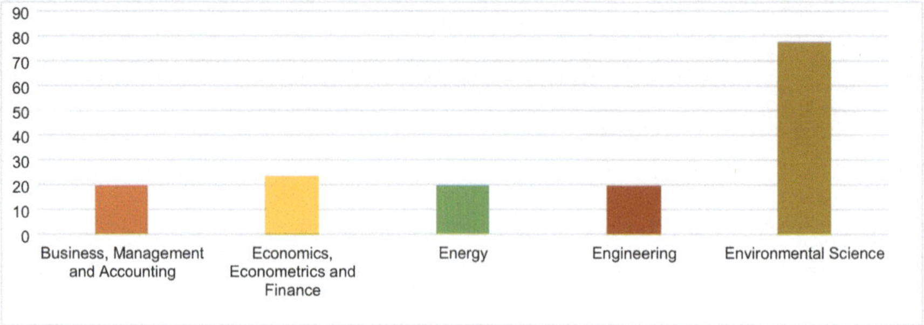

Fig. 3.7 Subject area of the journals

Fig. 3.8 Word frequency query (search in nodes) word cloud

3.4.1 *What Are the Current CWM Practices?*

This section looks at the standard CWM practices in the construction sector and considers the justification and effects of its implementation. The construction sector is under pressure to promote more effective CWM practices at the site level, guided by the 3Rs (Reduce, reuse, recycle) principle (Tam & Tam, 2006; Huang et al., 2018; Bakchan et al., 2019). In the same vein, Yang et al. (2017; as cited in Ghisellini et al., 2018) suggest the adoption of 4R principle (Reduce, reuse, recycle and recover) combined with adequate disposal to alleviate pressure in landfills.

The extensive studies on CW recommend the adoption of a waste management hierarchy that prioritises the 3Rs in descending order of environmental and economic preferences (Begum et al., 2007a; Yuan & Shen, 2011; Rodríguez et al., 2015). The studies highlight the need for devising of the 3R principle to manage the waste by firstly attempting to implement best practices with the least negative impacts on the sustainability dimensions of CWM. The CWM needs to be implemented as part of project management roles to effectively achieve waste reduction (Begum et al., 2007a).

3.4.1.1 Construction Waste Reduction/Minimisation (CWR) Practices

CWR helps to minimise waste generation and increases waste recycling and reuse rates (Won et al., 2016; Ding et al., 2016). There is a need for the construction industry to shift its focus to CWR practices to improve CWM, as the last-resort waste treatment of disposal cannot satisfy the sustainable development requirements of circular economy. Countries are transitioning to adopt the measures necessary to promote CWR as its benefits outweigh other waste management practices (Rodríguez et al., 2007). CWR is the priority as it is considered the most effective practice in reducing CW generated, preventing waste disposal and has the lowest adverse environmental impacts (Tam & Tam, 2006; Yuan & Shen, 2011; Al-Sari et al., 2012; Yuan, 2012; Wang et al., 2015; Ding et al., 2016; Wu et al., 2016; Huang et al., 2018).

Begum et al. (2007a) emphasise that improper CWM, excessive wastage of raw materials and low awareness of the waste reduction result in lack of consideration given to CWR measures during the planning and design phases. The practice offers benefits of minimising CW generation and cost-saving from cutting transportation, recycling and disposal fees. The CWR practices identified in the literature divide into two groups, which are source reduction and end of pipe waste reduction.

Source Reduction

Begum et al. (2007a, p. 91) define source reduction as "any activity that reduces or eliminates the generation of waste at the source, usually within a process". Source reduction practices are measures such as adopting low-waste technologies and on-site material management to minimise waste production before its generation (Ding et al., 2016; Ghisellini et al., 2018). Waste reduction at the source is recommended as the first step in avoiding waste generation (Won et al., 2016). Studies suggest the implementation of CWR at the start from the planning and design phase and source management during construction/deconstruction phase (Zhang et al., 2019). CWR implementation throughout the project lifecycle (design, planning, construction and deconstruction/demolition phases) is as follows:

During the Planning Phase

Estimation of the waste generated by construction activities is the most common waste reduction practice during the planning phase. The evaluation of the amount of CW that will be generated by construction activities is needed mainly for planning purposes (Lam et al., 2019). The construction projects generate continuous streams of waste that need systematic planning and management to enhance circular economy (Lu et al., 2016b). The estimations assist in the planning of ways of reducing waste generation by implementing alternative methods with less waste output.

The data on waste streams allows the construction companies to identify processes that generate significant streams of waste and help them adopt appropriate waste reduction technologies and methods (Li et al., 2016). The data also can help in deducing or estimating the impacts of a construction project on sustainable development. Further, the precise estimation of CW is necessary for the government to establish effective policies, guidelines, strategies and codes of practice towards circular economy (Li et al., 2016).

According to a survey executed to academics, lack of reliable data hinders them from making informed decisions in a bid to enhance CWM (Blaisi, 2019). The survey hints on the need for accurate estimates to draw up the relevant practices required to combat the vast CW generation. Developing an effective CWM plan needs identifying, quantifying and classification (quality) of CW streams generated and their appropriate destinations during different phases of construction (Bakshan et al., 2015; Dahlbo et al., 2015; Paz & Lafayette, 2016; Bakchan & Faust, 2019; Guerra et al., 2019). The forecasting of CW generation streams at different construction phases assist the project stakeholders in planning for best CW handling practices (Lu et al., 2015, 2016b; Bakchan & Faust, 2019).

The reliable data on generated CW estimation from construction sites is critical in improving CWM (Wang et al., 2019). Consequently, various ways of measuring and estimating CW generation involving different tools have emerged in the sector (Ajayi et al., 2015). Li et al. (2016) proposed a CW estimation model based on mass principle and work breakdown structure for building construction projects. Villoria Saez et al. (2011; as cited in Paz & Lafayette, 2016) developed an empirical model for estimating the waste generated, by creating several indicators relating to the amount of waste generated and the total constructed area of the building. Further, Lu et al. (2016b), used the S-curve model to forecast waste generation in construction projects. Other studies (Won et al., 2016; Guerra et al., 2019) show that there is the potential use of building information modelling (BIM) technology to reduce CW generation. The studies propose BIM uses like design validation, quantity take-off and prefabrication, which will, in turn, reduce design errors, ordering surplus material and offcuts waste.

In this respect, studies argue that existing CW estimation methods have limitations as some are time-consuming to implement, lack ways of verification or provide rough estimations, which are insufficient for developing informed CWM plan (lack of precision). Again, some rely on regional databases, which may not be compatible with studies from other regions (Bakchan & Faust, 2019; Guerra et al., 2019). Therefore, to improve the CW estimation, the construction companies need to have a reliable, user-friendly tool to estimate the quantity of CW generated during the different stages of a construction project (Gangolells et al., 2014).

On the other hand, Ajayi et al. (2015), state that the procurement stage is vital for CWM planning as it links with causes of CW such as improper material storage, double handling and packaging materials. Begum et al. (2007a) argue that the procurement of extra construction materials is due to the lack of consideration of CWR during the planning and design stage. Estimations of quantities of materials required are therefore vital as to minimise waste generation due to surplus materials, as well as damaged materials due to double handling and on-site storage. Accurate estimates of the number of materials are needed to prevent long periods of on-site storage (Gangolells et al., 2014). Likewise, a good logistics agreement with suppliers, such as just in time ordering, reduced packaging and appropriate ordering to minimise offcuts waste, could promote CWR during the planning phase (Ajayi et al., 2015; Vilventhan et al., 2019). Other studies (Tam & Tam, 2006; Wang et al., 2019) as well encourage the sector to return the packaging materials to suppliers for reuse.

During the Design Phase

Implementing CWR by design plays a vital role in avoiding CW production at the source (Li et al., 2015). Saez et al. (2013) argue that typically the practices adopted at the design stage of a construction project are aimed at appropriate CWM rather than targeting CWR. Nevertheless, the account of CWR at the design stage is the most sustainable practice. Ghisellini et al. (2018) highlight that the early stages of design in construction projects have the best chances of maximum CWR.

Some studies assert that the roles of designers towards CWR are not apparent under regulations, resulting in architects neglecting designs with waste minimisation (Wang et al., 2015). Ghisellini et al. (2018), for good measure, point out that previous studies focused on waste minimisation strategies and technologies, while there was an apparent omission of the designer's role consideration. However, a survey by Yuan (2012) concluded that poor CWR in Shenzhen is due to the designers' lack of awareness of its benefits. Another study by Osmani et al. (2008; as cited in Saez et al., 2013) indicated that architects presume CW to be mainly generated during the construction phase and barely during the design phase. However, Li et al. (2015), argue that the reduction of CW can be at both design and construction phases.

Del Río Merino et al. (2010) states that when designing infrastructure or building projects, the stakeholders should consider the potential CW generation in the project decisions they make. This consideration will ensure the reduction of the volume of CW generated by future demolition and enhance material reuse. Lam et al. (2019) contend that architects and engineers may appreciate the need for CWR if there is a presentation of the potential amount of CW to be generated by their designs in the BOQ (bill of quantities). This presentation can be useful in minimising CW produced by making waste-reducing changes before the actual construction begins. Similarly, a survey in Spain shows that construction companies consider the introduction of salvaged materials from old buildings in the design phase to minimise waste generation (Gangolells et al., 2014).

As stated by Innes (2004; as cited in Won et al., 2016), inappropriate design decision-making and design changes may increase the volume of CW generated up to 33%. Another study conducted by Saez et al. (2013) shows that one-third of CW generated is due to design decisions at the design stage. Likewise, Won et al. (2016) contend improper design and unexpected design changes as the major causes of CW generation. Additionally, Cheng et al. (2015; as cited in Won et al., 2016) propose that the use of integrated building design can avoid design changes and errors. Ajayi et al. (2015) recommend optimising designs to the industry's standards to facilitate the reuse of its removed materials in another optimised project. The design optimisation will significantly prevent waste production due to material offcuts. Moreover, Wang et al. (2015) found that the incorporation of fabricated components during design has the most substantial influence on waste reduction.

During the Construction/Deconstruction Phase

The green building practices are implemented during construction and demolition activities to help conserve resources and minimise the CW harm to the environment. USEPA (2018; as cited in Chethana et al., 2018, p. 722) defined green or sustainable building as "the practice of creating and using healthier and more resource-efficient models of construction, renovation, operation, maintenance and deconstruction". According to Chen et al. (2019), green building is the construction sector's mitigation measure to challenges such as exhaustion of non-renewable natural materials, water and air pollution, greenhouse gas emissions and human-induced global warming. Lu et al.'s (2018) study analysed the effects of green building on CWR using the Hong Kong accredited green building accrediting system. The results proved a 36.19% waste reduction in demolition activities.

Because of adopting more sustainable materials and methods, green building projects incur higher costs than traditional methods. Thereby the industry is less willing to have fewer profits due to adopting the green building methods. A study by Chen et al. (2019) confirms that green building costs are higher than conventional building by a range of 0.4–6% of the total cost. However, green building institutes disseminate that there may be a payoff of those higher prices in the future through improved environmental performance which results in lower utility bills and higher property value (Lu et al., 2018). Furthermore, the author suggests that in addition to social and environmental benefits of green building, there is a potential increase in construction and waste transport cost savings.

On the other hand, there is a proposal of implementing architectural technologies such as the use of prefabricated materials to assist in minimising the waste generated during construction (Saez et al., 2013; Wang et al., 2015; Won et al., 2016; Zhang et al., 2019). The study by Bakchan and Faust (2019) suggests that prefabrication, as a replacement of on-site production, is one of the most effective technologies in waste reduction. The prefabricated components include staircases, large panel formworks, thin panel walls, recycled scaffolding and hoardings (Wang et al., 2015; Ding et al., 2016). Accordingly, evidence shows that the use of precast or prefabricated materials promotes effective waste reduction as there is a minimisation of CW generation due to site mixing and offcuts (Ajayi et al., 2015; Lam et al., 2019). Jaillon et al. (2009; as cited in Won et al., 2016), found out that CW was reduced by 52% in Hong Kong, as an impact of using prefabricated materials. Huang et al. (2018) advocate for precast construction technologies as they reduce pollution during construction and demolition activities, and besides they enhance the recovery and reuse of materials without destructive dismantling of buildings. However, the industry struggles to promote prefabricated materials (mainly concrete) due to structural stability concerns (Wang et al., 2019). Thereupon a robust structural stability check should be put in place when manufacturing the prefabricated materials.

For further development, at the end of the lifespan of a construction project, the infrastructure or building has to be deconstructed or appropriately demolished to minimise waste production. Ajayi et al. (2015) contend that deconstruction differs from demolition, as it involves careful dismembering of the building components in an attempt to salvage them for reuse and recycling. Rodríguez et al. (2007) note that the demolition activities are not recommended for material recovery as they result in mixed wastes that are not suitable for reuse or recycle. Zhang et al. (2019) encourage the local authorities to enforce regulations for demolition licensing, which the contractor receives after notifying and getting approval to proceed with demolitions. The authorisation ensures necessary measures are put in place for the careful dismantling of buildings to promote reuse and recycle of materials.

End of Pipe Waste Reduction

The implementation of a detailed CWM plan requires a productive CW streams handling procedure and adoption of project management strategies to help in achieving the planned CWM objectives (Bakchan & Faust, 2019). After the unavoidable CW generation, there is the implementation of the end of pipe treatment strategy to reduce the amount of CW disposed of in landfills (Yuan & Shen, 2011). End of pipe waste reduction measures are sorting, reuse and recycle after the waste has been created (Ding et al., 2016; Ghisellini et al., 2018). The practices of recycling and reuse of CW are the best methods in mitigating CW problems at the end of the pipe and is the responsibility of practitioners in the field (Duan & Li, 2016).

Reuse

Reuse is salvaging construction materials to use them more than once. Yuan and Shen (2011) identify reuse as the best sustainable option after source reduction because it requires minimum processing and energy use. The energy savings by reusing salvaged materials is higher than recycling (da Rocha & Sattler, 2009). In like manner, reusing materials leads to a reduction in carbon emissions (Chen et al., 2019), has less adverse environmental impacts and might result in cost savings. The reuse of materials may include using the material either for the same function or as secondary aggregates (Ling & Leo, 2000; as cited in Yuan & Shen, 2011). The adoption of material reuse is a way of diverting CW from landfills (Ajayi et al., 2015).

The recovered materials can be reused for primary or secondary functions in the construction sector or other industries to reduce the CW produced into the environment. Paz and Lafayette (2016) argue that despite substantial amounts of CW are reusable, construction companies in Brazil improperly dispose the CW causing a series of economic, social and environmental problems. Paz et al. (2018) articulate that almost all CW generated can be reused; hence the industry should strive to produce the minimal possible CW. In the bargain, the construction buildings or infrastructure have to be carefully dismantled or deconstructed to enable the reuse of their components (Ajayi et al., 2015). Finally, the non-reusable waste material must be either recycled for new articles or disposed at landfills as a last resort if untreatable (Yuan & Shen, 2011).

Recycling

Recycling is defined as the recovery and reuse of what would otherwise be waste material (Begum et al., 2007a). The recovery of CW material involves processing it into new materials by using recycling technologies such as carbonisation of CW (Huang et al., 2018) and dry or wet recycling (Jain et al., 2020). The CW is recycled to produce derivative materials, which can replace virgin materials (Ajayi et al., 2015). Conservation of tons of natural materials is possible if a proper recycling system is adopted to produce quality-recycled materials (Kolaventi et al., 2019).

Lockrey et al. (2018) suggest that recycling can be treated as a process of CWM and material production and thus enhances sustainable construction. Again, Lockrey et al. point out that successful recycling depends on the generated revenue from the resale of recycled materials. Dahlbo et al. (2015) contend that an increase in recycling should be promoted by environmental and economic benefits to support Sustainable CWM. Nevertheless, there is weak competitiveness of recycled materials industry to that of virgin materials due to the unwillingness to use recycled materials and lack of supporting regulatory measures (Del Río Merino et al., 2010; Faleschini et al., 2016). Blaisi (2019) encourages governments to impose taxes on virgin materials to increase the industry use of recycled materials.

Jia et al. (2017) point out that of the waste generated in the Chinese construction industry, only approximately 5% is recycled. The low recycling rate is alarming, as China is the biggest waste producer (Wang et al., 2019). Saez et al. (2013) point out that the recycled material use is widely accepted in the construction sector, although only 8% of the surveyed construction agents in Spain implement it, due to quality standards. Therefore, to maintain competitive quality standards, according to many contemporary pieces of research the recommended replacement of natural aggregates is 50% (Kolaventi et al., 2019; Zhang et al., 2019). Moreover, to improve the quality of recycled materials, Dahlbo et al. (2015) counsel enhancing sorting, separation and processing of CW.

Sorting is separating the waste based on their characteristics, components for their reuse, recycling and disposal (Ghisellini et al., 2018). The recycling is beneficial when there is an efficient implementation of on-site sorting, and the transportation distance of CW from source to recycling facilities is under 30 km (Penteado & Rosado, 2016). Studies show that sorting is a useful measure for reducing the volume of CW to landfills by promoting higher rates of reuse and recycling (Ghisellini et al., 2018). Another study in Hong Kong found an increase in diversion of waste going to landfills as a result of on-site sorting practices (Dahlbo et al., 2015). Consequently, sorting might have barriers such as increased costs, lack of sorting space and lack of market for recycled materials (Gangolells et al., 2014; Ding et al., 2016).

According to Huang et al. (2013; as cited in Ghisellini et al., 2018), the hypothesis of a high recycling CWM system shows alleviation of virgin material use and generation of CW. Huang et al. further confirm the need to improve recycling and reduce material demand by prolonging construction project lifespan to reduce CW generation rates. The practice alleviates the environmental impacts caused by the rapid extraction of non-renewable natural materials (Del Río Merino et al., 2010; Kolaventi et al., 2019; Jain et al., 2020), greenhouse gas emissions (Zhang et al., 2019) and the hazards of disposing CW to landfills (Del Río Merino et al., 2010). Likewise, studies point out that recycling provides relief to landfills that are quickly filling up due to the massive CW production by the construction industry (Wang et al., 2019; Lockrey et al., 2018).

3.4.1.2 Landfill

Landfills are arguably the most common and traditional practice in CWM that require large areas of land. Although current research (circular economy) focuses towards a diversion of CW from landfills, however, the method is still widely used around the world (Ghaffar et al., 2020). Studies confirm that landfill space for CW disposal is scarce, particularly in and around continuously developing metropolitan urban areas (Yuan et al., 2011; Wu et al., 2016; Bakshan et al., 2015). Moreover, Jia et al. (2017) argued that due to the increased urban construction activities, there are challenges of finding sites for the disposal of huge volumes of CW. Again, Jin et al. (2019) point that the need to release pressure on landfills and promote waste diversion practices, to close the circular economy loop, is driving sustainable development in the construction sector.

In consequence, there are recommendations for hiking the dumping fees to enhance the landfill diversion rate and mitigate landfill impacts on the environment. The environmental impacts may include air and water pollution, leaching of non-inert waste and longer transportation distances (Del Río Merino et al., 2010; Ghisellini et al., 2018). Ding et al. (2016) note that the 2015 landslide accident in a landfill in Shenzhen proves the imbalance between the construction waste landfilling demand and limited landfill areas. Such incidents aid the adoption of CE practices. Likewise, due to the filling up of landfills, there is also a need to build engineered landfills which are more expensive. These economic impacts to the industry tend to support the diversion of CW to recycling plants and use of CW reduction practices (Wang et al., 2004; Zhang et al., 2019).

3.4.1.3 Illegal Dumping/Fly-Tipping

Illegal dumping refers to the intentional and unlawful disposal of waste in unauthorised areas to avoid landfill fees or to save time and costs of transportation to landfills (Lu et al., 2018). Extensive literature links illegal dumping to the landfill practice, as is triggered by lack of landfill space (Manowong, 2012), longer transportation routes to landfill sites (Blaisi, 2019) or higher landfill fees (Zhang et al., 2019). Again, Mahayuddin et al. (2008; as cited in Paz et al., 2018), argue that the lack of waste collection system at construction sites in Brazil leads to illegal dumping.

Consequently, the governments or regulatory boards should set penalties and improve the supervision of CW disposal along with an effective Sustainable CWM plan to minimise illegal dumping (Ghisellini et al., 2018). Chen et al. (2019) propose that unlawful dumping can be decreased more effectively under public participation with the government providing supervision hotlines and information platforms for convenient communications to report illegal dumping activities.

3.4.2 What Are the Factors That Contribute to the Implementation of Sustainable CWM Best Practices?

CWM practices towards CE have increasingly gained awareness within the construction industry and different practices and policies to enhance the system have been implemented. Thereupon this category discusses the factors, found in the literature, that drive the industry towards sustainable practices and its promotion. Table 3.4 shows an overview of the effects of country context that determines their stance on CWM practices. According to Li et al. (2018), the most significant measures to promote CWM practices are training contractors, enforcing effective regulations and increasing public awareness. These elements are grouped into legislation and regulations, and training and education.

3.4.2.1 Legislation and Regulations

The CWM depends on policies and regulations, which impose economic or administrative measures that promote the overall implementation, and effectiveness of CWR practices (Wang et al., 2015; Ding et al., 2016; Blaisi, 2019). points out the lack of precise and detailed CWM regulations for the construction industry in China hinders the adoption of the best practices. Regulation can enhance a positive attitude that focuses on sustainable best practices in CWM (Blaisi, 2019). The CWM professionals view the role of the government legislations and support of CE as the most critical contribution towards CWR (Ghisellini et al., 2018; Ghaffar et al., 2020). Saez et al. (2013) point out that some countries are developing laws to establish a

Table 3.4 Effects of country context on CWM Practices

Country/ Region	Effects	Reference
Kingdom of Saudi Arabia	Lack of institutional collaboration Limited coordination between regulators and generators Lack of motives, awareness and incentives to manage CW Lacking enforcement of law and regulations to prohibit illegal dumping	Blaisi (2019)
China	Policies to minimise waste production Government imposes strict penalties to illegal dumping Land reclamation from sea by dumping waste	Duan and Li (2016) Wu et al. (2017)
Canada	Government pledges to divert all CW away from landfills by 2030	Moraes et al. (2020)
European Union	Member states obliged to commit to a 70% reduction in CW generation by 2020	Moraes et al. (2020)
Brazil	As a developing country the rapid growth of construction has led to more waste generation and CWM is more challenging	Moraes et al. (2020)
United Kingdom	Government has developed initiatives to reduce waste, promote prefabricated materials and training The legislations and regulations drive the CWM	Ajayi et al. (2015); Ghaffar et al. (2020)
India	Imminent need to import raw materials such as sand	Jain et al. (2020)

framework for CWM and production to promote the 3R principle. Similarly, other studies suggest the government should take an active approach by providing practical guidelines, policies and laws that support the construction industry in implementing CE practices (Begum et al., 2007a, b; Lu et al., 2016a; Zhang et al., 2019).

For example, the EU has fully committed to achieving a goal of 70% reduction in total CW generation by 2020 (Moraes et al., 2020). Moreover, in the UK, there is a legislative framework that requires the development of a mandatory solid waste management plan (SWMP), before the commencement of the actual construction works, for construction projects valued above £300,000 (Ajayi et al., 2015). The SWMP includes statements of strategies used before construction to ensure waste minimisation and waste management practices during and after the construction phase. However, Zhang et al. (2019), encourages the government to adopt the balanced stakeholder approach to ensure the construction stakeholders' views and interests consideration in formulating and implementing of CWM policies. The incorporation of stakeholders' opinions in the decision-making process will assist in coming up with the best acceptable strategies within the industry (Blaisi, 2019).

Zhang et al. (2019) argue that recent studies suggest improvement of company policies towards sustainability than legislation enforcement. However, for the company policies to change, there should be an economic benefit, hence legislation to incentivise waste reduction is necessary to enhance the improvement of company policies. The law triggers companies to assume a reactive approach (end of pipe practices) to CWM. At the same time, incentives will encourage companies to adopt proactive (source reduction) policies to CW reduction. Therefore, improvement of company policies should be considered on top of government legislations.

The studies point that the government or local authorities enforcing legislation and regulations in the construction sector such as penalties for illegal dumping with strict supervision (Lu et al., 2015; Huang et al., 2018; Blaisi, 2019), progressive taxation for waste disposal (Chen et al., 2019; Wang et al., 2019), economic incentives for waste reduction (Ghaffar et al., 2020) and motivating their attitudes and behaviours towards effective CWM (Begum et al. 2007a) are the primary drivers towards CE model. The legislation and regulations in this study divide into four subsections which are the enactment of a mandatory standard CWM mode, incentivising the CWR practices, penalising illegal activities such as fly-tipping and finally taxation of virgin materials through the increase of waste disposal charges.

3.4.2.2 CWM Plan

CWM plan is a guide to the industry on how to handle different waste streams as well as minimising waste production. Existing studies show that an effective CWM plan, in the form of strategies to reduce waste production and raw materials extraction, results in substantial economic benefits to the construction sector (Ajayi et al., 2015; Wu et al., 2019). Blaisi (2019) advocates for CWM guideline documents to provide information on how the construction sector should comply with regulations as well as include best practices. Furthermore, Tam and Tam (2006), vouches for the mandatory enactment of a CWM plan so that all stakeholders can share the same goal of adopting circular economy. Notably due to the high number of construction activities, there is a need for a compulsory CWM plan that will propel the whole sector to CE practices. Rodríguez et al. (2015) highlight that the main aim of a CWM plan is to set apart the direct link between economic growth and the increase in the volume of CW generated to mitigate the associated environmental impacts.

The lack of CWM plan for every construction phase impedes the continuous improvement of best practices (Paz & Lafayette, 2016). Moreover, research by Vilventhan et al. (2019) observed the lack of on-site CWM planning as one of the primary causes of waste generation. Therefore, for example in Spain, a waste management model includes a waste management report developed during the design phase and a waste management plan developed during the construction work planning phase (Saez et al., 2013). These documents include descriptions of CWM best practice measures of the 3Rs and final disposal of waste. Jin et al. (2019) state that currently there are still needs of developing a more comprehensive performance measurement mechanism for C and D waste management or a suitable C and D waste management guide adopted in a particular organisational or local context.

Bakchan et al. (2019) recommend the development of a CWM plan that also enhances the efficiency of meeting the planned project management objectives. The author suggests the alignment of the joint project objectives with sustainability objectives at the planning phase to bridge the gap between project priorities and CE. The integration of project objectives and sustainable objectives is termed sustainable project management (SPM) (Sabini et al., 2019). Silvius and Schipper (2014a; as cited in Sabini et al., 2019, p. 822) define SPM as "the planning, monitoring and controlling of project delivery and support processes, with consideration of environmental, economic and social aspects of the life-cycle of the project's resources, processes, deliverables and effects, aimed at realising benefits for stakeholders, and performed in a transparent, fair and ethical way that includes proactive stakeholder participation".

3.4.2.3 Incentives

The studies show that the introduction of economic incentives stimulates a CW diversion from landfills by subsidising waste reduction, recycling and reuse practices (Manowong, 2012; Jia et al., 2017; Huang et al., 2018; Wang et al., 2019). Tax incentives given to contractors conforming to the CE practices are critical in driving the construction industry towards sustainability best practices (Mak et al., 2019). Begum et al. (2007a) vouch for the rewarding of construction stakeholders to urge the implementation of 3Rs to improve CWM. According to a survey carried out in mainland China (Wu et al., 2016), the participants agreed that economic benefits together with governmental supervision play a critical role in the implementation of CWM best practices.

A pilot program of using recycled materials in Shenzhen construction projects recommends the development of appropriate assessment systems to frequently measure the performance of companies in using recycled materials (Yuan, 2012). Thereupon rewards are for companies with better returns; as a result, this can work as a mechanism to push the construction sector into implementing recycling practices and the use of recycled materials in their projects. Gangolells et al. (2014) indicate the noticeable significant improvements in Spain due to introduced regulations that promote on-site sorting of CW and allows the use of recycled materials.

The Hong Kong government imposed contractual clauses to provide financial incentives for CW reduction (Lu et al., 2015). The incentivising increases competition across the sector, which may result in a significant increase in recycled material use and minimise waste generation. Correspondingly, it boosts the innovativeness of companies by pushing them to further research into sustainable practices to meet the waste targets set by law (Ghaffar et al., 2020). Huang et al. (2018) argue that most of the economic incentives in China are for CW treatment companies. Therefore, in the case of China, construction companies are not motivated to use waste minimisation technologies and techniques.

3.4.2.4 Fines/Penalties

The failure to comply with the imposed regulations is an offence punishable by law (Lu et al., 2016a), usually by enacted fines or penalties. Penalties are punishments enforced to effectively reduce illegal dumping of CW and encourage waste reduction practices (Huang et al., 2018). Accordingly, the research by Jia et al. (2017) indicates that imposing penalties will see a significant decrease in illegal dumping. In like manner, Chen et al. (2019) found that putting charges for unlawful dumping is insufficient, thus urges governments to improve the supervisory intensity and management level. Likewise, Blaisi (2019) indicates that fines and penalties may increase the recycling of CW but may alike lead to illegal dumping if there is a lack of enforcement and intense supervision.

3.4.2.5 Waste Disposal Charges/Landfill Taxes

Waste disposal charge increase may be adopted to promote the practice of recycling and other CW reduction strategies in the sector (Stenis, 2005; Begum et al., 2007a; Lam et al., 2019). Tax measures have assisted governments across many nations in diverting large volumes of CW from landfills to end of pipe treatment methods (Ajayi et al., 2015). The studies suggest that CW disposal charging scheme may reduce landfill area, decrease CW generation, enhance waste recycling practices by taxing virgin raw materials, increasing landfill fees and subsidising processing of recycled materials (Dahlbo et al., 2015; Ding et al., 2016; Wang et al., 2019; Blaisi, 2019). For instance, to avoid landfill taxes in Hong Kong, the contractors tend to focus on CW reduction, reuse and recycling (Lu et al., 2015). Likewise, Blaisi (2019), states that the United Kingdom (UK) has seen a decrease in landfill waste due to the rise in landfill taxes. Similarly, high landfill taxes in Denmark and the Netherlands have highlighted an increase in waste recovery (Mak et al., 2019).

Huang et al. (2018), also recommends introducing incentives that will push the construction sector to engage in CWR practices as well as increasing the disposal cost of CW. However, studies suggest that regulations such as progressive taxation for waste disposal might unintendedly promote illegal dumping (Stenis, 2005; Blaisi, 2019). For example, Germany experienced a surge in illegal dumping because of increasing landfill charges beyond market viability for the construction sector (Ried, 2000; as cited in Blaisi, 2019). As a result, the inflation of the taxation fees has to be constrained to avoid inspiring illegal dumping practices (Tam et al., 2014). Moreover, Wang et al. (2019) argue that the implementation of the waste disposal charges has not yet triggered subcontractors to change towards CW reduction methods. Despite that, some evidence suggests that waste disposal charges are useful largely. According to Yuan (2011; as cited in Gangolells et al., 2014), the increase in waste disposal fees leads to CWM benefits such as a decrease in illegal dumping, discounts in waste treatment fees due to better on-site sorting, use of more sustainable materials, practices and construction methods. Ultimately, there has to be a balance in the waste disposal fees to avoid triggering fly tipping.

3.4.2.6 Training and Education

Training sessions are provided to stakeholders to raise awareness towards appropriate CWM practices in minimising waste generation (Bakchan & Faust, 2019). Several studies imply that a lack of knowledge regarding the impacts of CW and awareness of the benefits of CWR makes the construction industry less interested in CWM (Del Río Merino et al., 2010; Yuan, 2012; Bakshan et al., 2017; Ding et al., 2016). A study conducted by Li et al. (2018) reveals that knowledge has a great deal of influence over the behaviour of the construction sector towards CW reduction.

Del Río Merino et al. (2010) state that in the case of Spain, no initiatives have been launched for education and awareness programmes for stakeholders in the sector. Again, a survey on the awareness of CE concept in the Chinese construction sector indicates that clients, designers and subcontractors are the least informed (Huang et al., 2018). The evidence above shows the need for educational intervention within the industry to raise awareness and promote best practices. Furthermore, Ghaffar et al. (2020) add that the challenge of adopting cleaner production is in changing the mind-sets of stakeholders and overcoming technical issues. Thereupon, efforts should be put in view of training and educating stakeholders of the sector to promote proactive measures in CWM. Zhang et al. (2019) suggest that more efforts are towards training and education, high level of engagement and supervision of stakeholders.

Begum et al. (2007a) highlight that the need to consider on-site staff training is vital to trigger environmental awareness and thus, promote CWM best practices. Consequently, the results of a survey amongst local contractors in Malaysia show that employees with a higher level of CW education can influence the contractor's attitude towards Sustainable CWM (Begum et al., 2009). In addition, Li et al. (2015), recommend visual demonstration of the adverse impacts of large volumes of CW generation or successful CWR projects to motivate decision-makers towards the adoption of Sustainable CWM best practices. For good measure, Gangolells et al. (2014) contend the support of efforts for sharing information on how to adopt the best practices at minimum cost. Sabini et al. (2017) pointed out the importance of the critical role played by professional institutions in developing and disseminating sustainable best practices. Likewise, Ghaffar et al. (2020) recommend the need for the industry stakeholders to adopt more robust dissemination strategy to share their practices.

3.4.3 What Are the Factors That Impede the Adoption of Sustainable CWM Best Practices?

The section looks at the factors that hinder the implementation of CE in waste management found in the literature review research. The study by Saez et al. (2013) classified drawbacks to the use of CWM best practices as economic (extra costs for technologies and techniques required for application), time (more time devoted to CWM), on-site space (lack of space for sorting and separation of CW) and extra paperwork (inspection reports and control forms). Furthermore, recent studies highlight that stakeholders identified the main barriers to adopting CWM as the government regulations, poor recycling infrastructure, lack of education, the perception that CWM impedes new construction progress, lack of consideration of CWM at planning and design stage, and logistical bottlenecks (Lockrey et al., 2018; Bakchan et al., 2019). The study splits barriers into four categories, namely: technical, behavioural, legal and market restrictions, corresponding to Zhang et al. (2019).

3.4.3.1 Technical Barriers

According to Zhang et al. (2019), technical barriers refer to a lack of technical expertise, standards and guidelines to promote the CWM plan. Again, Mahpour (2018) recognised barriers to waste reduction as a lack of appropriate indicators to measure source reduction and reuse, poor employee engagement and zero awareness of circular economic initiatives. On the other hand, Ghisellini et al. (2018) point out that lack of prefabrication technology, higher costs of handling prefabricated components, lack of skill for design and construction of prefabricated buildings as the technical barriers to adopting prefabrication. In brief, without the proper technology and proper skills, it is impossible to achieve a CE. For that reason, the findings by Oliveira et al. (2019) accept that low technology sites fail in implementing measures towards waste reduction and recycling as seen in Southern Brazil. Thus, governments are encouraged to assist the sector with proper technology to promote sustainable practices (Zhang et al., 2019).

Yuan (2012) adds that longer transportation distances of CW to treatment facilities result in higher costs of CWM and ultimately diminish the contractor willingness attitude to diverting CW from landfills. However, Zhang et al. (2019) encourage decentralisation of recycling facilities and markets to reduce the distance travelled by contractors and consequently, reducing the CWM costs. Contractors in Shenzhen identified the poor on-site sorting of waste as the primary technical issue leading to the low efficiency of reuse and recycling practices (Yuan, 2013). The major hindrances to on-site sorting may be due to limited space, health and safety restrictions, extra costs related to sorting and equipment required and its interference with project schedule (Ghaffar et al., 2020). Additionally, Huang et al. (2018) discovered the key barriers to reuse of CW as lack of guidance for effective sorting and collection, lack of standards and competitive market, and inappropriate urban planning demolition.

According to Dahlbo et al. (2015), the main barrier for recycling CW is the readily available virgin raw materials at low costs. Additionally, it is claimed that designers barely consider recycled materials during specifications, hence the low acceptance (Ajayi et al., 2015). Equivalently, Manowong (2012) adds that the process of reuse or recycling is considered to be costlier than buying raw materials. Furthermore, the ineffective CWM system, lack of advanced technology and weak recycled materials market are other major barriers (Huang et al., 2018). In like manner, the quality of recycled materials may be deficient relative to virgin materials. Some studies also hint high uncertainties or mistrust for the use of recycled materials for structural use (Faleschini et al., 2016; Huang et al., 2018; Zhang et al., 2019). It is recommended to test for the quality of recycled materials to ensure they comply with the construction standards. The examination of recycled materials can, however, be expensive. Thus, it may potentially offset the finances saved from recycling and reuse of CW (Ghaffar et al., 2020). The lack of profits might impede the industry's adoption of CWM best practices.

Ultimately, the recycled materials may fail to meet the required standards for reuse. Thereupon a need for further research into the quality of recycled waste is vital. According to a survey conducted in the UK with relevant stakeholders of the construction industry by Ghaffar et al. (2020), the recycled materials may not be used due to structural concerns or somewhat restricted for non-structural usage (Zhang et al., 2019). Again, the testing may require expensive advanced technologies and be time-consuming, affecting the progress of other construction activities. As a result, there is limited use of recycled materials, and the CW is preferably disposed of in landfills to avoid economic loss.

The recycled materials should demonstrate on top of the quality aspect, the economic viability (cheaper than virgin materials), social acceptability and environmental friendliness (Lockrey et al., 2018; Duan & Li, 2016). Therefore, as far as current recycling is concerned, Lockrey et al. (2018) noticed that concrete waste is used in temporary roads and then later transferred to landfills. Zhang et al. (2019) contend that the recent studies confirm the suitability of recycled waste material as base and sub-base materials for road pavements, and fit for precast construction and high-performance concrete.

3.4.3.2 Attitudes and Behaviour Barriers

Begum et al. (2009, p. 322) define attitude as "a positive or negative feeling toward specific objects", hence it exerts influence on behaviour (Al-Sari et al., 2012). According to Zhang et al. (2019, p. 14), behaviour barriers refer to "the attitudes and psychological perspectives which dictates human behaviour". Some scholars have pointed out that human factors such as stakeholders' attitude and behaviour influence the implementation of Sustainable CWM practices (Jia et al., 2017). Teo and Loosemore (2001; as cited in Begum et al., 2009) found that attitudes toward waste reduction are one of the major hindrances to Sustainable CWM. The most imperative behavioural barriers are the preference of CW disposal over recycling, poor quality perception of recycled materials, the spread of wrong information about recycled materials, lack of confidence in recycled materials and economic benefits of sustainable practices (Al-Sari et al., 2012; Zhang et al., 2019). Begum et al. (2009) state that contractor size, CWR problems, staff participation in training programs and waste disposal methods are vital factors that affect contractor attitudes in the sector.

Wu et al. (2017; as cited in Chen et al., 2019) point out government departments (legislation), construction contractors (competitors), economic viability and government supervision (regulations) as the main factors affecting contractors' behaviour towards Sustainable CWM. Yuan and Shen (2011) suggest that among other reasons affecting the effectiveness of CWM is the different views between stakeholders in the sector. The stakeholders can be divided into two categories, the first being authorities, the general public and NGOs who have positive attitudes towards diverting waste from landfills. Then the second category of stakeholders is identified as the project main contractor and subcontractor who tend to focus on the economic benefits and overlook the negative impacts of CW generation. Appreciating and optimising the construction stakeholder's behaviours is vital for improving CWR as the study by Li et al. (2018) revealed that knowledge had the most significant effect on CWM behaviour. Additionally, Loosemore et al. (2002; as cited in Ding et al., 2016) indicated the critical impact of human factors in CWR and contended that changing attitudes can minimise CW generation.

The studies imply that the enforcement of government guidelines should motivate the contractor's attitudes and behaviours towards effective CWM (Begum et al., 2007a). For instance, in China, the policies have a function of regulating and controlling actions of stakeholders in the transition towards sustainable practices (Ghisellini et al., 2018). Some studies, (Al-Sari et al., 2012; Chen et al., 2019), observed that without governmental regulations and supervision the attitudes and behaviours of the construction sector is driven by the occurrence of a direct economic benefit. However, the studies that focus on the decision-making behaviours in CWM are still limited (Chen et al., 2019). For example, there are no currently available studies that consider the attention and functioning of the construction professionals and workforce towards CWM in India (Kolaventi et al., 2019).

The study conducted by Jia et al. (2017) revealed that knowledge had the most considerable influence on CWM behaviour. Thus it is crucial to understand contractor employees' attitudes and actions to reduce waste generation effectively. Therefore, Al-Sari et al. (2012) point out that contractors with high numbers of unskilled workers have a negative influence on CWM. Bakshan et al. (2015) argued that there is an 83% probability of adopting CWM practices on-site when workers have a positive attitude to CWM. Kolaventi et al. (2019) highlight that most of the construction site workers are uneducated about sustainable practices, hence there is an urgent need for training them.

Consequently, the availability of training and awareness sessions is required to achieve positive attitudes of workers. Moreover, Chen et al. (2019) highlight the supervision intensity and costs, penalties, increased waste disposal costs and revenues from illegal dumping as significant factors influencing the behaviour of the construction industry. Although, Mak et al. (2019) suggest the implementation of waste disposal charges would motivate society's behavioural change to consider and become aware of the benefits of adopting Sustainable CWM best practices.

On top of the attitudes and behaviours, poor communication within the construction industry is a common barrier that hinders the stakeholders from sharing ideas and methods towards CWM best practices. Jin et al. (2017; as cited in Zhang et al.,

2019) highlight that improving communication amongst construction stakeholders concerning CWM will help in overcoming behavioural barriers. The results of Blaisi (2019) survey with academics showed a lack of data about the total amount of CW generated and methods of its management, which hinders their ability to make informed decisions towards waste reduction and disposal practices. Finally, there is a unique need for including all stakeholders in CWM decision-making by regulators or the government to influence their attitudes and behaviours by harmonising with their interests. As the study by Sabini et al. (2017) concluded that influences from other stakeholders are fundamental in integrating sustainability in project management.

3.4.3.3 Legal/Legislation Barriers

Legal barriers refer to the lack or absence of active policies, regulations and legislation to promote CWM best practices (Zhang et al., 2019). Ineffective government regulations and supervision prevent the implementation of waste reduction practices (Ghisellini et al., 2018). Yuan (2012) argues that, particularly in developing countries, project decision-makers focus on traditional project objectives, hence the weak regulatory environment and low implementation of CWM practices. Zhang et al. (2019) indicate that most of the current policies, specifically in China, tend to promote landfilling and down cycling but fail to address CWR in a sustainable approach.

Kolaventi et al. (2019) confirm that the Indian construction industry is well aware of the regulations for CWM. Despite that, still, there is no extensive use of the practices due to their weak enforcement and complexities of integrating them within the traditional systems. The government regulations are not strict and rigid enough to support waste reduction and environmentally sustainable CWM (Duan & Li, 2016). A study conducted by Ghisellini et al. (2018) in Shenzhen identifies the main weakness of practising Sustainable CWM as the lack of clarity on government regulations enhancing CW reduction and recycling, and responsibility of government departments in enforcing legislations.

3.4.3.4 Market/Economic Viability Barriers

Ghisellini et al. (2018) argue that the sector is profit-driven and economic viability is critical in the CWM decision making. Ghisellini et al. added that construction contractors instead pay attention to the cost, quality and schedule of the project as compared to CWR concerns. Studies by Yuan (2012) suggest the construction sector decision-makers emphasise more on cost, time, quality and safety (of the society) objectives, and less on the environment. Furthermore, Lu et al. (2015) point out that specifically the private construction sector emphasises more on time and cost-efficiency. Notably, the main motive of the construction industry is to generate economic benefits. As a result, the industry is reluctant to change as, in their view, sustainable approaches are time-consuming and tend to reduce profit margins.

In construction projects arguably clients do not allocate a budget for a proper CWM plan and hence there is a lack of market-driven benefits in adopting the best practices (Del Río Merino et al., 2010; Ding et al., 2016). Zhang et al. (2019) identify the high costs of CWM as one of the critical barriers to recycling. A survey among different construction stakeholders indicated that the return on investment (ROI) in diverting CW to recycling facilities is low (Jin et al., 2017; as cited in Ghisellini et al., 2018). The immature recycled materials market hinders the CWR practices, as the companies will not be able to compete in the market (Yuan, 2012; Ghaffar et al., 2020). Other studies identify the adoption of expensive cleaner technologies at the source, extra costs of recycling and the quality of recycled materials as the major hindrances to recycling practices (Yuan & Shen, 2011; Huang et al., 2018; Ghaffar et al., 2020).

Studies argue that from an economic standpoint recycling is implemented when the recycled materials are competitive to virgin materials in terms of cost, quantity and quality (Ghaffar et al., 2020; Blaisi, 2019). This competitiveness primarily occurs in areas with limited landfill space and sites where virgin materials are not available locally. For example, Jain et al. (2020) highlight that many regions of India will have to import sand from other countries if there is poor implementation of CW diversion methods. In this regard, the construction sector strives to focus on recycling and other CWR practices, as importing might result in high costs due to international tax and more extended transportation.

Ghisellini et al. (2018) contend that the initial cost of sustainable buildings is higher than conventional buildings, which discourages clients from investing in sustainable practices. Consequently, this brings about a reluctance to adopting sustainable practices as the main drivers of the sector is the profits (Bakchan et al., 2019). Wu et al. (2017; as cited in Chen et al., 2019) found that the construction sector's behaviours are affected by economic viability. Ultimately, as long as the CWR solutions are not as economically viable as the traditional ones, we will not witness the adoption of CE best practices on a large scale (Ghaffar et al., 2020).

The adoption of sustainable best practices may benefit the industry economically due to cost-saving, mainly from the reuse of salvaged materials from demolition activities and the use of recycled waste materials in place of virgin materials. Bakchan and Faust (2019) contend that the lack of Sustainable CWM practices proves to cause budget overruns on top of poor environmental performance. Evidence points out that 5% of construction waste reduction can save up to £130 million in the UK construction sector (BRE 2003; as cited in Ajayi et al., 2015). Manowong (2012) identifies minimisation of raw materials use, waste reduction, fewer greenhouse emissions and cost saving from reuse or recycling of waste materials. Likewise, Wu et al. (2016) note that effective CWM is considered vital in "achieving the visions of landfill space conservation, environmental impact reduction, job opportunity creation, and project expense reduction (USEPA, 2013)".

3.5 Conclusions

Waste management in construction has been a global issue that has brought to attention the need to come up with strategies to alleviate the environmental impacts of waste generation. Undoubtedly, the construction sector needs to adopt a system that will strive to consider all the three elements of the sustainability in harmony. In this literature review, Sustainable CWM was conceptualised through two main categories: triggers and barriers. Feasibility of the current and proposed strategies in Sustainable CWM is discussed.

The research provides evidence of global awareness of the 3R (reduce, reuse and recycle) principle as the significant CWM practice and finally landfilling as a last resort. The literature recommends the adoption of 3R in descending order of sustainability performance. The CW reduction principle is the eco-friendliest approach as it sought to reduce the production of CW at the source. The practice of estimating CW generation during the planning phase of construction projects is essential for developing a CWM plan that can reduce the waste generation. Various tools of CW estimation have emerged, although they lack reliability. Studies recommend the development and adoption of universally reliable and user-friendly estimation tools/methods, which can assist further in effective government legislation on minimising CW production.

Studies analysed indicate that when considered in the "design phase" there are the best chances to reduce waste. However, evidence from the very same research points that designers or architects lack awareness of the potential waste reduction at the design stage. When considered at that stage, the reduction in waste generated can be up for general construction projects. Another approach consists of the standardization of project design to facilitate reuse and recycling of material from old projects or buildings.

The introduction of precast or prefabricated materials during the construction stage promotes the CWM by minimising waste due to offcuts and site mixing. Similarly, the careful dismantling of old construction projects is advised, during the demolition phase, to reduce CW by furthering recovery of materials for reuse or recycle. More notable, the green building practices help significantly in reducing the waste produced during demolition activities. However, the green building practices are not highly implemented as they are considered more expensive than traditional building practices. There is a need to research on green building practices that are less expensive to trigger its implementation. Furthermore, a demolition-licensing scheme is encouraged to ensure proper regulations are in place during the demolition phase to maximise the salvaging of materials.

The production of CW is unavoidable, consequently, the end of pipe methods are adopted to divert waste from landfills. These end of pipe methods are mainly, reuse and recycle of CW. The reuse of salvaged CW is the priority as it is cost saving and has less environmental impact just like recycling. However, the literature suggest a lack of implementation in construction recycling practices (mainly for untrusted quality of recycled materials, poor technology and a lack of competitive market).

Another point emerged is the abundance of virgin materials, which hinders the sector from adopting recycling practices. Therefore, government regulations, such as imposing a tax on virgin materials, are encouraged to promote the recycled materials market. Ultimately, to facilitate reuse and recycling practices, proper on-site or offsite sorting practices must be adopted after careful demolition of infrastructure or buildings.

As a last resort, after all the waste reduction strategies have been implemented, the untreatable CW is disposed of in landfills. Landfills are the traditional and most widely used practice of disposing of CW. Studies recommend the diversion of CW from landfills as it has negative impacts on the environment and landfill space is filling up due to high CW production. Hiking of dumping fees has been put in place by governments to relieve the filling up of landfills (particularly in urban areas) and promote reuse or recycling of waste. On the other hand, the dumping fees should be monitored to avoid triggering the sector to illegal dumping as a way of escaping high landfill fees. However, the government supervision of CW disposal and setting of penalties will assist to decrease the illegal activities. Ultimately, for the legislations and regulations to be effective, the government has to involve the construction stakeholders' views and interests in decision-making.

The construction sector is reluctant to adopt the Sustainable CWM as they can be time-consuming, and the profit margins tend to be small as compared to traditional methods. The government policies and regulations are the major factors contributing to implementing or neglecting of sustainable best practices. The studies identified a lack of government regulations and support as the major drawback to the use of CWM best practices. Thus, the extensive literature recommends enforcing robust regulations to support and direct the sector towards the sustainability best practices.

The governments must incentivise waste reduction to increase competition towards CE in the sector. These economic incentives have shown to improve the CW diversion from landfills. However, the lack of CWM plan before construction hinders effective waste management. The literature stresses the need for the mandatory requirement in CWM to be approved by local authorities before construction starts. The failure to comply with the government regulations results in fines or penalties. These fines have played a role in reducing illegal dumping. On the contrary, other studies suggest that if there is a lack of enforcement and supervision, the fines may lead to an increase in illegal dumping activities. Therefore, the government need to intensify the supervision of the sector as well as monitor the fines.

Moreover, technical barriers such as low technologies, long CW hauling distances to treatment sites and the quality of recycled materials hinder the Sustainable CWM practices. The companies without advanced prefabrication or recycling technology fail to integrate sustainable practices in their construction activities. In conclusion, the government is encouraged to support the industry by providing financial support for high-cost technologies. Similarly, governments should decentralise recycling facilities and markets for reducing the distance for waste transportation. Therefore, to improve the quality of recycled materials, effective sorting practices should be implemented and research suggests a maximum of 50% mix ratio of recycled materials with virgin materials. However, further research on improving the quality of recycled materials for structural usage is recommended as the studies reference its use mainly on non-structural elements.

The research highlighted that the behaviour and attitudes of the construction stakeholders and employees play a critical role for or against sustainable practices. The main factors, which influence their behaviours, are government legislation, economic benefits and supervision. The enforcement of regulations and intense supervision can drive the sector towards effective CWM. The studies also conclude that there is a need for education and raising awareness in the construction industry on the benefits (mainly economic) of Sustainable CWM. The literature suggest that sustainable best practices have better economic benefits to conventional methods in the end. It is noticeable that the construction sector is reluctant to adopt them due to the high initial costs. Thus, the literature recommends thorough research on the economic benefits and how they can lure the industry towards sustainability.

References

Ajayi, S. O., Oyedele, L. O., Bilal, M., Akinade, O. O., Alaka, H. A., Owolabi, H. A., & Kadiri, K. O. (2015). Waste effectiveness of the construction industry: Understanding the impediments and requisites for improvements. *Resources, Conservation and Recycling, 102*, 101–112. https://doi.org/10.1016/j.resconrec.2015.06.001

Al-Sari, M. I., Al-Khatib, I. A., Avraamides, M., & Fatta-Kassinos, D. (2012). A study on the attitudes and behavioural influence of construction waste management in occupied Palestinian territory. *Waste Management and Research, 30*(2), 122–136. https://doi.org/10.1177/0734242X11423066

Bakchan, A., & Faust, K. M. (2019). Construction waste generation estimates of institutional building projects: Leveraging waste hauling tickets. *Waste Management, 87*, 301–312. https://doi.org/10.1016/j.wasman.2019.02.024

Bakchan, A., Faust, K. M., & Leite, F. (2019). Seven-dimensional automated construction waste quantification and management framework: Integration with project and site planning. *Resources, Conservation and Recycling, 146*, 462–474. https://doi.org/10.1016/j.resconrec.2019.02.020

Bakshan, A., Srour, I., Chehab, G., & El-Fadel, M. (2015). A field based methodology for estimating waste generation rates at various stages of construction projects. *Resources, Conservation and Recycling, 100*, 70–80. https://doi.org/10.1016/j.resconrec.2015.04.002

Bakshan, A., Srour, I., Chehab, G., El-Fadel, M., & Karaziwan, J. (2017). Behavioral determinants towards enhancing construction waste management: A bayesian network analysis. *Resources, Conservation and Recycling, 117*, 274–284. https://doi.org/10.1016/j.resconrec.2016.10.006

Begum, R. A., Siwar, C., Pereira, J. J., & Jaafar, A. H. (2007a). Factors and values of willingness to pay for improved construction waste management – A perspective of Malaysian contractors. Waste Management, 27(12), 1902–1909. https://doi.org/10.1016/j.wasman.2006.08.013

Begum, R. A., Siwar, C., Pereira, J. J., & Jaafar, A. H. (2007b). Implementation of waste management and minimisation in the construction industry of Malaysia. *Resources, Conservation and Recycling, 51*(1), 190–202. https://doi.org/10.1016/j.resconrec.2006.09.004

Begum, R. A., Siwar, C., Pereira, J. J., & Jaafar, A. H. (2009). Attitude and behavioral factors in waste management in the construction industry of Malaysia. *Resources, Conservation and Recycling, 53*(6), 321–328. https://doi.org/10.1016/j.resconrec.2009.01.005

Blaisi, N. I. (2019). Construction and demolition waste management in Saudi Arabia: Current practice and roadmap for sustainable management. *Journal of Cleaner Production, 221*, 167–175. https://doi.org/10.1016/j.jclepro.2019.02.264

Chen, J., Hua, C., & Liu, C. (2019). Considerations for better construction and demolition waste management: Identifying the decision behaviors of contractors and government departments through a game theory decision-making model. *Journal of Cleaner Production, 212*, 190–199. https://doi.org/10.1016/j.jclepro.2018.11.262

da Rocha, C. G., & Sattler, M. A. (2009). A discussion on the reuse of building components in Brazil: An analysis of major social, economical and legal factors. *Resources, Conservation and Recycling, 54*(2), 104–112. https://doi.org/10.1016/j.resconrec.2009.07.004

Dahlbo, H., Bachér, J., Lähtinen, K., Jouttijärvi, T., Suoheimo, P., Mattila, T., et al. (2015). Construction and demolition waste management – A holistic evaluation of environmental performance. *Journal of Cleaner Production, 107*, 333–341. https://doi.org/10.1016/j.jclepro.2015.02.073

Del Río Merino, M., Gracia, P. I., & Azevedo, I. S. W. (2010). Sustainable construction: Construction and demolition waste reconsidered. *Waste Management and Research, 28*(2), 118–129. https://doi.org/10.1177/0734242X09103841

Ding, Z., Yi, G., Tam, V. W. Y., & Huang, T. (2016). A system dynamics-based environmental performance simulation of construction waste reduction management in China. *Waste Management, 51*, 130–141. https://doi.org/10.1016/j.wasman.2016.03.001

Duan, H., & Li, J. (2016). Construction and demolition waste management: China's lessons. *Waste Management and Research, 34*(5), 397–398. https://doi.org/10.1177/0734242X16647603

Faleschini, F., Zanini, M. A., Pellegrino, C., & Pasinato, S. (2016). Sustainable management and supply of natural and recycled aggregates in a medium-size integrated plant. *Waste Management, 49*, 146–155. https://doi.org/10.1016/j.wasman.2016.01.013

Gangolells, M., Casals, M., Forcada, N., & Macarulla, M. (2014). Analysis of the implementation of effective waste management practices in construction projects and sites. *Resources, Conservation and Recycling, 93*, 99–111. https://doi.org/10.1016/j.resconrec.2014.10.006

Ghaffar, S. H., Burman, M., & Braimah, N. (2020). Pathways to circular construction: An integrated management of construction and demolition waste for resource recovery. *Journal of Cleaner Production, 244*, 118710. https://doi.org/10.1016/j.jclepro.2019.118710

Ghisellini, P., Ji, X., Liu, G., & Ulgiati, S. (2018). Evaluating the transition towards cleaner production in the construction and demolition sector of China: A review. *Journal of Cleaner Production, 195*, 418–434. https://doi.org/10.1016/j.jclepro.2018.05.084

Guerra, B. C., Bakchan, A., Leite, F., & Faust, K. M. (2019). BIM-based automated construction waste estimation algorithms: The case of concrete and drywall waste streams. *Waste Management, 87*, 825–832. https://doi.org/10.1016/j.wasman.2019.03.010

Huang, B., Wang, X., Kua, H., Geng, Y., Bleischwitz, R., & Ren, J. (2018). Construction and demolition waste management in China through the 3R principle. *Resources, Conservation and Recycling, 129*, 36–44. https://doi.org/10.1016/j.resconrec.2017.09.029

Jain, S., Singhal, S., & Pandey, S. (2020). Environmental life cycle assessment of construction and demolition waste recycling: A case of urban India. *Resources, Conservation and Recycling, 155*, 104642. https://doi.org/10.1016/j.resconrec.2019.104642

Jia, S., Yan, G., Shen, A., & Zheng, J. (2017). Dynamic simulation analysis of a construction and demolition waste management model under penalty and subsidy mechanisms. *Journal of Cleaner Production, 147*, 531–545. https://doi.org/10.1016/j.jclepro.2017.01.143

Jin, R., Yuan, H., & Chen, Q. (2019). Science mapping approach to assisting the review of construction and demolition waste management research published between 2009 and 2018. *Resources, Conservation and Recycling, 140*, 175–188. https://doi.org/10.1016/j.resconrec.2018.09.029

Kitchenham, B. A., & Charters, S. (2007) *Procedures for performing systematic literature review in software engineering*. EBSE Technical Report version 2.3, EBSE-2007-01, Software Eng. Group.

Kolaventi, S. S., Tezeswi, T. P., & Siva Kumar, M. V. N. (2019). An assessment of construction waste management in India: A statistical approach. *Waste Management and Research, 38*, 444. https://doi.org/10.1177/0734242X19867754

Lam, P. T. I., Yu, A. T. W., Wu, Z., & Poon, C. S. (2019). Methodology for upstream estimation of construction waste for new building projects. *Journal of Cleaner Production, 230*, 1003–1012. https://doi.org/10.1016/j.jclepro.2019.04.183

Li, J., Tam, V. W. Y., Zuo, J., & Zhu, J. (2015). Designers' attitude and behaviour towards construction waste minimization by design: A study in Shenzhen, China. *Resources, Conservation and Recycling, 105*, 29–35. https://doi.org/10.1016/j.resconrec.2015.10.009

Li, Y., Zhang, X., Ding, G., & Feng, Z. (2016). Developing a quantitative construction waste estimation model for building construction projects. *Resources, Conservation and Recycling, 106*, 9–20. https://doi.org/10.1016/j.resconrec.2015.11.001

Li, J., Zuo, J., Cai, H., & Zillante, G. (2018). Construction waste reduction behavior of contractor employees: An extended theory of planned behavior model approach. *Journal of Cleaner Production, 172*, 1399–1408. https://doi.org/10.1016/j.jclepro.2017.10.138

Lockrey, S., Verghese, K., Crossin, E., & Nguyen, H. (2018). Concrete recycling life cycle flows and performance from construction and demolition waste in Hanoi. *Journal of Cleaner Production, 179*, 593–604. https://doi.org/10.1016/j.jclepro.2017.12.271

Lu, W., Chen, X., Peng, Y., & Shen, L. (2015). Benchmarking construction waste management performance using big data. *Resources, Conservation and Recycling, 105*, 49–58. https://doi.org/10.1016/j.resconrec.2015.10.013

Lu, W., Chen, X., Ho, D. C. W., & Wang, H. (2016a). Analysis of the construction waste management performance in Hong Kong: The public and private sectors compared using big data. *Journal of Cleaner Production, 112*, 521–531. https://doi.org/10.1016/j.jclepro.2015.06.106

Lu, W., Peng, Y., Chen, X., Skitmore, M., & Zhang, X. (2016b). The S-curve for forecasting waste generation in construction projects. *Waste Management, 56*, 23–34. https://doi.org/10.1016/j.wasman.2016.07.039

Lu, W., Chen, X., Peng, Y., & Liu, X. (2018). The effects of green building on construction waste minimization: Triangulating 'big data' with 'thick data'. *Waste Management, 79*, 142–152. https://doi.org/10.1016/j.wasman.2018.07.030

Mahpour, A. (2018). Prioritizing barriers to adopt circular economy in construction and demolition waste management. *Resources, Conservation and Recycling, 134*, 216–227. https://doi.org/10.1016/j.resconrec.2018.01.026

Mak, T. M. W., Chen, P. C., Wang, L., Tsang, D. C. W., Hsu, S. C., & Poon, C. S. (2019). A system dynamics approach to determine construction waste disposal charge in Hong Kong. *Journal of Cleaner Production, 241*, 118309. https://doi.org/10.1016/j.jclepro.2019.118309

Manowong, E. (2012). Investigating factors influencing construction waste management efforts in developing countries: An experience from Thailand. *Waste Management and Research, 30*(1), 56–71. https://doi.org/10.1177/0734242X10387012

Moraes, F. T. F., Gonçalves, A. T. T., Lima, J. P., & Lima, R. D. S. (2020). An assessment tool for municipal construction waste management in brazilian municipalities. *Waste Management and Research, 38*(7), 762–772. https://doi.org/10.1177/0734242X20906886

Nunes, K. R. A., & Mahler, C. F. (2020). Comparison of construction and demolition waste management between Brazil, european union and USA. *Waste Management and Research, 38*(4), 415–422. https://doi.org/10.1177/0734242X20902814

Oliveira, M. L. S., Izquierdo, M., Querol, X., Lieberman, R. N., Saikia, B. K., & Silva, L. F. O. (2019). Nanoparticles from construction wastes: A problem to health and the environment. *Journal of Cleaner Production, 219*, 236–243. https://doi.org/10.1016/j.jclepro.2019.02.096

Paz, D. H. F., & Lafayette, K. P. V. (2016). Forecasting of construction and demolition waste in Brazil. *Waste Management and Research, 34*(8), 708–716. https://doi.org/10.1177/0734242X16644680

Paz, D. H. F., Lafayette, K. P. V., & Sobral, M. C. (2018). GIS-based planning system for managing the flow of construction and demolition waste in Brazil. *Waste Management and Research, 36*(6), 541–549. https://doi.org/10.1177/0734242X18772096

Penteado, C. S. G., & Rosado, L. P. (2016). Comparison of scenarios for the integrated management of construction and demolition waste by life cycle assessment: A case study in Brazil. *Waste Management and Research, 34*(10), 1026–1035. https://doi.org/10.1177/0734242X16657605

Rodríguez, G., Alegre, F. J., & Martínez, G. (2007). The contribution of environmental management systems to the management of construction and demolition waste: The case of the autonomous community of Madrid (Spain). *Resources, Conservation and Recycling, 50*(3), 334–349. https://doi.org/10.1016/j.resconrec.2006.06.008

Rodríguez, G., Medina, C., Alegre, F. J., Asensio, E., & De Sánchez Rojas, M. I. (2015). Assessment of construction and demolition waste plant management in Spain: In pursuit of sustainability and eco-efficiency. *Journal of Cleaner Production, 90*, 16–24. https://doi.org/10.1016/j.jclepro.2014.11.067

Sabini, L. (2016). Project management and sustainability. *Brief for GSDR*. https://sustainabledevelopment.un.org/content/documents/998449_Sabini_Project Management and Sustainability.pdf

Sabini, L., Muzio, D., & Alderman, N. (2017). *Integrating sustainability into project management practice: The Perspective of Professional Institutions*. IRNOP, April, 1–17. https://doi.org/10.5130/pmrp.irnop2017.5661

Sabini, L., Muzio, D., & Alderman, N. (2019). 25 years of 'sustainable projects'. What we know and what the literature says. *International Journal of Project Management, 37*(6), 820–838. https://doi.org/10.1016/J.IJPROMAN.2019.05.002

Saez, P. V., Del Río Merino, M., San-Antonio González, A., & Porras-Amores, C. (2013). Best practice measures assessment for construction and demolition waste management in building constructions. *Resources, Conservation and Recycling, 75*, 52–62. https://doi.org/10.1016/j.resconrec.2013.03.009

Stenis, J. (2005). Construction waste management based on industrial management models: A swedish case study. *Waste Management and Research, 23*(1), 13–19. https://doi.org/10.1177/0734242X05050184

Tam, V. W. Y., & Tam, C. M. (2006). A review on the viable technology for construction waste recycling. *Resources, Conservation and Recycling, 47*(3), 209–221. https://doi.org/10.1016/j.resconrec.2005.12.002

Tam, V. W., Li, J., & Cai, H. (2014). System dynamic modeling on construction waste management in Shenzhen, China. *Waste Management and Research, 32*(5), 441–453. https://doi.org/10.1177/0734242X14527636

U.S. EPA. Report on the 2013 U.S. Environmental Protection Agency (EPA) International Decontamination Research and Development Conference. Research Triangle Park, NC, November 05-07, 2013. U.S. Environmental Protection Agency, Washington, DC, EPA/600/R-14/210, 2014.

Vilventhan, A., Ram, V. G., & Sugumaran, S. (2019). Value stream mapping for identification and assessment of material waste in construction: A case study. *Waste Management and Research, 37*(8), 815–825. https://doi.org/10.1177/0734242X19855429

Wang, J. Y., Touran, A., Christoforou, C., & Fadlalla, H. (2004). A systems analysis tool for construction and demolition wastes management. *Waste Management, 24*(10), 989–997. https://doi.org/10.1016/j.wasman.2004.07.010

Wang, J., Li, Z., & Tam, V. W. Y. (2015). Identifying best design strategies for construction waste minimization. *Journal of Cleaner Production, 92*, 237–247. https://doi.org/10.1016/j.jclepro.2014.12.076

Wang, J., Wu, H., Tam, V. W. Y., & Zuo, J. (2019). Considering life-cycle environmental impacts and society's willingness for optimizing construction and demolition waste management fee: An empirical study of China. *Journal of Cleaner Production, 206*, 1004–1014. https://doi.org/10.1016/j.jclepro.2018.09.170

Won, J., Cheng, J. C. P., & Lee, G. (2016). Quantification of construction waste prevented by BIM-based design validation: Case studies in South Korea. *Waste Management, 49*, 170–180. https://doi.org/10.1016/j.wasman.2015.12.026

Wu, Z., Shen, L., Yu, A. T. W., & Zhang, X. (2016). A comparative analysis of waste management requirements between five green building rating systems for new residential buildings. *Journal of Cleaner Production, 112*, 895–902. https://doi.org/10.1016/j.jclepro.2015.05.073

Wu, Z., Yu, A. T. W., & Shen, L. (2017). Investigating the determinants of contractor's construction and demolition waste management behavior in mainland China. *Waste Management, 60*, 290–300. https://doi.org/10.1016/j.wasman.2016.09.001

Wu, H., Zuo, J., Yuan, H., Zillante, G., & Wang, J. (2019). A review of performance assessment methods for construction and demolition waste management. *Resources, Conservation and Recycling, 150*, 104407. https://doi.org/10.1016/j.resconrec.2019.104407

Yuan, H. (2012). A model for evaluating the social performance of construction waste management. *Waste Management, 32*(6), 1218–1228. https://doi.org/10.1016/j.wasman.2012.01.028

Yuan, H. (2013). A SWOT analysis of successful construction waste management. *Journal of Cleaner Production, 39*, 1–8. https://doi.org/10.1016/j.jclepro.2012.08.016

Yuan, H., & Shen, L. (2011). Trend of the research on construction and demolition waste management. *Waste Management, 31*(4), 670–679. https://doi.org/10.1016/j.wasman.2010.10.030

Yuan, H. P., Shen, L. Y., Hao, J. J. L., & Lu, W. S. (2011). A model for cost-benefit analysis of construction and demolition waste management throughout the waste chain. *Resources, Conservation and Recycling, 55*(6), 604–612. https://doi.org/10.1016/j.resconrec.2010.06.004

Zhang, L. W., Sojobi, A. O., Kodur, V. K. R., & Liew, K. M. (2019). Effective utilization and recycling of mixed recycled aggregates for a greener environment. *Journal of Cleaner Production, 236*, 117600. https://doi.org/10.1016/j.jclepro.2019.07.075

Part II
CE Projects and Sustainable Business Change

Chapter 4
Responsible Project Management Tensions in a Tier 1 UK Infrastructure Organization

Charles Spooner and Nigel L. Williams

Abstract Researchers have previously examined the tensions between individual actions and organizational structures. In the domain of Responsible Management, these tensions may manifest in the conceptualization of an organization's corporate responsibility and the individual agency of managers. In this paper, we examine these tensions using the case of a large UK infrastructure firm. To identify these tensions, organizational documents and interviews of professionals were conducted and analysed. Results indicate the presence of tensions including Cognitive Organisational-Individual Responsibility Dissonance, External role tension and Visibility of values. Future research could examine the role of smaller incremental organizational actions which demonstrate values internally on overall organisational sustainability behaviours.

Keywords Responsible project management · Procurement · Tensions · Sustainability

4.1 Introduction

The debate about the nature and extent of the responsibility of professionals, shareholders and stakeholders has been a prominent subject in management research. Corporate social responsibility can be considered a visible commitment to humanistic values predicated on an exchange basis where organisations attempt to secure future financial or non-financial returns (Kitzmueller & Shimshack, 2012). More recently, responsible management shifted the focus from the organisation to the

---

C. Spooner (✉)
J. Murphy & Sons Limited, London, UK
e-mail: charlesspooner@murphygroup.co.uk

N. L. Williams
University of Portsmouth, Faculty of Business and Law, Portsmouth, UK
e-mail: nigel.williams@port.ac.uk

V. Obradović (ed.), *Sustainable Business Change*,
https://doi.org/10.1007/978-3-031-23543-6_4

beliefs, behaviours and practices of managers of sustainability and social responsibility (Laasch & Conaway, 2015). This individual basis is currently explored in Project Management by the Responsible Project Management movement with current activities at www.ResponsiblePM.com (Thompson & Williams, 2018). For this paper, we define Responsible Management in Projects as the actions of people involved in the management of projects (Project Managers, Project Sponsors, Project Team Members) who seek to deliver beneficial environmental outcomes as part of their roles and are not dedicated to sustainability or social value professionals. These differing dimensions, organisational and individual, may result in tensions or paradoxes as organisations attempt to reduce their environmental impact.

Project Management is currently configured around a linear economic model in which resources are extracted, processed, assembled and disposed of at the end of useful life (Ghisellini et al., 2016). An alternative economic approach has been provided in the form of the circular economy that takes a closed-loop approach. The goal is to minimise the number of materials that are disposed of. In contrast, another economic model that has been gaining attention in the last decades is the Circular Economy, defined as a regenerative approach that aims to keep resources in a closed loop at their highest value (Within, 2015). This economic paradigm is opposed to the current linear take-make-dispose resource model that generates a significant amount of waste.

In the Circular Economy models, the end of life materials should be reused and their components and parts deconstructed to act as material banks for new initiatives, keeping the components and materials in a closed-loop. Firms have begun incorporating ideas from the circular economy to find useful ways to reduce, reuse or recycle (Andrews, 2015). These factors aim to find the best possible opportunities to eliminate waste materials (Prendeville, 2014) by implementing sustainable design into company processes (Esa, 2017). However, this idea is still developing in the project domain (Pomponi & Moncaster, 2017). To accomplish this paradigm shift requires project stakeholders to hold a shared conceptualisation of environmental responsibilities (Ghaffar et al., 2020). While corporate social responsibility requires organisations to respond to material environmental requirements, responsible project management suggests that personal values may drive the Project Manager's implementation of circular economy principles and practices. This sets up a tension between responsible management (focused on the individual manager in daily practice) and corporate social responsibility, which is focused on organisational level representations and processes of enacting societal and community responsibility. For example, the RPM principle of stakeholder engagement may be at odds with organisational interests in responding to entities that have been defined in CSR frameworks. There is, therefore, a need to examine the differences between organisations and project procurement employees' conceptualisations of environmental responsibility to identify tensions that can shape the responsible management activities of managers involved in projects. The research question is, therefore:

> What are the tensions between organisational and individual conceptualisations of responsibility for circular economy activities in the procurement function of Project Organisations?

This study was conducted in the UK, where projects produced 131 million tonnes of solid waste in 2014, just over half of the country's municipal solid waste (DEFRA,

2019). A significant amount of this waste is generated by tier 1 construction companies, defined as the top 100 UK building and infrastructure contractors (Institute of Civil Engineers, 2018). The study focuses on the procurement function of a company that is involved in the buying of goods and services that enable an organisation to operate profitably and ethically responsible manner (CIPS, 2019). Project procurement consists of the acquisition of resources from outside of the project environment. Depending on the project, this could include materials, machinery, tradespeople, consultants and a host of other goods and services (Pheng, 2018). With procurement being recognised as strategically significant (Humphreys, 2001) and some organisations spending nearly two-thirds of their revenue within the procurement function, they hold a significant influence and responsibility to source responsibly (CIPS, 2020).

4.2 Literature Review

As a significant amount of resources are used in projects, efforts have been made to encourage resource efficiency and reuse. In the project domain, the circular economy perspective integrates institutional and individual perspectives (Pomponi & Moncaster, 2017) where organisations develop processes to design, maintain and regenerate outputs (Mahpour, 2018). A significant amount of current research identifies the use of circular economy approaches at the front end of projects where decisions for subsequent development and delivery activities are made. Related research examines the use of life cycle analysis as an approach to compare design options for reducing the environmental impact of projects. This research implicitly takes an economic approach where organisations translate different types of sustainability value (environmental and social) into financial costs and benefits. This aligns with the usage of tools such as Life Cycle analysis or the Triple bottom line, which fits within existing financial reporting frameworks (Lake et al., 2015). In this way, they can be integrated within organisational improvement and development initiatives.

In addition to internal goals, the adoption of these circular economy practices reflects the need to respond to external pressures. Circular economy approaches provide a signal to external shareholders that the firm seeks to maximise long term value as waste is perceived as a financial loss in energy, effort and materials (Gray, 2006). In this way, the adoption of the approaches supports the acquisition of repeat business from customers who increasingly seek organisations that share their values. They are also used by organisations to obtain legitimacy or a "social licence to operate" from external stakeholders in the geographical communities in which project organisations operate.

In addition to regulations and standards, environmental sustainability in infrastructure projects is dependent on the beliefs and knowledge of stakeholders who participate in activities. While the circular economy may be planned/not planned into significant infrastructure projects based on regulatory and voluntary commitments, the realisation of these outcomes can be influenced by the knowledge and beliefs of procurement personnel. Specifically, the level of awareness of the circular economy, along with professionals' response uncertainty to the implementation of circular practices, can actively influence the nature and extent of implementation of circular economy practices (Sparrevik et al., 2021). These gaps may be linked to the systems characteristics of projects with the ambiguity of boundaries, negotiated responsibilities and temporal dynamism (Martin et al., 2013). It is not uncommon for these stakeholders to have conflicts of interest about sustainability within a network of a project (Lin et al., 2017). Project procurement professionals often have to make a personal decision on how to balance different stakeholders' wishes (Mok et al., 2015). While organisations have attempted to implement holistic sustainability frameworks, these may be more effective on internal rather than external stakeholders. For example, internal competency development by organisations does not only develop capability but also supports the development of sustainability mindsets (Silvius et al., 2017). Previous research has also examined the perception, awareness and sustainability behaviour of organisations and professionals (Masi et al., 2018). Other research has examined gaps in each of these dimensions, such as the gap between sustainability awareness and behaviour (Diófási-Kovács & Valkó, 2017).

These differing conceptualisations, organisational and legal requirements (standards and regulations) and individual beliefs and knowledge suggest a dialectical scenario in which the organisational entity has pluralistic interacting conceptualisations. Organisations can treat these tensions as a conflict engagement and seek to "win" by suppressing dissenting views, establishing the company sustainability narrative (Neale & Northcraft, 1989). Others may seek a creative synthesis that can be formalised into organisational routines. This approach recognises that environmental stances can be contradictory, resulting in tensions and paradoxes that can influence and change the overall stance of sustainability over time (Das & Teng, 2000). For project activities where there is a high degree of uncertainty, the latter approach may predominate as individual managers need to address emergent issues that cannot be entirely resolved via organisational processes (De Rond & Bouchikhi, 2004). This research, therefore, seeks to examine these competing conceptualisations to identify their possible influence on the implementation of project circular economy practices.

4.3 Methodology

The research was conducted as a case study of the procurement function of a large UK infrastructure project organisation using a process perspective. Process perspectives assume that context, organisation and individuals are linked, enabling the examination of complex relationships among entities, including tensions and paradoxes in a case study setting (Eisenhardt, 1989). Data collection was conducted via semi-structured interviews with eight procurement employees of a project organisation along with analysis of organisational policy and process documents related to

Table 4.1 Example of categories

Participant	Sub-category	Findings
All	Knowledge	Buyers had varying levels of understanding regarding sustainability
2		Some had a detailed understanding of job-specific requirements, e.g. sustainable wood
All All		
All	Personal opinions	Sustainability is considered by all participants to some degree
2,3,5		Some feel it should be considered in every action a buyer makes
1,6		Others feel it should be driven from the top
3,4		The older participants have seen it become more prominent over the years
All		All noted that it isn't as important to the company as commercial goals – business comes first
4		One participant had negative views on HS2 regardless of JMS' involvement in the project

environmental sustainability. As this research involves capturing the knowledge of procurement personnel, semi-structured interviews were used to examine sustainability beliefs and experiences. Analysis of sensemaking was done using coding to identify concepts and themes. Coding is defined as a label for assigning meaning to a segment of text which was then summarised into themes. An excerpt of categories derived from the interviews codes is provided in Table 4.1.

To analyse the organisational representation, company policy and related documents were analysed and compared with the interview findings. In the case of this research, any documents provided were compared to interview responses and findings of both interviews and secondary data were triangulated. This enabled the discovery of any differing perspectives between individuals and organisations that might occur at the company.

The benefit of this approach was not only the capture of different perspectives but the creation of new knowledge via triangulation. An example is presented in Table 4.2.

Sensemaking or theorising from these codes was done during a two-stage process of Narratives and Visual Mapping. Narratives were used to present a description of the emergent tensions in sustainability between company and individual. Visual mapping was then used to provide data expansion (making new connections between concepts), transformation (converting data into meaningful units) and reconceptualisation (rethinking theoretical associations) (Langley, 1999).

4.4 Findings and Analysis

The analysis identified four tensions in perspectives that can influence responsible management actions in project procurement: (1) Cognitive Tension, (2) Sustainability Direction Dissonance, (3) Stakeholder Role tension and (4) Visible Symbolism.

Table 4.2 Organizational perspective and individual perspective comparison

Source	Document content	Comparison to interviews
Social responsibility and sustainability policy	It aims to protect the environment, communities and stakeholders Ensuring the business and supply chain are governed correctly	Supports what was said in the interviews, specific policies that govern how the buyers operate
Environmental sustainability strategy	3+ year strategy It aims to include IS014001 Both an internal and external strategy	Buyers didn't know as much about IS14001 as shown in the policy Supports consideration of aligning external and internal strategies
Procurement sustainability strategy	Aims to meet government legislation in the area of net zero/circular economy Considers clients, suppliers and environment	Buyers had very limited knowledge of legislation Supports consideration of sustainability/circular economy across all levels of the procurement strategy
CP6 procurement strategy	Explains partnership with supply chain sustainability school. Through participation in the school, the company will partner with the government and competitors to reduce environmental impact	Supports that the client leads how sustainable a project can be Buyers showed a lot of support and interest in the supply chain school
Supplier sustainability questionnaire	Ensures suppliers can operate to the required sustainability level Considers emissions of vehicles, sustainability products and waste management	Supports buyers as indicate suppliers are crucial in reducing impact as they cover a large portion of supply chain

4.4.1 Cognitive Tension

Buyers had knowledge gaps in the scope of standards for sustainability (Table 4.2). Research has found that industry standards that support the circular economy, like ISO14001, are most effectively used when integrated with suppliers and clients (Weingarten et al., 2013), this in contrast to what the buyers said as they only considered ISO as an internal tool for sustainability. When asked about ISO14001, the buyers had minimal knowledge of the capabilities of the standards. This lack of understanding may have led them to disregard it's capabilities to provide guidance to not only them but also to suppliers and clients. This indicates that potentially this industry-standard may have supported alignment across both perspectives, supporting a consistent drive for sustainability (Weingarten et al., 2013). The existence and impact of cognitive tensions in the form of knowledge gaps have been identified in previous research (Diófási-Kovács & Valkó, 2017). In this study, this tension could have led to individuals' perceptions that the organisation was not making sufficient efforts towards sustainability which is in contrast to the organisation's financial and resource commitment to the circular economy via these standards.

4.4.2 Organisational–Individual Responsibility Dissonance

In this tension, the direction from which sustainability initiatives should be implemented has clashing perspectives that are not easily harmonised, hence dissonance. The procurement personnel's perspective is that more guidance could be top-down while the organisation indicated that top-down and bottom-up perspectives could be accommodated. This is in contrast to both responsible management perspectives that indicates that individual values can guide circular economy behaviour, and CSR perspectives that indicate that indirect or direct financial views are the primary drivers of behaviours. This dissonance occurs as the buyers' beliefs are sometimes challenged by the formal signals in the form of KPIs (Key Performance Indicators) that have been presented by senior management. The situation was further challenged as individual managers may have different views on undesirable consequences, therefore, will consider, for instance, commercial over sustainability. Procurement personnel buyers focused more on aligning sustainability with companywide policies but held the belief that the company perceived that commercial is more important than sustainability. Sustainability was considered to some extent by the buyers; they indicated that company KPIs did not weigh this area as heavily as commercial or health and safety. As one person stated:

> And one of the elements within the tender process is you know about sustainability. How sustainable are they? .... And we'll ask for examples of reduced plastics, carbon emissions, vehicle fleet... things in that space. Now that is weighted when you get your tender return. The sustainability element will have a weighting. That weighting won't be as prominent as commercial or health and safety.

The organisation saw sustainability as having top-down and bottom-up dimensions. The organisation saw itself as a major contractor and felt a responsibility to act sustainably. However, buyers believed that their opinion may often be overlooked as they are required to achieve certain commercial goals regardless of the impact on sustainability which contradicts the organisations' idea that sustainability should always be considered and that sustainability was part of their differentiator and responsibility. This tension may influence these buyers perceived responsibility for environmental matters. The buyers believed that the organisation should create the framework for responsibility; however in projects, due to the high level of uncertainty, the organisation sought to rely on the professional's expertise and judgement in procurement matters.

This tension created dissonance about incorporating external input. As projects have unclear boundaries, capabilities based on externally supplied knowledge can be built. However, without a shared conceptualisation of the organisation's stance, it may be difficult to determine the type of knowledge to be utilised, and which entity (organisational or individual) has responsibility for integration. For example, ideas were also being offered from outside the organisation from suppliers in the areas of reducing environmental damage and carbon offset. The buyers took the initiative to incorporate some of these methods, including low emission vehicles, carbon offset schemes and sustainable products.

4.4.3 External Stakeholder Role Tension

For large infrastructure projects, the government is both the major client and the creator of the circular economy institutional framework. This creates challenges as the government may mandate both environmental sustainability and the use of locally produced materials that may not necessarily meet the former requirement. Industry transformations towards sustainability are facilitated by government support but may cause challenges for individual organisations via contrasting policy dictates (Pitt et al., 2009). As a commercial entity, the organisation and the procurement personnel saw the client as the key stakeholder who, in many of their projects, was not only the negotiator of commercial terms but was the manager of the framework in which commercial transactions are negotiated. This challenge is exacerbated with local projects. As one respondent asked:

> How do Network Rail requirements influence our procurement processes? So at the beginning of a major project for a local government, our procurement point of contact will have to write a procurement strategy about how they're going to procure goods and services and subcontractors throughout the project life cycle to meet those requirements and local needs.

A related issue was the contrast between organisational and project specific policies. Requirements for the circular economy from the client perspective were focused specifically on projects (Zainul-Abidin, 2008) while the procurement personnel talked about sustainable policies as a companywide entity, supported by organisational policies including Environmental Sustainability Policy and Procurement Sustainability Policy. It was not entirely clear whether the effect of sustainable goals was achievable when policies are implemented companywide and not just restricted to a specific project. Although stakeholders such as pressure groups and activists may be heavily opinionated about certain projects, they may not actually have a direct influence on outcomes in current institutional framework (Xiong et al., 2015). For the procurement professionals in this study, this tension is another influence on the perceived scope of action. As the main client (the state) is also the shaper of institutional rules, then input from smaller stakeholders will necessarily be ignored. For buyers seeking to responsibly procure products, this adds uncertainty to their scope of action.

4.4.4 Visibility of Values

The case organisation made a significant amount of investment in formal systems to manage sustainability as it saw sustainability as a differentiator among other competing organisations. In addition to ISO14000, the organisation engaged in industrywide initiatives such as membership in training schemes, voluntary carbon reduction and circular economy initiatives such as materials reuse and recycle. The organisation also invested in internal initiatives such as workshops that would bring together people from different aspects of the organisation in order to work on and implement sustainability. However, from the individual professional's perspective their belief was that the values of sustainability were not embedded at the individual level. The organization needs to encourage the adoption of visible micro practices that while they acknowledge would not achieve significant carbon savings, were a demonstrator that the institutional character was embedded in daily practice of organisational employees. One respondent stated:

> It shouldn't have to be an initiative that this month we are having low plastics month. I want it to be how people live.

The organisation focused on the larger picture of carbon and emissions and did not consider the smaller incremental changes to environmental impact reduction, for example office plastic reduction. From the interviewee's perspective they wanted a greater consistency between the values practiced at the organisational level and the daily activities of managers. This perspective that these visible micro practices could signal a deeper commitment to sustainability may set up another implicit tension at the peer level where these managers who do not have responsibility for sustainability are attempting to shape lifestyle actions of organisational members. While there is some support for this in the literature, the organisation attempted to provide support for attitude change via events like hub days to reach a larger proportion of staff (Ehrenfeld, 2008). Figure 4.1 summarises these tensions below.

The outcome is an emergent responsibility stance that has a static and dynamic character. The static character is based on the organisations' external commitments in the form of standards, industry partnerships and supplier activities (Table 4.2). The dynamic nature is based on the internal uncertainty of procurement professionals as they seek to translate these ideas into daily practice.

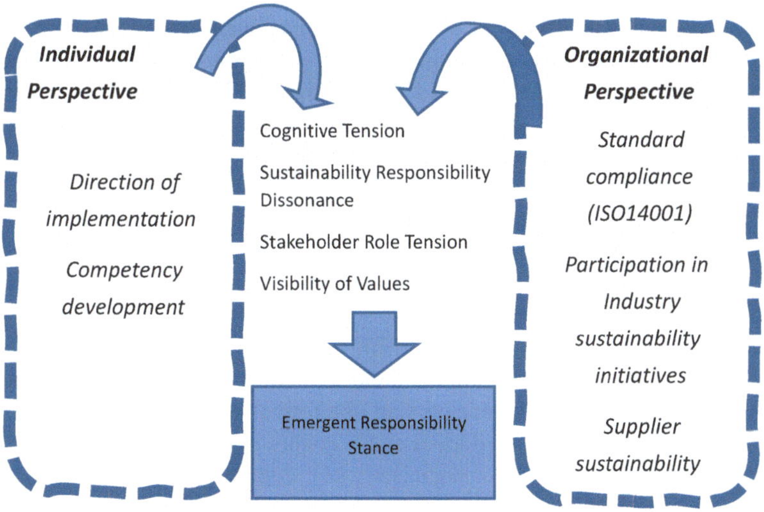

Fig. 4.1 Project procurement professional responsibility stance. (Source: The Authors)

4.5 Discussion and Conclusion: Tensions in the Project Procurement Professional Environmental Responsibility Stance

The debate over the responsibilities of managers and management dates back to the early formulation and recognition of management as a separate discipline. Research discussions have moved from notions of shareholder to stakeholder to societal responsibility, however in projects due to their high degree of uncertainty this extent of responsibility is still very much a debate among practitioners and professionals (Derakhshan et al., 2019). The identification in tensions based on sensemaking by stakeholders of different organisations has been identified in sustainability research (Chen et al., 2021). To date, however, there has been little exploration of tensions in circular economy responsibility stances within a given organisation. In this case study, we identified four tensions that can shape the emergent responsibility stance or perceived environmental responsibility role of procurement professionals in an organisation involved in large infrastructural projects. These tensions emerge between the beliefs of project procurement professionals and official organisational representations and the outcome has its own inherent contradictions. For example, the belief from practitioners that sustainability should be top-down is in contrast with their belief that visible micro practices shape organisational behaviour. In the former organisational rules may result in a particular set of behaviours while engaging in company sanctioned activities (Ahn, 2013) but will probably not extend to personal and individual actions outside of the company. However, the view of Procurement professionals in this study was that they not only wanted sustainability to be led from the top-down but also wanted individual organisational to engage in practices that extended beyond the boundaries of the workplace into their personal lives. This is not consistent with a top-down view of sustainability in projects. It is also not consistent with a Responsible Project Management perspective that focuses on the individual activities of managers as they attempt to encourage, not impose environmentally sustainable practices on the organisation. Responsible management seeks to increase the agency of managers, encouraging them to broaden their scope of action, and additional research could identify feasible pathways for procurement professionals to successfully enact change in their organisations.

Addressing of this tension may require joint sensemaking between the senior members of the organisation and the procurement team. This can take the approach of participatory workshops such as roadmapping sessions to establish shared sustainability values. These initiatives can be supported with shared KPI development to ensure that internal signals are congruent with organisational intent and participant values. Further, to further embed an agreed shared responsibility stance, the organisation can consider the development of a formalised circular economy workplace learning program to embed ideas and to ensure that internal participants have a shared understanding of organisational commitments (Andrianova & Antonacopoulou, 2020).

Further, the client was both user of the outputs and creator of the environmental regulations. These goals may result in stakeholder conflict of interest (Lin et al., 2017). The situation could become complex in mega projects where the central government seeks the economic benefits from the project, while local politicians and activists may wish for the project to be stopped altogether for environmental reasons. This may increase the uncertainty experienced by procurement professionals as they negotiate the differing perspective of client and community. Further research into the circular economy in infrastructure projects may need to explicitly recognise the monopsony or buyer controlled nature of this market. To date, little research has examined the impact of these tensions on internal organisational participants. In the case of this research, it may have created uncertainty in procurement participants. Industry participants may wish to work with joint initiatives such as the sustainability school (Table 4.2) in order to create industry frameworks that detail the responsibility for central and local government circular economy responses (Pinto & Williams, 2012).

The dynamic interplay among tensions will result in changes to responsibility stance over time, in some cases, dramatically. Specifically, large-scale rapid changes in personal circumstances such as the impact of COVID19 may result in the members of the organisation either seeking a more significant amount of assurance from official organisational values or may embolden professionals to provide stronger signals of personal values. These radical changes are consistent with a dialectical perspective of interacting organisational/individual perspectives.

These tensions and paradoxes are not easily resolvable at the organisational and industry level and will require additional research as well as action by the project community as a whole. Future research could examine the role of these smaller incremental changes that will have an impact on organisational sustainability behaviours. The emergence of the Responsible Project Management community (www.ResponsiblePM.com) could work alongside existing professional associations to support organisations and professionals as they navigate these tensions to embed circular economy principles in projects.

References

Ahn, C. (2013). Integrated framework for estimating, benchmarking, and monitoring pollutant emissions of construction operations. *Journal of Construction Engineering and Management, 139*.

Andrews, D. (2015). The circular economy, design thinking and education for sustainability. *Local Economy, 30*, 305–215.

Andrianova, O., & Antonacopoulou, E. (2020). Responsible managers workplace learning. In *Research handbook of responsible management*. Edward Elgar Publishing.

Chen, M. S., Eweje, G., & Kennedy, J. C. (2021). Managerial sensemaking of tensions in sustainability: Empirical evidendavice from Chinese and New Zealand business partnerships. *Journal of Cleaner Production, 128699*.

CIPS. (2019). *What is procurement*. Retrieved from CIPS: https://www.cips.org/en-GB/cips-for-individuals/what-is-procurement/

CIPS. (2020, March 31). *Why procurement matters*. Retrieved from CIPS: https://www.cips.org/cips-for-business/why-procurement-matters/

Das, T. K., & Teng, B. S. (2000). A resource-based theory of strategic alliances. *Journal of Management, 26*(1), 31–61.

DEFRA. (2019, March 5). *UK statistics on waste*. Retrieved from DEFRA: https://bit.ly/2ygeDr7

Derakhshan, R., Turner, R., & Mancini, M. (2019). Project governance and stakeholders: A literature review. *International Journal of Project Management, 37*(1), 98–116.

De Rond, M., & Bouchikhi, H. (2004). On the dialectics of strategic alliances. *Organization Science, 15*(1), 56–69.

Diófási-Kovács, O., & Valkó, L. (2017). Transition to green economy: Green procurement implementation strategies experiences from Hungarian public and private organisations. In *Sustainable economic development* (pp. 243–257). Springer.

Ehrenfeld, J. R. (2008). Sustainability needs to be attained, not managed. *Sustainability: Science, Practice and Policy, 4*(2), 1–3.

Eisenhardt, K. (1989). Building theories from case study research. *The Academy of Management Review, 14*, 532–550.

Esa, M. (2017). Developing strategies for managing construction and demolition wastes in Malaysia based on the concept of circular economy. *Cycles Waste Management, 19*, 1144.

Ghaffar, S. H., Burman, M., & Braimah, N. (2020). Pathways to circular construction: An integrated management of construction and demolition waste for resource recovery. *Journal of Cleaner Production, 244*, 118710.

Ghisellini, P., Cialani, C., & Ulgiati, S. (2016). A review on circular economy: The expected transition to a balanced interplay of environmental and economic systems. *Journal of Cleaner Production, 114*, 11–32.

Gray, R. (2006). Does sustainability reporting improve corporate behaviour? Wrong question? Right time? *Account and Business Research, 36*, 65–88.

Humphreys, P. (2001). Designing a management development programme for Procurement executives. *Journal of Management Development, 20*, 604–626.

Institute of Civil Engineers. (2018, January). *Where do Tier 1 contractors make their money?* Retrieved from https://www.ice.org.uk/news-and-insight/policy/where-do-tier-1-contractors-make-money

Kitzmueller, M., & Shimshack, J. (2012). Economic perspectives on corporate social responsibility. *Journal of Economic Literature, 50*(1), 51–84.

Laasch, O., & Conaway, R. (2015). *Principles of responsible management: Glocal sustainability, responsibility, and ethics*. Mason: Cengage.

Lake, A., Acquaye, A., Genovese, A., Kumar, N., & Koh, S. C. L. (2015). An application of hybrid life cycle assessment as a decision support framework for green supply chains. *International Journal of Production Research, 53*(21), 6495–6521.

Langley, A. (1999). *Strategies for theorizing from process data. Academy of Management Review, 24*(4), 691–710.

Lin, X., Ho, C., & Shen, G. (2017). Who should take the responsibility? Stakeholders' power over social responsibility issues in construction projects. *Journal of Cleaner Production, 154*, 318–329.

Mahpour, A. (2018). Prioritising barriers to adopt circular economy in construction and demolition waste management. *Resources, Conservation and Recycling, 134*, 216–227.

Martin, K., Legg, S., & Brown, C. (2013). Designing for sustainability: Ergonomics–carpe diem. *Ergonomics, 56*(3), 365–388.

Masi, D., Kumar, V., Garza-Reyes, J. A., & Godsell, J. (2018). Towards a more circular economy: Exploring the awareness, practices, and barriers from a focal firm perspective. *Production Planning and Control, 29*(6), 539–550.

Mok, K., Shen, G., & Yang, J. (2015). Stakeholder management studies in mega construction projects: A review and future directions. *International Journal of Project Management, 33*(2), 446–457.

Neale, M. A., & Northcraft, G. B. (1989). Experience, expertise, and decision bias in negotiation: The role of strategic conceptualization. In B. Sheppard, M. Bazerman, & R. Lewicki (Eds.), *Research on negotiations in organizations, Vol. 2*. Greenwich, CT: JAI Press.

Pheng, L. (2018). *Project procurement management*. Springer.

Pinto, J. Â., & Williams, N. (2012). *Country project management maturity*. Project Management Institute.

Pitt, M., Tucker, M., Riley, M., & Longden, J. (2009). Towards sustainable construction: Promotion and best practices. *Construction Innovation: Information, Process Management, 9*, 201–223.

Pomponi, F., & Moncaster, A. (2017). Circular economy for the built environment: A research framework. *Journal of Cleaner Production, 143*, 710–718.

Prendeville, S. (2014). *Circular economy: Is it enough?* EDCW.

Silvius, G., Kampinga, M., Pangiagua, S., & Mooi, H. (2017). Considering sustainability in project management decision making; an investigation using Q-methodology. *International Journal of Project Management, 35*, 1133–1150.

Sparrevik, M., de Boer, L., Michelsen, O., Skaar, C., Knudson, H., & Fet, A. M. (2021). Circular economy in the construction sector: Advancing environmental performance through systemic and holistic thinking. *Environment Systems and Decisions, 41*, 1–9.

Thompson, K. M., & Williams, N. (2018). *A guide to responsible project management.*

Weingarten, F., Pagell, M., & Fynes, B. (2013). ISO 14000 certification and investments in environmental supply chain management practices: Identifying differences in motivation and adoption levels between Western European and north American companies. *Journal of Clean Production, 56*, 18–28.

Within, G. (2015). *A circular economy vision for a competitive Europe* (pp. 1–98). Ellen Macarthur Foundation.

Xiong, W., Yuan, J., Li, Q., & Skibniewski, M. (2015). Performance objective-based dynamic adjustment model to balance the stakeholders' satisfaction in PPP projects. *Journal of Civil Engineering and Management, 21*(5), 539–547.

Zainul-Abidin, N. (2008). Achieving sustainability through value management: A passing opportunity? *International Journal of Construction Management, 8*, 79–91.

Chapter 5
Corporate Social Responsibility, Circular Economy and Sustainable Development: Business Changes and Implications in Project-Oriented Companies

Isidora Milošević, Sanela Arsić, and Anđelka Stojanović

Abstract By the end of the last century, the only goal of the company's businesses was to make a profit and paid very little attention to other aspects that were outside the obligations prescribed by law. However, constant social, environmental, and economic changes have affected the dynamics of changes in the business environment, which contributed to companies starting to apply the new concepts as Corporate Social Responsibility, Circular Economy, and Sustainability. Hence, this research aims to investigate the implementation of Circular Economy through Corporate Social Responsibility practice to achieve sustainable development in project-oriented companies in various industries. For this purpose, an adequate measuring scale was developed for assessing the respondents' opinions. The respondents were the employees of all levels and the management structures in companies from Serbia, Russia, and Bulgaria. For analyzing of results, the Shenon Entropy Method was used for estimating criteria weights. For the final ranking of five types of industries in surveyed countries, the TOPSIS method was used. Given that does not exist a research framework that is systematically dealing with the analysis of the Corporate Social Responsibility and Circular Economy in order to sustainability development in project-oriented companies, this research is contributed to theoretical and practical implication.

Keywords Corporate Social Responsibility · Circular Economy · Sustainability · Project-oriented organizations

I. Milošević (✉) · S. Arsić · A. Stojanović
University of Belgrade, Technical faculty in Bor, Bor, Republic of Serbia
e-mail: imilosevic@tfbor.bg.ac.rs; saarsic@tfbor.bg.ac.rs; anstojanovic@tfbor.bg.ac.rs

V. Obradović (ed.), *Sustainable Business Change*,
https://doi.org/10.1007/978-3-031-23543-6_5

5.1 Introduction

Modern capitalism and globalization should establish and implement universal business principles, which will contribute to creating a sustainable world. Corporate Social Responsibility (CSR) is a business practice that emerged first in developed countries, originally in the USA, while more recently, CSR and Sustainability have begun to gain attention in other emerging and developing markets (Doh et al., 2015; Stojanović et al., 2020). European business ground had a substantial contribution to the development of the CSR concept. However, within the continent, given that not all countries are at the same economic and social development level as well as historical and cultural oriented, it should be expected that CSR has been understood and implemented in various ways.

Corporate Social Responsibility is established as an instrument to manage organizations facing different economic and ecological environment issues created by the linear economy model. However, the linear economic model on which the previous industrial development was based has shown that the sustainability of economic prosperity and environmental protection requires a new model that will eliminate the weaknesses of the previous model. This model is called in the literature and practice the model of the Circular Economy (CE). Therefore, in line with the contemporary development of organizations, Corporate Social Responsibility is being modernized by a Circular Economy.

CSR is a broadly adopted social phenomenon, especially in the resources, financial, and services sectors, where the business activities create substantial benefits for stakeholders (Xia et al., 2018; Stojanović et al., 2020). Given the various business drivers of CSR, there is a broad spectrum of industry and organization features that influence whether and how CSR practice is adopted in order to reach sustainability (Doh et al., 2015; Đorđević et al., 2020). Sustainability arises from the idea of managing an organization without compromising the economic, ecological, and social environment while meeting current and future society demands (Crane & Matten, 2007). Sustainability has become one of the central matters and essential indicators of business success (Goni et al., 2017). The more financially stable and profitable an organization is, the more likely it is to invest in CSR activities (Doh et al., 2015). Corporate operations should be managed at a strategic level and performed with measurable impact on objectives in mind if the organization wants to be successful and sustainable (Plessis & Grobler, 2014).

Many countries have a long history of developing CSR culture and Sustainability. Others, which have passed the transition process, proactively accept the responsible business concept and ensure that companies behave socially acceptable.

Considering CSR in terms of Circular Economy and Sustainability, the gap in academic literature dealing with the elements of Circular Economy in various industries among different countries research is spotted. Hence, this paper aims to investigate the application of Circular Economy elements through CSR practice in order to achieve sustainable development in companies in various industries in Serbia, Russia, and Bulgaria. In this study, five types of industries according to the elements of the Circular Economy were ranked by integrated Entropy-TOPSIS methodology.

The research is presented as follows. The first section is the introduction. The second section gives an overview of the relevant scientific literature of Circular Economy, the third section deals with the connection between Corporate Social Responsibility and Circular Economy. The fourth section gives highlights of the application of CSR and CE in project management, while the fifth section is given the application of Corporate Social Responsibility and Circular Economy in various industries. The sixth section presents the basics of the methodology as well as data analysis and discussion of results. Finally, the conclusion section with recommendations and limitations are present in the last section.

5.2 Circular Economy

The industrial evolution has been dominated by a linear (one-way) model of production and consumption called take (from nature), make (in the process of production), use, and discard (waste). However, due to the increase of the global economy and the growing consumption of resources, there was a need for an innovative economic model. That has started companies to explore new ways to reuse products or their components and restore more of their precious material, energy, and labor inputs. As an alternative to an unsustainable linear production-consumption system, the Circular Economy has emerged. The Circular Economy is an umbrella concept that addresses how humanity produces and consumes goods and services (Schöggl et al., 2020; Suárez-Eiroa et al., 2021). The main aim of the Circular Economy is to extend the useful life of products, components, and materials in circulation and without loss of value and reduce waste as much as possible. From that point of view, the Circular Economy represents a regenerative economic system based on business models that continue the end of life of materials through a maintenance process, recycling, and reuse (Machado & Morioka, 2021). The basic premise of the Circular Economy is that achieving sustainable development at the global level. It does not mean a change in people's quality of life, nor a decline in production and profits on the part of producers, but that the circular model has to be profitable such as the linear model, and that consumers enjoy in products and services. The Circular Economy concept changes economic logic because it replaces production with sufficiency: reuse what you can, recycle what cannot be reused, repair what is broken, remanufacture what cannot be repaired (Stahel, 2016).

The Circular Economy business models are realizing in two directions. The first approach is pointed at a business model that fosters reuse and extends service life through repair, remanufactures upgrades, and retrofits. The second business model focused on turn old goods into new resources by recycling materials (Stahel, 2016).

The concept of Circular Economy is not limited to company size, specific industries, or regions, already for its successful implementation, the underlying logic and patterns of business must be suitably understood. This concept can be implemented in companies in different ways, such as introducing internal structural changes in the company, business model adaptations, product development and design, or changing framework conditions (Zhu et al., 2014; Moreno et al., 2016; Lewandowski, 2016; Leal Filho et al., 2019; Stumpf et al., 2021).

The first country in the world to adopt a law for the Circular Economy was China in 2008. Later, in 2012, the European Commission published a document called the "Manifesto for a Resource-Efficient Europe," in which it is clearly emphasized growing pressure in the world, due to the lack of more natural resources, hence the European Union has no choice but to move to a resource-efficient and ultimately regenerative model of the Circular Economy (https://ec.europa.eu/commission/presscorner/detail/en/MEMO_12_989). The application of this concept is to enhance the sustainability of production and consumption and to provide to achieving "a cleaner and more competitive Europe" (European Commission, 2020). The European action plan about the Circular Economy is based on promoting the changes that lead to the strengthening of the Circular Economy (Stumpf et al., 2021). Korhonen et al. (2018) highlighted that the European Commission estimated that implementing the CE model of business could create 600 billion euros annual economic gains for the EU manufacturing sector alone. For example, McKinsey (2014) and Finland's Independence Celebration Fund (FICF, SITRA) considered that Finland, through a Circular Economy, can get annual gains of 2.5 billion euros. Next to the EU national governments, the CE concept is used in the UK, Japan, China, Canada, and some advanced international companies worldwide.

The Circular Economic concept has gained momentum among scholars and practitioners and has become an important academic research field with a steep increase in the publication number (Lewandowski, 2016; Kirchherr et al., 2017; Geissdoerfer et al., 2017; Govindan & Hasanagic, 2018; Daú et al., 2019). Findings by Kirchherr et al. (2017) indicate that the Circular Economy is most frequently depicted as a combination of reducing, reuse, and recycling activities, whereas it is often not highlighted that CE necessitates a systemic shift. The previous studies indicate that the CE concept has some barriers that vary considerably concerning industrial focus and level of implementation. However, in their systematic review research, Govindan and Hasanagic (2018) indicate that cross-sectoral and empirical studies relating to CE barriers are still relatively scarce.

The Circular Economy is recommended as an approach to economic growth that aligns with sustainable environmental and economic development (Korhonen et al., 2018; Suárez-Eiroa et al., 2021). Sustainability refers to benefiting the environment, the economy, and society (Elkington, 1997), while the primary beneficiaries of the CE are the economic actors that implement this concept (Geissdoerfer et al., 2017). In this rapidly changing economic and social environment, thinking about sustainability and social and environmental management is a way for companies to positioning and thriving (Stoyanova, 2019). Therefore, the strategic approach of Corporate Social Responsibility, which focuses on Circular Economy, is becoming increasingly crucial for the competitiveness of companies. The benefits are multiple and contribute to cost reduction, human resources management, risk management, customer relations, and innovation capacity.

5.3 Corporate Social Responsibility and Circular Economy Towards Sustainability

The concept of CSR has been developing as a business practice for years, while the Circular Economy appears as a concept in the contemporary literature, all under the umbrella of Sustainability.

There are many pending questions concerning the impact of CSR activities on Sustainability. CSR and Sustainability enter all areas of business. Those are supported by the fact that they are gaining more importance in companies running in various industries. The possibility for a process or a certain situation, to be maintained at a high level without resource restrictions is a goal to be achieved. The existing dominant technologies that companies use to perform business functions to meet their needs are slowly but surely exhausted. In that sense, the direction of thinking and eventual solution of the problem of preservation of the natural environment is the notion of sustainable development, i.e., Sustainability as one of the basic concepts of the economics of natural resources and environment (Silvius, 2017). It can say that many human activities are directed and maintained through Sustainability. People's way of life creates a complex set of values, goals, and activities and certainly implies social, economic, and environmental dimensions (Elmualim, 2017). Nowadays, Sustainability is one of the most significant challenges of modern companies, in which project management has a vital role in the realization of more sustainable business practices.

Sustainable development represents an ethical standard according to the World Commission on Environment and Development (Brundtland, 1987). Therefore, the sustainability scientists defined the framework for strategic development (FSSD), which defines four general sustainability principles (Robèrt et al., 2002; Baumgartner, 2014). These principles refer to a sustainable society such as (1) in a sustainable society, nature is not subject to systematically increasing concentrations of substances extracted from the Earth's crust; (2) nature is not subject to concentrations of substances produced by society; (3) nature is not subject to degradation by physical means; (4) in that society, people are not subject to conditions that systematically undermine the efforts to meet their needs (Brundtland, 1987).

Sustainable development can reflect on the innovation, profitability, and success of the companies. To achieve sustainability, companies need a framework they can implement on in order to identify internal and external factors of the business and improve corporate sustainability strategies to be more successful on the market. Hence, sustainability programs can be divided into internal and external dimensions. Internal sustainability programs are primarily directed at environmental concerns that are taken care of by implementing measures within the company, while the external sustainability programs are referred to as supply chains and invest the collaborative efforts to address environmental issues (Gimenez et al., 2012).

The global economy has been created with the idea of a high level of welfare, justice, and equality. However some of the instruments used for achieving those goals are failed and short-term interest prevailed, bringing with them products and innovations that are not environmentally friendly, societies that are not ruled by democracy, human rights violations, the free market illusion, and business actors that are not willing partners for establishing sustainable development. As a result of the problems and inequalities, the public's conflicting feelings towards business emerged. On the one hand, companies provided people with the necessary goods, services, and jobs. On the other hand, people believe that companies conduct their operations with the sole purpose of increasing their profits without taking into account the needs of others in the vicinity. In order to be able to balance the needs of the environment and achieve long-term sustainability of their business, companies were forced to mitigate the confrontation with society by introducing a new business model of Corporate Social Responsibility in their business strategies.

CSR in the global economy is an inevitable issue. Considerable development of society and information technologies cause greater attention directed towards improving social and environmental conditions, human rights, and significant pressure on companies to balance people, planet, and profit elements in their business (Kanji & Chopra, 2010).

Many concepts, definitions, and elements have been proposed to define corporate behavior that complies with the needs of the environment in which a company operates. However, none of them are generally accepted and comprehensive, therefore in literature can be found terms such as Sustainable Development (SD), Corporate Responsibility (CR), Corporate Sustainability (CS), Corporate Citizenship (CC) (Martens & Carvalho, 2016). Therefore, Corporate Social Responsibility is relevant for achieving sustainable development goals (Xia et al., 2018).

Despite the different views, Corporate Social Responsibility is considered the company's involvement in minimizing the unfavorable effects on the environment and society, going beyond minimum legal requirements, therefore undertaken voluntarily. Therefore, many definitions emphasized the voluntary moment as a qualification whether specific company behavior can be characterized as CSR.

Carroll (1979) introduced a model of four company responsibilities: economic, legal, ethical, and discretionary (philanthropic) responsibility. They form a well-known Pyramid of Corporate Social Responsibility which served as a base for many proposed models (Carroll, 1991). The base of the Pyramid is economic responsibility, where the role of business is fundamental, creating value for owners and shareholders. The next level is legal, where it is assumed that some issues encompassed in the frame of CSR can be imposed with regulations, but the moral implication of companies in that way can be omitted. Ethical responsibility comes as a consequence of embedded high values and norms in a company's business and significantly exceeds the legal level. At this point, the activities are being undertaken to prevent any social harm. Carrol's third level of the Pyramid of Corporate Social Responsibility, named ethical responsibility, precisely defines the voluntariness in applying CSR as a high standard of social involvement (Carroll, 2016). The most elevated position in the Pyramid belongs to altruism. This discretionary responsibility encompasses voluntary actions that are not directly connected to business and even not expected to deal with it. However, companies are guided with a desire to take an active role in a dynamic social environment. If helping society at this level is missing, that is not considered unethical behavior.

Authors often argue that economic responsibility is cardinal, especially in business survival, and goes even to the attitude that this responsibility is the only one that matters (Stojanović et al., 2021a). This rigorous understanding of the purpose of the business began with Friedman, in which opinion the company's sole obligation is to generate profits for owners and shareholders (Friedman, 1970). Carroll (2016)) pointed out that it maybe seems unusual to perceive economic responsibility as Corporate Social Responsibility. However, society expects that a company provides products and services at a certain quantity and quality level. Consequently, the company is expected to ensure financial effectiveness even more in today's global business, where long-term sustainability is becoming prerogative (Carroll, 2016). Companies that fail in fulfilling economic responsibility will not be able to fulfill any other responsibilities. Therefore this is the prerequisite for all others.

Managers and shareholders of the company determine the level of CSR involvement in the company. Therefore, CSR can be the way to fine adjust companies' activities to gain some additional benefits, such as special political benefits, social license to operate, comparative advantage, etc., and economic gains (Milne, 2002).

It should be pointed out that social-oriented activities are not mandatory and set by any law; instead, they come from the ethical sense of the companies that should return something to the society in which they operate. Those aspirations can be expressed in various forms. For example, the company can donate financial incentives to local sports, educational, or civil organizations to enforce community development. Also, under the same frame, the company can allow and encourage employees to participate in socially engaged activities. Managers should understand the broader social environment and decide and act according to morals, ethics, and base values that do not depend on the nation, state, or religion (Freeman et al., 2010). Given that social responsibility is discretionary, it is closely connected to the voluntariness aspect.

The stakeholder aspect of Corporate Social Responsibility presupposes responsibilities for the needs of wider stakeholders, not only shareholders and customers. For companies, the first responsibility to stakeholders deals with satisfying the clients' needs following ethical methods in performing business operations (Hanzaee & Rahpeima, 2013). By going beyond, companies built even stronger relations with stakeholders directly involved in the business (suppliers, customers, business partners, financial institutions). By promoting ethical attitudes and values, they spread good business practices even wider. Stakeholder theory becomes a central concept around which a new way of organizing the fulfillment of social responsibilities is formed. However, attempts to integrate responsibilities to different stakeholders into business operations can be very difficult in practice. It must be borne in mind that stakeholder demands are often conflicting, and resources that the company has to allocate for CSR are limited, so procedures to determine the priority of certain claims over others need to be established. Therefore, managers need to understand relationships with their stakeholders and make decisions that will not satisfy only one group of stakeholders while cause harm to another. Careful selection of stakeholders according to their power, legitimacy, or urgency is not uncommon, especially in developing countries where resource constraints are even more expressed, competitive pressures are intensified, and the CSR context is less developed (Jamali, 2008). Devotion to managing stakeholders is voluntarily continuously seeking solutions to be more responsive to stakeholders, rather than leaving it to the regulatory institutions (Freeman et al., 2010).

In stakeholder theory, the business is seen as an alliance among stakeholders where the common goal is achieving certain values. In that sense, the company management faces numerous challenges embodied in the diversity of stakeholder requirements locally and globally.

However, viewed from the company's strategic position, it is necessary to achieve positive economic results, and strategically, the focus of CSR is shifting towards finding common ground with economic results. In that sense, the dependence of the companies and the environment is established. But, on the other hand, the values and strategic goals of the company are harmonized with the needs of society, which achieves a win-win situation for business and society (Porter & Kramer, 2006).

The aspects that most often come out in CSR models and definitions are environmental responsibility, social responsibility, and economic responsibility. Increased productivity and development followed by enormous consumption brought welfare in many countries. On the other hand, growing industrialization meant the use of local resources and their devastation, which resulted in the degradation of the environment and endangering human health. This caused more and more attention being paid to global warming, CO_2 emission, overconsumption and depletion of natural resources, severe consequences of natural disasters, environmental accidents, and major corporate scandals. Recently, renewable energy, clean air, decreasing emission, recycling, and water protection have attained more prominence and have become inevitable topics for policymakers and scientists. The role of the companies, especially big multinational companies in the mentioned problems couldn't be neglected. Companies are forced to take responsibility for the environmental changes they cause by their actions, and on that occasion, they usually obey local regulations or customs in business practice. Regulations usually prescribe restrictions, i.e. safe limits, but the problem of global environmental degradation remains. Steffen et al. (2015) proposed a "planetary boundaries" framework which provides "science-based analysis of the risk that human perturbations will destabilize the Earth system at the planetary scale." Two core boundaries are climate change and biosphere integrity and an additional seven others, which need to be embedded in sustainable development goals to provide clean energy and sufficient food supply for upcoming generations.

The world's expectation is focused on the governments of the most developed countries and the management of large multinational companies to embed in the emerging economic system a new change that will integrate economic growth and environmental protection and strive to a Circular Economy (Vazquez-Burst et al., 2014).

This process requires investments in resource conservation, green management, and a focus on sustainable development to reduce the destruction of nature and produce environmentally friendly products. Environmental behavior expected from companies means control of environmental aspects through the whole product or service life cycle (Hrbáčková et al., 2019). It is supposed that investments in environmental protection have been made by companies in order to fit in the contemporary model of the Circular Economy. The ability of the company to incorporate in business operation constraints concerning natural resources is based on three connected environmental strategies: pollution prevention, environmental management system, and sustainable development (Hart, 1995; Hrbáčková et al., 2019). On the other hand, the raising question is whether this has financial benefits for the companies. The answers are double-sided, but by following environmental strategy companies develop assets and human resources capable of green innovations, product differentiation, and long-term sustainability (Porter & van der Linde, 1995; Clarkson et al., 2011).

In recent years, awareness of the need for industrialization to be carried out in an environmentally sustainable way has been developed, rejecting the linear model that assumes production, use, and disposal of used products (Kirchherr et al., 2018). As a result, the influential concept such as Circular Economy gains importance, striving to redesign business activities so that used products turn into resources for other industries. In this way, a unique approach to the connection between the economy, the environment, and society is defined. Research using indicators to monitor the Circular Economy indicates that the focus is on resource conservation through recycling programs, monitoring and redesigning of processes through the so-called Life Cycle Thinking (LCT) approach, programs based on sharing platforms, plans for surplus products, and multifunctionality of product, upcycling, etc. (Moraga et al., 2019).

5.4 Corporate Social Responsibility and Circular Economy in Project Management

Project management includes planning, organizing resources, monitoring and collaborating with stakeholders, and motivating employee teams. Project management planning has to focus on harmonizing team members in employing Corporate Social Responsibility strategies. It embraces being socially aware during all project phases in order to deliver ongoing ecological and economic sustainability. Sustainability should be embedded into project management methods and concepts to support the organizations in achieving competitive advantage, known as sustainable project management (Chofreh et al., 2019). Silvius and Schipper (2014) defined sustainable project management as the practices of "ensuring profitable, fair, transparent, safe, ethical, and environmentally friendly project delivery aiming at a project deliverable that is socially and environmentally acceptable throughout its lifecycle." Silvius and Schipper (2014) identified several opportunities for taking sustainability principles into project management. The four-dimensional framework in project management that includes sustainability in the triple bottom line perspective (environment, society, and economy) was defined by Marcelino-Sádaba et al. (2015). Hwang and Ng (2013) consider that the project manager needs to satisfy the traditional roles of project management and operate with the project in order to achieve effective and efficient sustainability business.

One of the main challenges for project managers is integrating a Corporate Social Responsibility strategy into sustainability project management to ensure that all projects comply with global CSR recommendations (Schieg, 2009). In the project, the project manager has a central position and can affect many aspects during the project realization (Silvius, 2016). Therefore, project managers should implement Corporate Social Responsibility and ethical orientation for all company stakeholders to provide corporate governance during all business and project actions. Also, the project manager has to affect the application of sustainability principles during the realization of the project (Goedknegt, 2013; Maltzman & Shirley, 2013).

Corporate Social Responsibility strategy in project management must be focused on ensuring that the project is carried out by meeting the wider social community's social, economic, and environmental interests. This means integrating Corporate Social Responsibility activities into all aspects and phases of the project activities in order to comply with all international guidelines. The basic elements of the project management concept are time, resources, and costs on the one hand and planning, monitoring, and control of individual project phases on the other hand. The success of the functioning of projects related to sustainable development as a system depends on how the defined goals are realized and achieved their purpose in a dynamic environment (Zhang et al., 2017). Corporate Social Responsibility guidelines need to be included in all projects in order to ensure compliance with the laws because the implementation of the Corporate Social Responsibility strategy becomes mandatory throughout to provide global long-term sustainable development. The guidelines and legislation of Corporate Social Responsibility sustainable development present an overall framework for project managers to ensure that they are dedicated to sustainable development and responsibility. These initiatives can enable managers in project-oriented organizations to be committed to raising their organizations' ethical standards and sustainability practices. Incorporating norms socially responsible business in project management operations and processes ensures the execution of and monitoring of social responsibility through all phases of operation of the project-oriented organization.

The perception of project-oriented companies as a factor in society is constantly changing to reflect the changing societal expectations represented by the community groups, governments, and other stakeholders (Plessis & Grobler, 2014). In traditional business, organizations are seen as responsible only for delivering products and services to the marketplace, contributing employments and workers' security, complying with legislation, rewarding investors with profit, and paying taxes to the government. In addition, the assignment of project management is to recognize environmental systems, identify the internal and external dimensions of Corporate Social Responsibility, and examine defined CSR standards for their use in various projects. Therefore, CSR in project-oriented companies contributes to establishing values such as integrity, reputation, and credibility. For the prosperous performance of Corporate Social Responsibility activities, it is necessary to harmonize the dedication of the project-oriented organization to its objectives and business operations.

The integration of Corporate Social Responsibility into strategic project plans should provide a range of social, economic, and environmental guidelines that can help project-oriented organizations meet all stakeholders' needs. An essential idea of project-oriented organizations is to connect competencies and different stakeholders to solve and overcome obstacles and problems better and more quickly. Therefore, the idea of knowledge combination and cooperation of key stakeholders, as an important dimension of the Corporate Social Responsibility concept, is central to project management (Hou et al., 2010; Stojanovic et al., 2021a). A partnership between project managers, hierarchical levels, and external stakeholders should be a primary value (Beringer et al., 2013). It is required that the organization views itself as project-oriented and all stakeholders have clearly defined key values of a project-oriented organization (Gareis & Huemann, 2000; Eskerod et al., 2015; Gemünden et al., 2018). It is, therefore, necessary that all project plans have a clear and detailed explanation of how each phase of the project will comply with global Corporate Social Responsibility regulations in order to inform all stakeholders, including customers, suppliers, distributors, and partners, about implementing a sustainability strategy for project management. Then everyone would understand how much an organization is socially aware, which will significantly help them become and remain sustainable in the future. This will set a positive example for all other market participants who will follow the positive examples of social sustainability and encourage the implementation of Corporate Social Responsibility strategy in all project-oriented organizations. In addition to the orientation towards social responsibility, the project management culture should also contain an additional value, such as the orientation towards sustainable development as well as towards Circular Economy (Huemann, 2015).

Today, integrating the Corporate Social Responsibility concept in project management is one of the most prominent global project management trends (Alvarez-Dionisi et al., 2016). This relationship indicates that project-oriented organizations assume responsibility for their actions to increase social impact, leading to changes in the organization's products, processes, services, practices, and resources (Magano et al., 2021). It is very important to make project management sustainable. In project-oriented organizations, employees are assigned to projects, processes are changed and adjusted, new stakeholders are hired, all to monitor changes in the environment. These changes are especially related to the Circular Economy, which is not short-term but has a long-term character (Magano et al., 2021).

Therefore, for project management, the Circular Economy is important because the principles of the CE must be a fundamental part of the project management process. These principles include product recovery management, life cycle assessment, adaptability, product design to be easier to use and later for recycling, which leads to sustainability (Sanchez & Haas, 2018). Furthermore, considering that the Circular Economy is recognized as a high-impact strategy helping society be aware of the limits of economic growth (Leipold & Petit-Boix, 2018), the project-oriented organization has to adopt this practice. However, they require a high initial investment. Therefore, in project-oriented organizations, managers should respect the principles of Circular Economy and be socially responsible for achieving the set goals, and have a responsibility towards society and the environment (Bănacu et al., 2016).

5.5 Application of Corporate Social Responsibility and Circular Economy in Various Industries

Sustainable development can be a source of success, innovation, and profitability for companies. Hence, the network between sustainability and project management is intensively developed. An increasing number of scholars and professionals are dealing with this topic which is still a challenge in the field of project management (PM) (Silvius et al., 2013; Martens & Carvalho, 2016; Martens & Carvalho, 2017; Carvalho & Rabechini, 2017). According to Marcelino-Sádaba et al. (2015), many unresolved questions still exist related to project management and sustainability. Chofreh et al. (2019) analyzed the literature in sustainability, project management, and sustainable project management. Carvalho and Rabechini (2017) considered the relations between project sustainability management and project success and discovered a low degree of commitment to social and environmental aspects in the surveyed projects. Also, the significant positive connections between project sustainability management and project success in reducing social and environmental negative impact were determined. Therefore, sustainability principles need to be incorporated in the project management concept by controlling various projects, programs, and portfolios (Chofreh et al., 2019). The application of Corporate Social Responsibility activities is widespread in many industries that base their business on projects, as evidenced by numerous research studies around the world.

Corporate Social Responsibility in the mining industry is increasingly represented because the mining industry is very important in the economic development of countries (Velasquez, 2012). In the world, over 20 million people depend on exploiting mineral resources as a basis for their living. Therefore, Narula et al. (2017) suggested a three-stage model that provided an innovative framework for sustainable rural livelihoods in mining areas in the context of the changing CSR regime.

The implementation of Corporate Social Responsibility also is prevalent in the construction industry. Analyzing the state of the art in the construction projects, Xia et al. (2018) identified four research topics of Corporate Social Responsibility comprising CSR perception, CSR dimensions, CSR implementation, and CSR performance. Corporate Social Responsibility is more complex in the construction industry because of its project-based nature. The construction industry has some contradictions, at the same time is building a new environment but having an adverse effect on the environment (Wang et al., 2016). Based on that, Corporate Social Responsibility activities are more flexible and dynamic than in other industries (Evangelinos, 2016; Loosemore & Lim, 2017; Xia et al., 2018).

Corporate Social Responsibility in the Information Technology (IT) sector is formed as voluntary engagement of the companies and is supported by public policies and regulations (Martinuzzi et al., 2011). Institutional regulatory initiatives such as EU directives are aimed at ecological problems such as waste creation, disposal and recycling, hazardous substances, and chemicals. The role of CSR is to provide the locus for the European Information and Communication Technology (ICT) sector on the global market (MacGillivray et al., 2006). The authors argue that the European ICT companies need to put environmental and social concerns at the forefront of their business in order to gain a competitive advantage in the market. Dhanesh (2014) dealt with Corporate Social Responsibility as a possible strategy for managing to strengthen relationships between companies and their employees. Results revealed strong, significant, and positive associations between Corporate Social Responsibility and organization-employee relationships.

The service industry has more important environmental effects than companies realize. The problem with environmental impacts in the service industry is that ecological issues are generated across many different places. The application of CSR requires changes in many other business areas and establishing sustainability teams. Implementation of the CSR concept, in such a way, may increase the value of the companies' products or services for its clients (Camilleri, 2009; Bello et al., 2017; Fandos-Roig et al., 2020). Also, the CSR concept is positively linked to consumer trust and loyalty (Choi & La, 2013). Considering the services are intangible and used at the time of purchase, a high level of confidence in the suppliers is required in order to perform the buying process effectively (Wu et al., 2018). Fandos-Roig et al. (2020) investigated how service companies, through their CSR actions, can improve their client's loyalty and determined that CSR becomes a critical strategic asset for determining trust and loyalty among consumers. For companies that are oriented on services, it is harder to evaluate CSR efforts. Casado-Díaz et al. (2014) state that service company owners need to react more positively to CSR activities because the nature of services is specific.

As production requires a high degree of interaction with the environment, whether it is the acquisition of resources, labor, or target market, CSR practices need to focus on environmental issues in manufacturing industries (Handayani et al., 2017). That requires actively building and developing relations with the environment and establishing environmental management practices. A wide range of resources is used in production activities, so it is necessary to put sustainable development at the center when defining business strategies. According to Nagyová et al. (2016), the embracing of Sustainability and Circular Economy is important because it creates a special relationship between the manufacturing industry and the environment. Those relations are essential for continuous supply with high-quality raw materials and, on the other hand, contributing to the preservation of limited natural resources. Cherian et al. (2019) also suggested that monitoring the manufacturing industry is of vital importance compared to other industries since manufacturing cause notable devastation to nature, resulting in great environmental pollution. However, implementing environmental practices can generate numerous benefits for the industry, such as green innovation, high-quality products, greater customer satisfaction, lower production cost, and in the frame of Circular Economy, energy-saving, product reusing, and recycling. Sardana et al. (2020) examined the impact of environmental responsibility and supplier sustainability practices on companies' performance within the manufacturing industry. The study revealed that environmental sustainability had a direct influence on companies' performance. The effect of supplier sustainability on the performance of an organization was indicated to be positively moderated by organization ability. Xia et al. (2018) suggested that small and medium manufacturing enterprises also need to change the traditional business practices. They have to improve attitudes related to environmental responsibility and increase CSR implementation in their production activities to effectively be involved in the Circular Economy wave.

5.6 Methodology

The methodology used in this research consists of several steps and represents in Fig. 5.1. First, defining the research problem and analyzing previous literature research, a survey on Corporate Social Responsibility was conducted regarding the

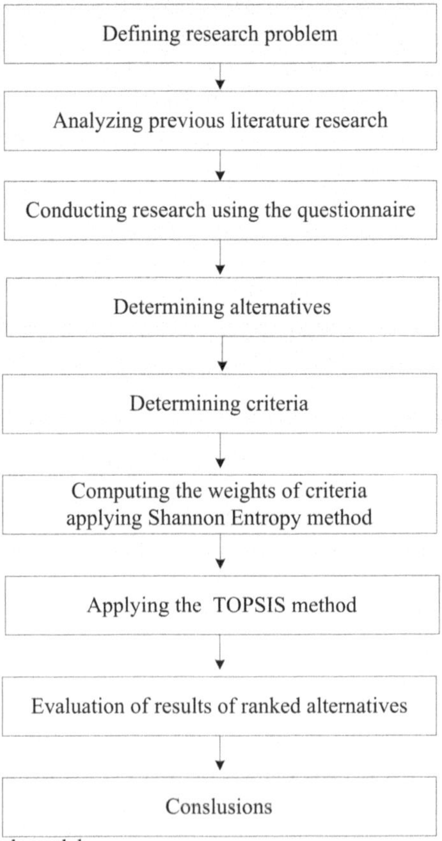

Fig. 5.1 Defined research model

Table 5.1 The statements that were evaluated

	Item
Q1	Energy saving
Q2	Waste recycling
Q3	Mobility management
Q4	Sustainable packaging
Q5	Develop of environmental friendly product
Q6	Life cycle assessment processes
Q7	Management of environmental system
Q8	Use of renewable resources

environmental aspect of sustainability. Then, a multi-criteria model was defined to determine the level of environmental impact practiced by project-oriented companies from different business sectors. The alternatives represent the combination of industry sectors and countries in the specified model, while the criteria are defined through selected questions about the Circular Economy. The next step is to determine the weights of the proposed criteria using the Shenon Entropy method. Finally, the ranking of the proposed alternatives using the TOPSIS methodology was performed.

The survey method, used for data collecting, enables gathering standardized data towards understanding the application of measures to reduce the environmental impact in project-oriented companies. In order to measure corporate environmental responsibility, eight questions from the questionnaire were used (Table 5.1), which referred to the Circular Economy thought activities carried out in order to reduce company's environmental impact. These statements were evaluated by employees of project-oriented companies operating in various industries such as Mining, Construction, Manufacturing, Services, and IT. The research was conducted in Serbia, Russia, and Bulgaria. Respondents rated the acceptance of specific environmentally responsible measures by their company on a scale of 1 to 5 (Likert scale) where a value of 1 meant "the measure is never implemented in the company I work for" while a value of 5 indicated "a measure is always implemented in the company I work for."

5.6.1 Entropy Method

Up to now in scientific research, many methods have been proposed for determining criterion weights in multicriteria models, including subjective as well as objective ones. Subjective weight methods, for example, AHP (Analytic Hierarchy Process) and the Delphi method, depend on experts' experience and judgments (Du & Gao, 2020). This can be considered as an advantage as well as a drawback. According to Li et al. (2011), evaluation of the weights of criteria using subjective methods such as due to subjective factors could deviate the criteria weights. One person's opinion is highly reliable, and weights are correctly estimated if the person is a real expert in the field. However, the opinion of experts can often be biased or under the influence of some circumstances. Also, in situations where fewer experts participate in defining the importance of criteria, group decision-making and fuzzy methods can be employed (Lamata et al., 2016). Objective methods reflect the weights that are contained in data itself. They are especially useful in situations where it is difficult to accurately determine the respondents' preferences, or the number of collected data is significantly large.

In this research, the Shannon Entropy method determines the magnitude of the diversity found in the data (Hamsayeh, 2019; Arsić et al., 2021). This means that if the entropy is low, the weight of the criteria will be higher because of the greater amount of information the data carries and vice versa (Du & Gao, 2020; Stojanović et al., 2021b).

Shannon (1948) was the first who developed the concept of the entropy of the system in his theory of communicators and proposed the function of entropy represented by the Eq. (5.1):

$$H(p_1, p_2, \ldots, p_n) = -K\sum_{i=1}^{n} p_i \log(p_i) \tag{5.1}$$

where (p_1, p_2 ..., p_n) represents the probability of random variables calculated from the probability function P a K and represents a positive constant.

Calculating the weight of the criteria $W = \{w_1, w_2 \ldots w_n\}$ using the entropy, where $C = \{C_1, C_2 \ldots, C_n\}$ is the criteria, $A = \{A_1, A_2 \ldots, A_m\}$ alternatives of the decision matrix, and x_{ij} indicators of the alternative value according to criterion j, can be done in a few steps (Hafezalkotob & Hafezalkotob, 2016). First, the normalization of the value of the decision matrix is determined using Eq. (5.2):

$$p_{ij} = \frac{x_{ij}}{\sum_{i=1}^{m} x_{ij}} \tag{5.2}$$

Then, the measure of entropy is calculated using the Eq. (5.3):

$$E_j = -k\sum_{i=1}^{m} p_{ij} \ln(p_{ij}) \tag{5.3}$$

where $k = 1/\ln(m)$. The objective values of the weight of the criterion are obtained by Eq. (5.4):

$$w_j = \frac{1 - E_j}{\sum_{j=1}^{n} (1 - E_j)} \tag{5.4}$$

The lower the entropy value is the lower is the degree of disorder of the system, which indicates that if the difference in the value between the evaluated object for the same criteria is high, the criteria will provide more useful information (Zhang et al., 2014; Arsić et al., 2021).

5.6.2 TOPSIS Method

TOPSIS (Technique for Order Preference by Similarity to Ideal Solution) was proposed by Hwang and Yoon (1981), enabling the determination of the positive ideal solution (A^+) and negative ideal solution (A^-). Based on this, alternatives can be ranked by estimating the shortest distance from the positive ideal solution and the farthest from the negative ideal solution. This method is one of the most popular methods in MCDM (The Multi-criteria Decision Making) (Dymova et al., 2013). For further improvement, it is often integrated with fuzzy logic or other MCDM methods to reduce biases in results (Cato, 2009).

The first step in the implementation of the TOPSIS methodology is the normalization of the initial matrix, using Eq. (5.5):

$$r_{ij} = \frac{x_{ij}}{\sqrt{\sum_{i=1}^{m} x_{ij}^2}} \tag{5.5}$$

Each element of the normalized matrix is multiplied by the corresponding weight criteria w_j, and thus, the elements v_{ij} of the weight matrix V are obtained, using Eq. (5.6):

$$V = v_{ij} = W_j^* r_{ij}, \quad \sum_{i=1}^{n} w_j = 1 \tag{5.6}$$

The next step is to form an ideal positive and an ideal negative solution. For each alternative A_i, the components A^+ of the positive ideal solution and A^- of the negative ideal solution are determined by the Eqs. (5.7) and (5.8):

$$A^+ = \left\{ \left(\begin{matrix} \max \\ i \end{matrix} v_{ij} \mid j \in J' \right) \text{and} \left(\begin{matrix} \min \\ i \end{matrix} v_{ij} \mid j \in J'' \right) \right\} = \left\{ v_1^+, v_2^+, \ldots, v_j^+, \ldots, v_n^+ \right\}, i = 1, 2, \ldots, m \tag{5.7}$$

$$A^- = \left\{ \left(\begin{matrix} \min \\ i \end{matrix} v_{ij} \mid j \in J' \right) \text{and} \left(\begin{matrix} \max \\ i \end{matrix} v_{ij} \mid j \in J'' \right) \right\} = \left\{ v_1^-, v_2^-, \ldots, v_j^-, \ldots, v_n^- \right\}, i = 1, 2, \ldots, m \tag{5.8}$$

where

J'⊆J→J' is a subset of the set J when it consists of max type criteria

J"⊆J→J" is a subset of the set J when it consists of min type criteria

Calculation of the separation measure (Euclidean distance) by using Eqs. (5.9) and (5.10):

$$S_i^+ = \sqrt{\sum_{j=1}^{n} \left(v_{ij} - v_j^+\right)^2} \quad (5.9)$$

$$S_i^- = \sqrt{\sum_{j=1}^{n} \left(v_{ij} - v_j^-\right)^2} \quad (5.10)$$

Finally, the relative closeness C_i to the ideal solution is determined using Eq. (5.11).

$$C_i = \frac{S_i^-}{S_i^- + S_i^+}, 0 \leq C_i \leq 1 \quad (5.11)$$

A larger index value C_i indicates a better position of the alternative.

5.6.3 *Results*

Primary data analysis was performed to obtain the sample's descriptive statistics using the SPSS v.25 software package.

Regarding the involvement of the company in CSR activities in general and a given management commitment, the question was asked at what level the activities of socially responsible behavior are implemented in project-oriented companies. The implementation of CSR and environmental activities in the company can be carried out from several levels. The top-down approach implies a focus on developing management strategies and initiatives that are spread through the company. In contrast, following the global economic and technological changes, employees are becoming more informed and aware of environmental problems and therefore want to engage in the direction of ensuring productivity, quality, and sustainability. This approach involves bottom-up initiation of environmental activities and behavior of employees following the principles of Circular Economy. Most answers were given to the Strategic/CEO level, 50.0% in Bulgaria, 34.5% in Russia, and 47.0% in Serbia, which implies that the engagement in socially responsible business is directed by the highest management level (Fig. 5.2).

The obtained results about considered industries are depicted in Fig. 5.3. In the Manufacturing industry, 48.5% of respondents think that planning and implementing CSR activities is the duty of higher management levels, 28.4% think it is the obligation of executive levels, while 22.4% of respondents consider that CSR activities are the bottom-up initiated. In the Construction industry top-down level is presented as the most important for CSR implementation with 48.4%, followed by

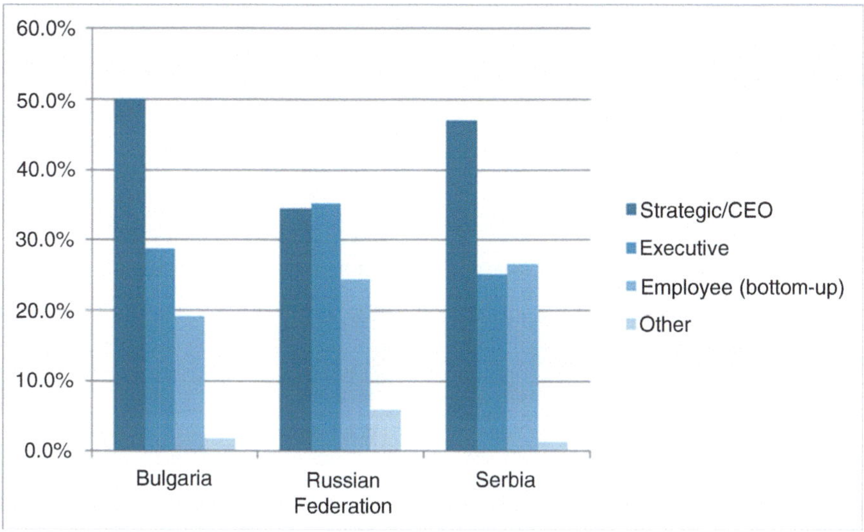

Fig. 5.2 Company's level on which CSR is managed, by countries

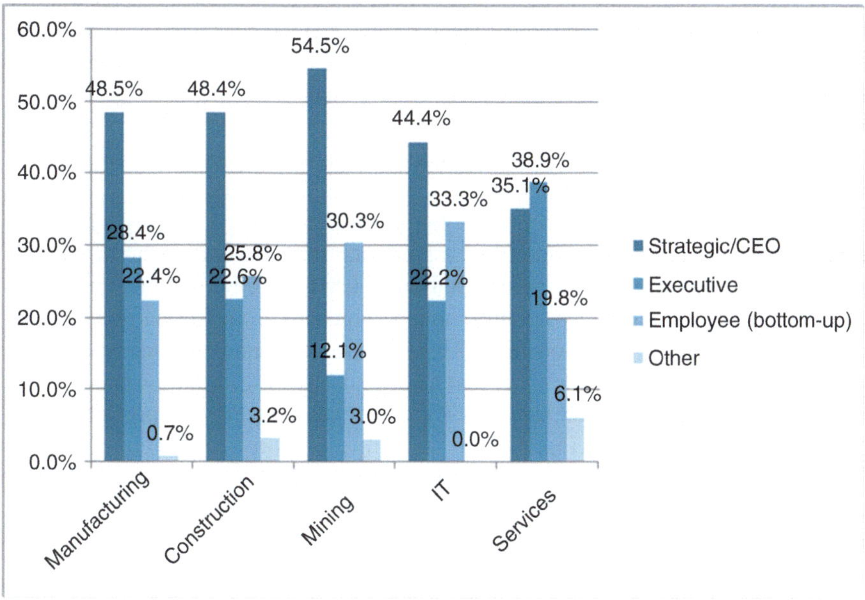

Fig. 5.3 The level of the company at which the CSR is managed

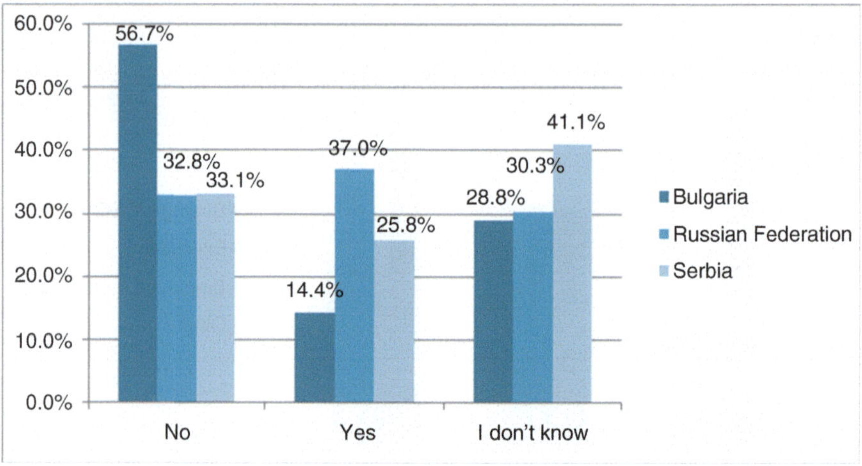

Fig. 5.4 Does the company have an in-charge unit/person for CSR within the organizational structure?

employee initiatives 25.8%, and executive-level 22.6%. CSR's highest level of obligation is on the Strategic/CEO level in the Mining industry, even 54.5%, then employee initiatives with 30.3% and executive-level with 12.1%. When observing the IT industry results, the highest management levels are the most responsible for CSR 44.4%, followed by employee level 33.3%, and executive-level 22.2%. The Service industry showed a little bit different results where the highest percent for managing CSR practice have executives 38.9% than Strategic/CEO 35.1%, and finally employees 19.8%.

At an organizational level, the respondents gave various answers to questions about how implemented CSR practices in companies they work.

Considering that project-oriented organizations have a specific organizational structure, the issue of the existence of a particular unit or person who deals with planning and monitoring the implementation of CSR and environmental activities was considered. Observed by country, only respondents from Russia mainly gave positive answers to this question with 37.0%. On the other hand, the majority answer of the respondents from Bulgaria is that a particularly obliged unit or person does not exist at the organizational level, 56.7%. In comparison, most respondents from Serbia do not know 41.1% (Fig. 5.4).

Also, an important issue for determining the place of CSR in a company is the existence of a defined CSR policy. This issue is vital for CSR communication through the company itself, regardless of whether it is project-oriented or not, and for communication with other stakeholders and through the entire supply chain. The results obtained from the analysis of the answers reveal that the answer "yes" is with the lowest percentage in all considered countries. Namely, in Russia, 33.6% of respondents answered positively to the question, in Bulgaria 24.0%, while the share of Serbian respondents is 21.4% (Fig. 5.5). Thus, responses that indicate that official CSR policies either do not exist or employees do not know about them prevail.

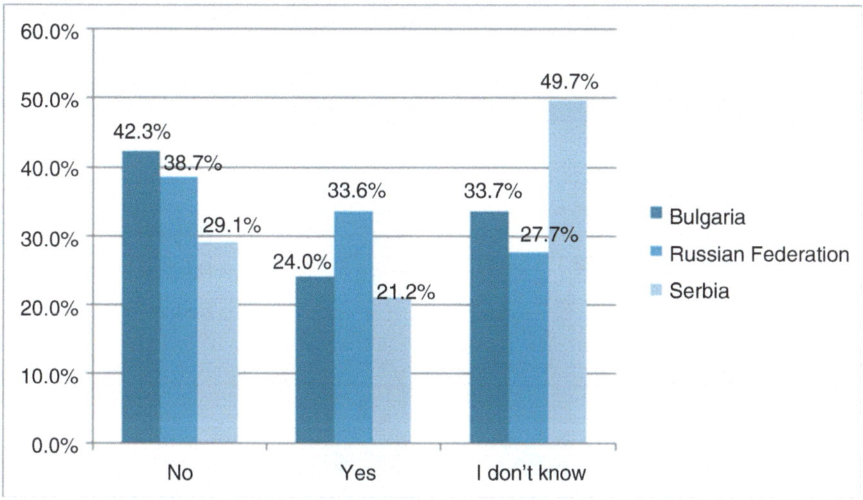

Fig. 5.5 Does the company have a CSR policy in place?

Corporate social reporting is not mandatory in many countries. Still, many companies have adopted this type of non-financial reporting as a well-established business practice and a way of communicating with stakeholders. In addition to CSR reports, sustainability reports and other forms of reporting are often used by companies. However, large companies are much more conditioned by regulations to submit environmental impact reports and sustainability and corporate responsibility reports, while small and medium-sized enterprises are, on rare occasions, conditioned by laws and are less bound by regulations (Rakić et al., 2021). Therefore, there are a large number of companies that probably don't publish this kind of report.

In the countries considered in the research, the results show a low percentage of companies reporting on their CSR practices when it comes to publishing CSR reports. In Russia, only 24.4% of respondents stated that reports on CSR practice and its impact are published in companies where they are employed. Bulgarian respondents answered in the affirmative in 22.1% of cases, while only 15.2% of respondents from Serbia answered in the affirmative (Fig. 5.6).

Earlier researches showed that managers often initiate CSR activities based on their personal beliefs, while in multinational companies, CSR activities are initiated based on directives from the company's headquarters (Jamali & Mirshak, 2007). What has also been observed is that there is no systematic measurement of social impact and that CSR reports are very rare. The general conclusion is that companies struggle with the basic elements of CSR and the impact of implemented activities and that few ask questions about the essential responsibilities that companies can take to influence changes in their environment.

By observing the results, it can be noted that there are many occurrences of answers "No" and "I don't know," when the place of systematic and organizational inclusion of CSR comes to questioning. These results impose the need to further

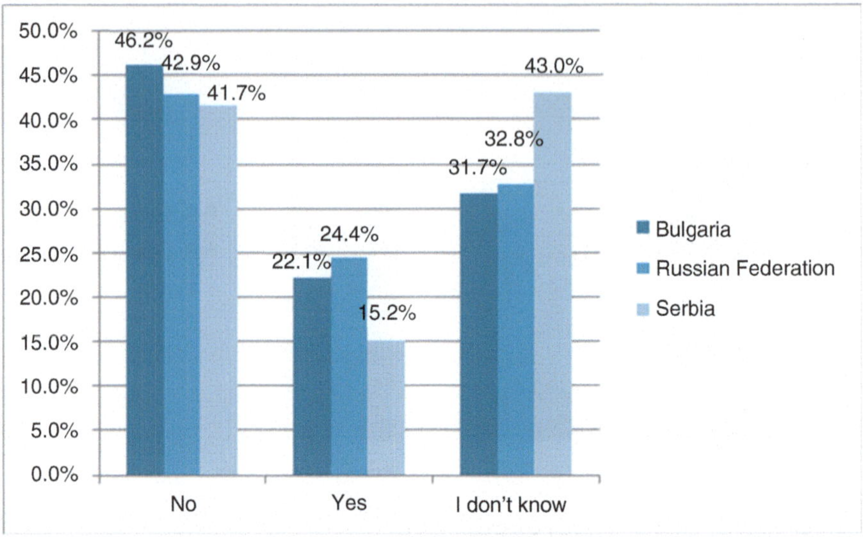

Fig. 5.6 Does the company publicly report CSR practices and their impact?

analyze the level of recognition of the companies' activities, especially in the ecological domain, which is the main interest of this research.

For this purpose, the model consisting of eight criteria and 13 alternatives is proposed, and initial data are presented in Table 5.2.

The weights obtained based on the Entropy method are objective from the information contained in the data itself. When evaluating weight (Eqs. 5.1, 5.2, 5.3, and 5.4), those criteria with lower entropy had a higher degree of variation in the data and, therefore, a higher value of the weighting coefficient. In contrast, those criteria with high entropy have a lower degree of variation expressed through the data and have a lower weight coefficient.

In the proposed model, the items "Sustainable Packaging" and "Development of Environmentally Friendly Products" had the highest values of weights, 0.22 and 0.17, respectively. High weights mean that the values obtained for Sustainable Packaging are significantly different between industries and states. Also, a similar conclusion can be made for the Production of Environmentally Friendly Products. On the other hand, the items "Waste Recycling" and "Energy Savings" are the only two items where the weight values are below 0.1 (0.07 and 0.08, respectively), which means that the attitudes of the respondents are more uniform in terms of applying these activities.

The overall result was obtained by integrating the TOPSIS methodology and entropy weights are depicted in Table 5.3. After calculating the distance of each alternative from the ideal solution (Eqs. 5.5, 5.6, 5.7, 5.8, 5.9, 5.10, and 5.11), a ranking was obtained for each observed industry/state combination.

The graphical interpretation of the obtained results is presented in Fig. 5.7.

Table 5.2 Initial data

	Q1	Q2	Q3	Q4	Q5	Q6	Q7	Q8
Manufacturing/Bulgaria	3.44	3.24	3.04	2.32	2.00	2.32	2.24	2.20
Manufacturing/Serbia	3.48	3.08	2.81	3.21	3.13	2.85	3.46	3.40
Construction/Bulgaria	3.00	3.35	3.38	2.92	2.58	2.54	2.73	2.73
Construction/Serbia	2.40	3.00	3.20	1.60	2.20	1.60	2.80	2.80
Mining/Bulgaria	4.00	4.13	3.80	3.53	3.73	3.40	3.60	3.53
Mining/Russian Federation	4.25	3.75	2.17	3.42	2.75	2.50	2.50	2.67
Mining/Serbia	3.17	3.83	3.50	2.67	2.50	2.83	3.50	2.50
IT/Bulgaria	3.63	2.75	2.75	1.63	2.38	2.38	2.50	3.00
IT/Russian Federation	3.81	3.11	2.26	2.67	3.00	3.15	3.15	3.22
IT/Serbia	2.50	2.50	2.10	1.50	1.40	1.80	2.70	2.20
Services/Bulgaria	2.90	2.50	2.23	2.00	1.83	1.77	1.70	1.87
Services/Russian Federation	3.44	3.51	2.70	2.96	2.96	2.87	3.10	3.20
Services/Serbia	3.14	2.73	2.82	2.50	2.27	2.50	2.64	1.95

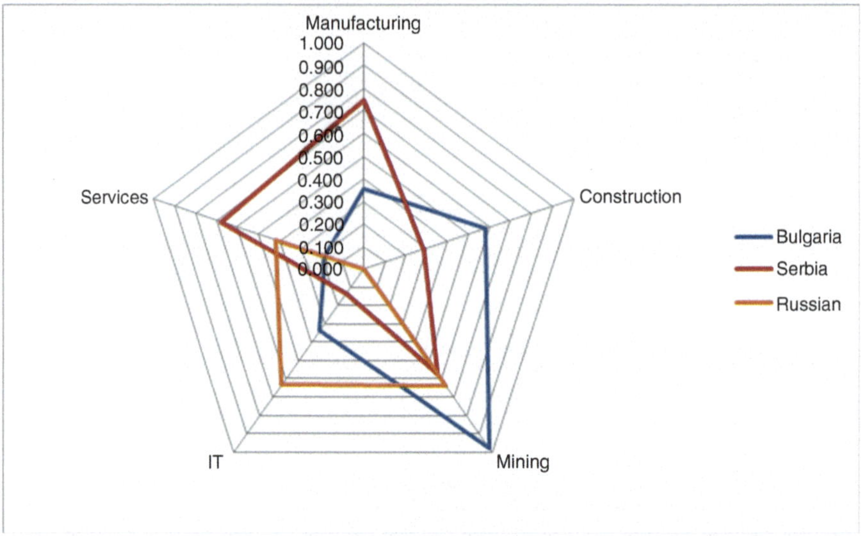

Fig. 5.7 Graphic representation of industry rankings by analyzed countries

Table 5.3 Ranking of industries by TOPSIS

	C_i	Rank
Mining/Bulgaria	0.980	1
Manufacturing/Serbia	0.749	2
Services/Russian Federation	0.680	3
Mining/Russian Federation	0.635	4
IT/Russian Federation	0.631	5
Construction/Bulgaria	0.586	6
Mining/Serbia	0.575	7
Services/Serbia	0.422	8
Manufacturing/Bulgaria	0.356	9
IT/Bulgaria	0.336	10
Construction/Serbia	0.284	11
Services/Bulgaria	0.183	12
IT/Serbia	0.136	13

Based on the obtained results presented in Table 5.3 and Fig. 5.7, it can be concluded that according to the level of implementation of environmental activities in the project-oriented companies, the best-ranked is the Mining industry from Bulgaria. At the same time, the worst position is the Services sector in the same country. When it comes to Serbia, the Manufacturing industry is the best ranked, while the IT industry took the last position. In Russia, the Services sector is the best-ranked industry, unlike the IT sector, which is the worst-ranked industry. In general, unlike other countries, all considered industries in Russia showed a higher level of implementation of environmental activities. Also, when looking at industries in general (Fig. 5.7), it can be seen that the Mining industry is highly positioned in all countries.

5.7 Discussion

For the purpose of research, the multi-criteria model was defined to determine the level of environmental impact practiced by project-oriented companies from various industries.

The obtained results indicate that the best ranking is the Mining industry. The Mining industry is under constant surveillance globally and locally, given the specific nature of performing mining activities, where a significant amount of natural and human resources are used. Therefore, mining companies are constantly expected to monitor and manage their environmental and social impact. In order to achieve the balance between harmful activities that deplete natural resources, and during extraction and processing cause various other environmental issues such as severe air and water pollution, endangering ecosystems, and adverse effects on human health, sustainable practices are becoming a necessity that must accompany the mining process (Govindan et al., 2014). On the other hand, as an answer to public pressures, a lot of legislation and regulations are being enacted to address mining issues. One of the answers given by the mining companies is that Corporate Social Responsibility is applied beyond required legislation with the purpose to achieve high overall performances at a socially acceptable cost. The management of the mining companies is aware of the significant hazards of mining and tries to facilitate local development in communities where they operate (Esteves, 2008). However, the research conducted in developing countries showed that mining industries were lagging in fulfilling Corporate Social Responsible strategies (Govindan et al., 2014). Considering the obtained results in this research, the Mining industry has a good position in the implementation of environmental corporate social responsibilities, hence it can be said that it is in alliance with the requirements of the Circular Economy. This can be explained by numerous initiatives and strict regulations enacted in the EU affecting the mining sector. Voluntary social engagement by the mining industry has also improved lately (Esteves, 2008). CSR programs developed by the management of mining companies are initiated as answers of strong interdependence with people in the vicinity of the mining places and on the other hand with weak cooperation with local governments and lack of serious planning of economy and society. Also, one of the prevailing arguments in the literature of CSR is that "if corporate interventions to address social problems are to be substantial and sustainable, they must also be profitable" (Esteves, 2008). This study is in line with previous research findings, where the mining industry is highly positioned in CSR implementation.

Huge funds have been invested in construction projects lately, yet sustainable construction in light of expectations that deliverables be socially acceptable is still lagging (Banihashemi et al., 2017). This research results showed the opposite, where the implementation of CSR in the Construction sector from Bulgaria is highly recognized. Unfortunately, this result cannot be generalized for all countries since the Construction industry in Serbia is positioned lower, whereas there is a lack of respondents from Russia.

In this research, the IT industry is the worst ranked industry in all considered countries. It can be explained by the fact that the implementation of environmental activities in the IT sector cannot be quantified and cannot be easily measured. However, besides products and services, the IT sector encompasses technological innovations that can help all other industries incorporate environmentally acceptable behavior in their practices. In this way, the Circular Economy makes it possible to identify the possibilities of the ongoing fourth industrial revolution and sustainable business practices while Corporate Social Responsibility is seen as the link between sustainability and the IT sector. Sustainability implies striving for business processes to be realized in such a way as to enable the conservation of natural resources by eliminating intensive spending and considering all used materials as potential resources for reuse. In addition to the commitment to redesign existing business processes, further transition to a Circular Economy requires the development of information and communication technology. This can be enabled by big data storage, information and communication systems with the goal to reduce costs, establish sustainable supply chains, and reduce energy and material consumption and management assistance for more efficient, smarter, and more responsive management (Daú et al., 2019).

There is diversity in activities and business models in the Services industry, and many are not covered under the circular economy. In Russia, the services sector is the most dominant when looking at the elements of the circular economy, while for the other two countries considered in the research, the results that the Services sector showed in the ranking are average. The need for direction toward circular and more sustainable economic models has become more evident lately. Therefore, it is expected for Services to embrace these trends.

The Manufacturing industry is the best ranked in Serbia, while in Russia and Bulgaria the ranks are lower. Nowadays, after numerous theoretical clarifications of Circular Economy, the missing link seems to be the practical solution for its implementation. The manufacturing industry is dealing with significant challenges when considering the increasing product demand and already severe shortage in raw materials and energy sources. With regard to this, the Manufacturing industry has to overcome serious barriers to transition towards the Circular Economy. The Manufacturing industry is very diverse, but each segment requires effective transition and opportunity that embraces significant economic benefit through decreasing environmental impact. By combining existing production processes and smart technologies with resource efficiency, the Manufacturing industry in integration with the Circular Economy can become a leading innovative business concept that contributes to global welfare. It should be emphasized that in order to further strengthen the Circular Economy, the manufacturing industry needs support in favoring products obtained from the chains of the Circular Economy and strategically planning production and supplying where everything and everyone is in the right place and work efficiently.

The Multi-criteria Decision Making (MCDM) methods can be a helpful instrument in the management and policy-making decisions. They could provide the flexibility and capacity to assess the opinions on the economic, environmental, social, stakeholders, and voluntariness context of CSR at the same time (Doukas et al., 2006). The managers can direct their activities to improve the perception of stakeholders and, as research proved (Reverte et al., 2015), to improve the overall results of their projects and companies.

5.8 Conclusion

The constant growth of population and consumption condition persistent depleting of natural resources. As the population grows, so does the need for products and services, and at the same time occurs an increase in waste. Thus, production and consumption activities and waste disposal significantly burden the economy and the environment. The possibility for scaling down of serious negative consequences can be seen in the model proposed by the Circular Economy. In order to reduce the deterioration of the planet, it is necessary to redesign the production and consumption cycle by carrying out certain activities aimed at achieving sustainability goals.

Sustainability, as a topic, results from raising considerations of environmental effects caused by business operations. At first, only environmental concerns are taken into account. Along with addressing ecological issues appeared the economic argumentation that companies play vital roles in the local and country economy but cannot solve all arising problems. Also, all this has reflected on relationships that companies built with their stakeholders, especially with the broader community. The Corporate Social Responsibility concept can be especially useful in implementing the new idea of a Circular Economy because it introduces a social, environmental, and ethnic dimension to a company's business. This concept offers improvement in business reflected in green business practices, greater innovation, a better reputation, and ultimately better economic performance while achieving sustainability.

Although the projects are seen as temporary ventures, project deliveries have a long-term impact on the social and natural environment, therefore, have obligation to balance between sustainability elements and primary project goals, time, scope, and costs. The research by Jovanović et al. (2019) showed that incorporating the environmental aspect in project management contributes to intangible benefits for project-oriented companies. Therefore, through better company image, the trust of stakeholders, and greater loyalty and engagement of employees, long-term sustainability can be achieved.

Considering the spotted gap in academic literature dealing with the measurement of environmental impact in project-oriented companies in different industries, the research was carried out on this topic from the position of CSR and Circular Economy in various countries. In this way, this research is contributed to fulfilling this literature gap.

The results showed a low level of implementation of CSR activities in some industries along with a low interest in Circular Economy and Sustainability. This research can have implications on the practitioners through the need to introduce some changes to inform and educate companies' actors on what those concepts mean and how they are important and valuable for business. Organizational and policy changes made in that direction shouldn't be dramatic for the company since project organizations are flexible and already accustomed to changes and learning. Although some research identifies specific issues for managing projects in a sustainable development manner (Swain, 2018), Sustainability incorporates enhancements of human living conditions in the sense of social well-being and ecological safety (Cato, 2009). Project management should meet emerging changes in global business.

This research has some limitations. The first limitation is based on the fact that research was conducted in three countries that share similar economic development in the previous 30 years, so the results cannot be generalized until compared to the results in some Western society economies. The second limitation is based on the use of only one MCDM methodological approach and will be overcome by applying the additional analytical techniques in future research.

Acknowledgement The research presented in this paper was done with the support of the Ministry of Education, Science and Technological Development of the Republic of Serbia, within the funding of the scientific research work at the University of Belgrade, Technical Faculty in Bor, according to the contract with registration number 451-03-68/2020-14/200131.

The research presented in this chapter is initiated through DAAD sponsored project "Corporate Social Responsibility (CSR) as the entrepreneurial activity," in the frame of the International Resita Network for Entrepreneurship and Innovation. The project coordinators were Professor Peter Schulte from INEA, Germany, and Professor Ivan Mihajlovic from the University of Belgrade, Faculty of Mechanical Engineering, Serbia.

References

Alvarez-Dionisi, L. E., Turner, R., & Mittra, M. (2016). Global project management trends. *International Journal of Information Technology Project Management, 7*, 54–73.

Arsić, S., Nikolić, D., & Jevtić, M. (2021). An investigation of the usability of image-based CAPTCHAs using PROMETHEE-GAIA method. *Multimedia Tools and Applications, 80*, 9393–9409.

Bănacu, C. S., Zecheru, V., & Olaru, B.G. (2016, November 3rd–4th). Project management in organic waste recycling. In *Proceedings of the 10th international management conference "challenges of modern management"*. Bucharest, Romania.

Banihashemi, S., Hosseini, M. R., Golizadeh, H., & Sankaran, S. (2017). Critical success factors (CSFs) for integration of sustainability into construction project management practices in developing countries. *International Journal of Project Management, 35*, 1103–1119.

Baumgartner, R. J. (2014). Managing corporate sustainability and CSR: A conceptual framework combining values, strategies and instruments contributing to sustainable development. *Corporate Social Responsibility and Environmental Management, 21*, 258–271.

Bello, F. G., Banda, W. J., & Kamanga, G. (2017). Corporate social responsibility (CSR) practices in the hospitality industry in Malawi. *African Journal of Hospitality, Tourism and Leisure, 6*, 1–21.

Beringer, C., Jonas, D., & Kock, A. (2013). Behavior of internal stakeholders in project portfolio management and its impact on success. *International Journal of Project Management, 31*, 830–846.

Brundtland, G. (1987). *Report of the world commission on environment and development: Our common future*. United Nations General Assembly document A/42/427. Available at: https://sustainabledevelopment.un.org/content/documents/5987our-common-future.pdf

Camilleri, M. A. (2009). *Creating shared value through strategic CSR in tourism* (Ph.D. Thesis). University of Edinburgh, Edinburgh, UK, (Unpublished).

Carroll, A. (1979). A three-dimensional conceptual model of corporate performance. *Academy of Management Review, 4*(4), 497–505.

Carroll, A. (1991). The pyramid of corporate social responsibility: Toward the moral management of organizational stakeholders. *Business Horizonts, 34*(4), 39–48.

Carroll, A. (2016). Carroll's pyramid of CSR: Taking another look. *Carroll International Journal of Corporate Social Responsibility, 1*, 3.

Carvalho, M. M., & Rabechini, R. (2017). Can project sustainability management impact project success? An empirical study applying a contingent approach. *International Journal of Project Management, 35*(6), 1120–1132.

Casado-Díaz, A. B., Nicolau-Gonzálbez, J. L., Ruiz-Moreno, F., & Sellers-Rubio, R. (2014). The differentiated effects of CSR actions in the service industry. *Journal of Services Marketing, 28*(7), 558–565.

Cato, M. S. (2009). *Green economics: An introduction to theory, policy and practice*. Earthscan.

Cherian, J., Umar, M., Thu, P. A., Nguyen-Trang, T., Sial, M. S., & Khuong, N. V. (2019). Does corporate social responsibility affect the financial performance of the manufacturing sector? Evidence from an emerging economy. *Sustainability, 11*(4), 1182.

Chofreh, A. G., Goni, F. A., Malik, M. N., Khan, H. H., & Klemes, J. J. (2019). The imperative and research directions of sustainable project management. *Journal of Cleaner Production, 238*, 117810.

Choi, B., & La, S. (2013). The impact of corporate social responsibility (CSR) and customer trust on the restoration of loyalty after service failure and recovery. *Journal of Services Marketing, 27*, 223–233.

Clarkson, P. M., Li, Y., Richardson, G. D., & Vasvari, F. P. (2011). Does it really pay to be green? Determinants and consequences of proactive environmental strategies. *Journal of Accounting and Public Policy, 30*(2), 122–144.

Crane, A., & Matten, D. (2007). *Business ethics* (2nd ed.). Oxford University Press.

Daú, G., Scavarda, A., Scavarda, L. F., & Portugal, V. J. T. (2019). The healthcare sustainable supply chain 4.0: The circular economy transition conceptual framework with the corporate social responsibility mirror. *Sustainability, 11*, 3259.

Dhanesh, G. S. (2014). CSR as organization–employee relationship management strategy. *Management Communication Quarterly, 28*(1), 130–149.

Doh, J., McGuire, S., & Ozaki, T. (2015). The Journal of World Business Special Issue: Global governance and international nonmarket strategies: Introduction to the special issue. *Journal of World Business, 50*(2), 256–261.

Đorđević, P., Panić, M., Arsić, S., & Živković, Ž. (2020). Impact of leadership on strategic planning of quality. *Total Quality Management & Business Excellence, 31*(5–6), 681–695.

Doukas, H., Patlitzianas, K.D., & Psarras, J. (2006) Supporting the sustainable electricity technologies in Greece using MCDM, *Resources Policy, 31*(2), 129–136. https://doi.org/10.1016/j.resourpol.2006.09.003

Du, Y.-W., & Gao, K. (2020). Ecological security evaluation of marine ranching with AHP-entropy-based TOPSIS: A case study of Yantai, China. *Marine Policy, 122*, 104223.
Dymova, L., Sevastjanov, P., & Tikhonenko, A. (2013). An approach to generalization of fuzzy TOPSIS method. *Information Sciences, 238*, 149–162.
Elkington, J. (1997). *Cannibals with forks: The triple bottom line of 21st century*. Capstone.
Elmualim, A. (2017). CSR and sustainability in FM: Evolving practices and an integrated index. *Procedia Engineering, 180*, 1577–1584.
Eskerod, P., Huemann, M., & Savage, G. (2015). Project stakeholder management - past and present. *Journal of Project Management, 46*(6), 6–14.
Esteves, A. M. (2008). Mining and social development: Refocusing community investment using multi-criteria decision analysis. *Resources Policy, 33*(1), 39–47.
European Commission. (2020). *Circular economy action plan – for a cleaner and more competitive Europe*. Publications Office of the European Union.
Evangelinos, K. (2016). Exploring the status of corporate social responsibility disclosure in the UK building and construction industry. *International Journal of Global Environmental Issues, 15*(4), 377–399.
Fandos-Roig, J. C., Sánchez-García, J., Tena-Monferrer, S., & Callarisa-Fiol, L. J. (2020). Does CSR help to retain customers in a service company? *Sustainability, 13*(1), 300.
Finland's Independence Celebration Fund (FICF, Sitra) and Mckinsey. (2014). *Kiertotalouden Mahdollisuudet Suomelle (the possibilities of circular economy for Finland)*. SITRA.
Freeman, R. E., Harrison, J., Hicks, A., Parmar, B., & de Colle, S. (2010). *Stakeholder theory – the state of the art*. Cambridge University Press.
Friedman, M. (1970). The social responsibility of business is to increase its profits. *The New York Times Magazine*, Section SM, 17.
Gareis, R., & Huemann, M. (2000). Project management competences in the project-oriented organisation. In: Turner, J.R., Simister, S.J. (Eds.), The Gower Handbook of Project Management. Gower, Aldershot, 709–721.
Geissdoerfer, M., Savaget, P., Bocken, N. M. P., & Hultink, E. J. (2017). The circular economy – A new sustainability paradigm? *Journal of Cleaner Production, 143*, 757–768.
Gemünden, H. G., Lehner, P., & Kock, A. (2018). The project-oriented organization and its contribution to innovation. *International Journal of Project Management, 36*(1), 147–160.
Gimenez, C., Sierra, V., & Rodon, J. (2012). Sustainable operations: Their impact on the triple bottom line. *International Journal of Production Economics, 140*, 149–159.
Goedknegt, D. (2013). Responsibility for adhering to sustainability in project management. In *7th Nordic conference on construction economics and organization* (pp. 145–154). Trondheim.
Goni, F. A., Chofreh, A. G., Mukhtar, M., Sahran, S., Shukor, S. A., & Klemeš, J. J. (2017). Strategic alignment between sustainability and information systems: A case analysis in Malaysian public higher education institutions. *Journal of Cleaner Production, 168*, 263–270.
Govindan, K., & Hasanagic, M. (2018). A systematic review on drivers, barriers, and practices towards circular economy: A supply chain perspective. *International Journal of Production Research, 56*, 278–311.
Govindan, K., Kannan, D., & Shankar, K. M. (2014). Evaluating the drivers of corporate social responsibility in the mining industry with multi-criteria approach: A multi-stakeholder perspective. *Journal of Cleaner Production, 84*, 214–232.
Hafezalkotob, A., & Hafezalkotob, A. (2016). Extended MULTIMOORA method based on Shannon entropy weight for materials selection. *Journal of Industrial Engineering International, 12*, 1–13.
Hamsayeh, E. (2019). Shannon entropy weight determination method implementation in MATLAB, 1–2. https://doi.org/10.13140/RG.2.2.27159.93604
Handayani, R., Wahyudi, S., & Suharnomo, S. (2017). The effects of corporate social responsibility on manufacturing industry performance: The mediating role of social collaboration and green innovation. *Verslas Teorija ir Praktika, 18*(1), 152–159.

Hanzaee, K., & Rahpeima, A. (2013). Corporate social responsibility (CSR): A scale development study in Iran. *Research Journal of Applied Sciences, Engineering and Technology, 6*(9), 1513–1522.
Hart, S. L. (1995). A natural-resource-based view of the firm. *Academy of Management Review, 20*(4), 986–1014.
Hou, S., Fu, W., & Li, X. (2010). Achieving sustainability with a stakeholder-based CSR assessment model for FIEs in China. *Journal of International Business Ethics, 3*(1), 41–55.
Hrbáčková, L., Stojanović, A., Tuček, D., & Hrušecká, D. (2019). Environmental aspects of product life cycle management and purchasing logistics: Current situation in large and medium-sized Czech manufacturing companies. *Acta Polytechnica Hungarica, 16*(7), 79–94.
Huemann, M. (2015). *Human resource management in the project-oriented organization. Towards a viable system for project personnel*. Gower.
Hwang, B. G., & Ng, W. J. (2013). Project management knowledge and skills for green construction: Overcoming challenges. *International Journal of Project Management, 31*(2), 272–284.
Hwang, C.-L., & Yoon, K. (1981). *Multiple attribute decision making: Methods and applications a state-of-the-art survey*. Springer-Verlag.
Jamali, D. (2008). A stakeholder approach to corporate social responsibility: A fresh perspective into theory and practice. *Journal of Business Ethics, 82*, 213–231.
Jamali, D., & Mirshak, R. (2007). Corporate social responsibility (CSR): Theory and practice in a developing country context. *Journal of Business Ethics, 72*, 243–262.
Jovanović, I., Stojanović, A., & Milijić, N. (2019). Influence of corporate social responsibility on intangible benefits in the project-based companies. In *Proceedings advances in economics, business and management research, 5th IPMA SENET conference*. Belgrade, Serbia.
Kanji, G., & Chopra, P. (2010). Corporate social responsibility in a global economy. *Total Quality Management, 21*(2), 119–143.
Kirchherr, J., Reike, D., & Hekkert, M. (2017). Conceptualizing the circular economy: An analysis of 114 definitions. *Resources, Conservation and Recycling, 127*, 221–232.
Kirchherr, J., Piscicelli, L., Bour, R., Kostense-Smit, E., Muller, J., Huibrechtse-Truijens, A., & Hekkert, M. (2018). Barriers to the circular economy: Evidence from the European Union (EU). *Ecological Economics, 150*, 264–272.
Korhonen, J., Honkasalo, A., & Seppälä, J. (2018). Circular economy: The concept and its limitations. *Ecological Economics, 143*, 37–46. https://doi.org/10.1016/j.ecolecon.2017.06.04
Lamata, M., Liern, V., & Pérez-Gladish, B. (2016). Doing good by doing well: A MCDM framework for evaluating corporate social responsibility attractiveness. *Annals of Operations Research*, 1–18.
Leal Filho, W., Tripathi, S.K., Guerra, J.A., & Gine-Garriga, R., (2019). Using the sustainable development goalstowards a better understanding of sustainability challenges. *International Journal of Sustainable Development &World Ecology 26*(2), 179–190.
Leipold, S., & Petit-Boix, A. (2018). The circular economy and the bio-based sector-perspectives of European and German stakeholders. *Journal of Cleaner Production, 201*, 1125–1137.
Lewandowski, M. (2016). Designing the business models for circular economy-towards the conceptual framework. *Sustainability, 8*, 43.
Li, X., Wang, K., Liu, L., Xin, J., Yang, H., & Gao, C. (2011). Application of the entropy weight and TOPSIS method in safety evaluation of coal mines. *Procedia Engineering, 26*, 2085–2091.
Loosemore, M., & Lim, B. T. H. (2017). Mapping corporate social responsibility strategies in the construction and engineering industry. *Construction Management and Economics, 36*(2), 1–16.
MacGillivray, A., Martens, H., Rudinger, K., Vilanova, M., Zollo, M., Begley, P., & Zadek, S. (2006). *Responsible competitiveness in Europe. Enhancing European competitiveness through corporate responsibility*. Institute of Social & Ethical Accountability.
Machado, N., & Morioka, S. N. (2021). Contributions of modularity to the circular economy: A systematic review of literature. *Journal of Building Engineering, 44*, 103322.
Magano, J., Vidal, D. G., Sousa, H. F. P. E., Dinis, M. A. P., & Leite, Â. (2021). Validation and psychometric properties of the Portuguese version of the coronavirus anxiety scale (CAS)

and fear of COVID-19 scale (FCV-19S) and associations with travel, tourism and hospitality. *International Journal of Environmental Research and Public Health 18*:E427. https://doi.org/10.3390/ijerph18020427

Maltzman, R., & Shirley, D. (2013). Project manager as a pivot point for implementing sustainability in an enterprise. In *Sustainability integration for effective project management* (pp. 261–278). IGI Global.

Marcelino-Sádaba, S., González-Jaen, L. F., & Pérez-Ezcurdia, A. (2015). Using project management as a way to sustainability. From a comprehensive review to a framework definition. *Journal of Cleaner Production, 99*, 1–16.

Martinuzzi, A., Kudlak, R., Faber, C., & Wiman, A. (2011). CSR Activities and Impacts of the ICT Sector. RIMAS Working Papers, 5.

Martens, M. M., & Carvalho, M. M. (2016). The challenge of introducing sustainability into project management function: Multiple-case studies. *Journal of Cleaner Production, 117*, 29–40.

Martens, M. M., & Carvalho, M. M. (2017). Key factors of sustainability in project management context: A survey exploring the project managers' perspective. *International Journal of Project Management, 35*(6), 1084–1102.

Milne, M. J. (2002). Positive accounting theory, political costs and social disclosure analysis: A critical look. *Critical Perspectives on Accounting, 13*(3), 369–395.

Moraga, G., Huysveld, S., Mathieux, F., Blengini, G. A., Alaerts, L., Van Acker, K., de Meester, S., & Dewulf, J. (2019). Circular economy indicators: What do they measure? *Resources, Conservation and Recycling, 146*, 452–461.

Moreno, M., De los Rios, C., Rowe, Z., & Charnley, F. (2016). A conceptual framework for circular design. *Sustainability, 8*, 937.

Nagyová, Ľ., Holienčinová, M., Košičiarová, I., & Holota, T. (2016). Corporate social responsibility in food manufacturing companies – Environmental dimensions. *Acta Universitatis Agriculturae et Silviculturae Mendelianae Brunensis, 64*(3), 1037–1043.

Narula, S. A., Magray, M. A., & Desore, A. (2017). A sustainable livelihood framework to implement CSR project in coal mining sector. *Journal of Sustainable Mining, 16*, 83–93.

Plessis, N., & Grobler, A. F. (2014). Achieving sustainability through strategically driven CSR in the South African retail sector. *Public Relations Review, 40*, 267–277.

Porter, M. E., & Kramer, M. R. (2006). Strategy and society the link between competitive advantage and corporate social responsibility. *Harvard Business Review, 84*(12), 78–92.

Porter, M. E., & van der Linde, C. (1995). Toward a new conception of the environment-competitiveness relationship. *The Journal of Economic Perspectives, 9*(4), 97–118.

Rakić, A., Milošević, I., & Filipović, J. (2021). Standards and standardization practices: Does organization size matter?. *Engineering Management Journal, 34*(2), 291–301.

Reverte, C., Gomez-Melero, E., & Cegarra-Navarro, J. G. (2015). The influence of corporate social responsibility practices on organizational performance: Evidence from Eco-Responsible Spanish firms. *Journal of Cleaner Production, 112*, 2870–2884.

Robert K -H, Schmidt-Bleek, B, Aloisi de Laderel, J., Basile, G., Jansen, J. L., & Kuehr, R. (2002). Strategicsustainable development - selection, design and synergies of applied tools. *Journal of Cleaner Production, 10*(3), 197–214.

Sanchez, B., & Haas, C. (2018). Capital project planning for a circular economy. *Construction Management and Economics, 36*(6), 303–312.

Sardana, D., Gupta, N., Kumar, V., & Terziovski, M. (2020). CSR 'sustainability' practices and firm performance in an emerging economy. *Journal of Cleaner Production, 258*, 120766.

Schieg, M. (2009). The model of corporate social responsibility in project management. *Verslas: Teorija ir praktika Business: Theory and Practice, 10*(4), 315–321.

Schöggl, J.-P., Stumpf, L., & Baumgartner, R. J. (2020). The narrative of sustainability and circular economy - a longitudinal review of two decades of research. *Resources, Conservation & Recycling, 163*, 105073.

Shannon, C. E. (1948). A mathematical theory of communication. *The Bell System Technical Journal, 27*, 379–423.

Silvius, A. J. G. (2016). Sustainability as a competence of project managers. *PM World Journal, 5*(9), 1–13.
Silvius, G. (2017). Sustainability as a new school of thought in project management. *Journal of Cleaner Production, 166*, 1479–1493.
Silvius, A. J. G., & Schipper, R. P. J. (2014). Sustainability in project management: A literature review and impact analysis. *Social Business, 4*(1), 63–96.
Silvius, A. J. G., Schipper, R., & Nedeski, S. (2013). Sustainability in project management: Reality bites 1. *Project Management World Journal, 2*(2), 1–14.
Stahel, W. R. (2016). The circular economy. *Nature, 531*, 435–438.
Steffen, W., Richardson, K., Rockström, J., Cornell, S. E., Fetzer, I., Bennett, E. M., Biggs, R., Carpenter, S. R., de Vries, W., de Wit, C. A., Folke, C., Gerten, D., Heinke, J., Mace, G. M., Persson, L. M., Ramanathan, V., Reyers, B., & Sörlin, S. (2015). Planetary boundaries: Guiding human development on a changing planet. *Science, 347*(6223), 1259855.
Stojanović, A., Milosević, I., Arsić, S., Urošević, S., & Mihaljović, I. (2020). Corporate social responsibility as a determinant of employee loyalty and business performance. *Journal of Competitiveness, 12*(2), 149–166.
Stojanović, A., Mihajlović, I., Safronova, N., Kunev, S., & Schulte, P. (2021a). The multi-criteria analysis of corporate social responsibility: A comparative study of Russia, Bulgaria and Serbia. *Journal of Management & Organization, 27*(4), 809–829.
Stojanović, A., Sofranova, N., Arsić, S., Milošević, I., & Mihajlović, I. (2021b). The effects of CSR activities on business according to employee perception. *European Review, 30*(5), 686–707.
Stoyanova, T. (2019). CSR strategies applied in terms of circular economy. *Economic Alternatives, 2*, 263–274.
Stumpf, L., Schöggl, J.-P., & Baumgartner, R. J. (2021). Climbing up the circularity ladder? – A mixed-methods analysis of circular economy in business practice. *Journal of Cleaner Production, 316*, 128158.
Suárez-Eiroa, B., Fernández, E., & Méndez, G. (2021). Integration of the circular economy paradigm under the just and safe operating space narrative: Twelve operational principles based on circularity, sustainability and resilience. *Journal of Cleaner Production, 322*(1), 129071.
Swain, R. B. (2018). A critical analysis of the sustainable development goals. In F. W. Leal (Ed.), *Handbook of sustainability science and research* (World sustainability series). Springer.
Velasquez, T. A. (2012). The science of corporate social responsibility (CSR): Contamination and conflict in a mining project in the southern Ecuadorian Andes. *Resources Policy, 37*, 233–240.
Vazquez-Burst, D., Smith, A., & Sarkis, J. (2014). Managing the transition to critical green growth: The green growth state. *Futures, 64*, 123–148.
Wang, H., Lu, W., Ye, M., Chau, K., & Zhang, X. (2016). The curvilinear relationship between corporate social performance and corporate financial performance: Evidence from the international construction industry. *Journal of Cleaner Production, 137*, 1313–1322.
Wu, W., Liu, Y., Chin, T., & Zhu, W. (2018). Will green CSR enhance innovation? A perspective of public visibility and firm transparency. *International Journal of Environmental Research and Public Health, 15*, 268.
Xia, B., Olanipekun, A., Chen, Q., Xie, L., & Liu, Y. (2018). Conceptualising the state of the art of corporate social responsibility (CSR) in the construction industry and its nexus to sustainable development. *Journal of Cleaner Production, 195*, 340–353.
Zhang, Y., Wang, Y., & Wang, J. (2014). Objective attributes weights determining based on Shannon information entropy in hesitant fuzzy multiple attribute decision making. *Mathematical Problems in Engineering*, 1–7.
Zhang, D., Morse, S., & Kambhampati, U. (2017). *Sustainable development and corporate social responsibility. London,* Routledge.
Zhu, Y. O. Siegal, M. L. Hall, D. W. & Petrov, D. A. (2014). Precise estimates of mutation rate and spectrum in yeast. *Proceedings of the National Academy of Sciences, 111*(22), E2310–E2318. https://doi.org/10.1073/pnas.1323011111

Chapter 6
Development and Review of Circular Economy Indicators: Evidence from European Union

Tijana Milanović, Aleksandar Jovović, and Nataša Petrović

Abstract The linear model of the economy, which is currently present, implies the way of production and consumption in enormous quantities, where the environment is exploited, the constant use of natural resources and energy to obtain certain products, which after use are discarded, i.e. end up in landfills and everywhere in the environment. The circular economy is one of the possible concepts that must be considered as a potential solution and a response to the challenge of sustainability. But the transition and application of the circular economy and its principles is a long process, a complex, not at all simple and quite serious task. In order to move from a linear model of economy, to a completely new and dynamic system, such as circular economy, it is necessary to develop a series of tools, indicators, measuring instruments, which would allow better and simpler monitoring of efficiency and effects of circular economy. This chapter will provide an overview of circular economy indicators at different levels, with a focus on the review of indicators as well as the evaluation system developed by the European Union, which would help decision-makers, both nationally and locally and at the enterprise level.

Keywords Circular economy · Circular economy indicators · European Union indicators

T. Milanović (✉)
Belgrade Academy of Business and Art Vocational Studies, Beograd, Republic of Serbia
e-mail: tijana.milanovic@bpa.edu.rs

A. Jovović
University of Belgrade, Faculty of Mechanical Engineering, Beograd, Republic of Serbia
e-mail: ajovovic@mas.bg.ac.rs

N. Petrović
University of Belgrade, Faculty of Organizational Sciences, Beograd, Republic of Serbia
e-mail: natasa.petrovic@fon.bg.ac.rs

V. Obradović (ed.), *Sustainable Business Change*,
https://doi.org/10.1007/978-3-031-23543-6_6

6.1 Introduction

The progress of society, technology, living standards and industrialisation entails serious consequences. This pace of life and the way of dealing with resources over time leads to their scarcity, which further leads to increasing demand and competition for these same products and raw materials, then to environmental degradation and market price volatility. "In parallel, an increase in the global population has led to an increase in inefficient consumption of natural resources" (Radaković et al., 2017; Smyth, 2006).

The deterioration of the environment around the world has resulted in the requirement for a new way of thinking, innovation and change in all segments of life. The economic model, known as the linear economy, has not been sustainable for a long time because the resources on the planet Earth are limited. The term sustainability in this contest represents "an idea, a process, a strategy and/or an objective that allows to address the current situation of concatenated ecological, social and economic crisis, labeled together as 'global change'" (Borojević et al., 2017; Hugé et al., 2016; Biggs et al., 2011).

This linear model of the economy, which is currently present, implies the way of production and consumption in enormous quantities, where the environment is exploited, the constant use of natural resources and energy to obtain certain products with consequential "a strong greenhouse gas effect" (Houshfar et al., 2012), which after use are discarded, i.e. end up in landfills and everywhere in the environment. The linear model of the economy implies intensive use of resources, which are unfortunately becoming less on the Planet, and does not guarantee prosperity for future generations, because "humans use more energy, create more waste, and produce more air pollutants than ever before" (Iannuzzi, 2017; Petrovic et al., 2016).

Further assumptions are that in the next 20 years, the middle class will grow from one to three billion people, and all this together will cause too much pressure on natural resources and survival on the Planet (Milanović et al., 2019). All this leads to thinking about how industrial systems must be designed, in order to achieve a balance between the resource capacities of the Earth and the environment and, on the other hand, the enormous consumer habits of society. There is a need for a primordial transformation in the behaviour of all, both producers and consumers, in increasing production efficiency, optimising the use of materials, as well as directing the system to use all possible resources with as little waste as possible. Such a system is necessary, which will be directed towards production from waste, as well as "waste reduction and possibilities for using waste materials" (Vukadinović et al., 2016). In more developed countries, reuse, repair and recycling, returning products to the re-production process, use of all materials that still have use-value, are becoming key activities in many sectors (Milanović et al., 2020a).

Therefore, the application of the circular economy model, which is a model of sustainable production and consumption, is increasingly recommended as a "suitable solution for achieving sustainable development goals" (Saidani et al., 2019), but also as a "completely new approach to solving global problems and challenges" (Banaitė, 2016). As already mentioned, the circular economy is a dynamic system and its establishment and implementation involve putting a lot of effort, as well as the development of new tools, measuring instruments and indicators. The transition to a circular economy is a complex, comprehensive and, above all, long-term process. It is a way to separate societies from unsustainable linear economies that result in the depletion of limited resources (Mihajlov et al., 2019).

Another important thing is the general agreement on what the application of circular economy means and what aspects of the transition process and the effects should be measured (Potting et al., 2017). It is necessary to define exactly what is it to be achieved by circular economy application, to decide which aspect should be measured. For example, the transition to the circular economy model contributes to the improvement of the environment. There are two situations here, one is that some companies environmental protection, understood as a form of improving the transition to circular economy, while many other actors are taking it to be an integral part of circular economy (Potting et al., 2017). It can be concluded from this that, depending on the way of accepting the application of circular economy, there is an enormous breadth of understanding, and thus a wide range of acceptance of indicators for measuring the transition of circular economy.

Around the world, academics, industrialists and politicians agree on the need to develop and apply a variety of measurement tools and indicators related to the circular economy, to facilitate governance and all that the transition to this model entails. In this context, a wide range of C-indicators of circularity, i.e. indicators, has been developed in recent years (Saidani et al., 2019). The need to develop indicators is crucial for monitoring of all changes related to the circular economy.

There is already enough knowledge and ways on how and in what way to measure the progress of the transition to the circular economy model. In particular, considering the sets of developed indicators and other data for their application, there is knowledge on how to measure the effects of the circular economy and its application. It can also be said that in some ways there is a difference of opinion of many scientists, rivalry and competition and that the degree of agreement between them is still low, which significantly slows down and complicates the application of circular economy indicators and its principles. Viewed in this way, some believe that the credibility of circular economy transition strategies may also be affected by the extent to which scientists and consultants consider their efforts to be recognised as part of the decision-making process for measuring circular economy transition indicators (Potting et al., 2017).

The authors Mitrović and Veselinov (2018) believe that a complex indicator of progress towards the circle is necessary to maximise discussions on priority measurements and uniformly initiate development to create an innovative and prosperous carbon-free circular economy.

There are a large number of indicators for different purposes and needs developed by various scientists, states, government agencies, NGOs, companies, as well as many others, which will be presented in this chapter.

6.2 The Concept of Circular Economy

The concept of circular economy has become a popular and global important issue, and this concept has gained increasing attention in recent years, worldwide, presenting an approach to solving the problems of dwindling resources and environmental pressures associated with the current linear economic model. This concept has gained increasing importance in politics in recent years, development on the international, European Union, and the national level of governance and in business practices and consumer behaviour (World Health Organization [WHO], 2018).

The very idea of a circular economy model has existed for a long time, but according to many, it represents a completely new concept (Murray et al., 2017). "The circular economy model is the opposite of the linear one and is based on closed loops such as the biological life cycle" (Smol et al., 2017). In recent years, circular economy has been a guideline by which various researchers, institutions, decision-makers, policymakers, scientists, companies, countries and many others are increasingly engaged and interested in providing opportunities to increase the sustainability of our economic system (Milanović et al., 2020a), and according to Staniškis (2012), circular economy strategies currently represent the latest concept for striving to achieve global sustainability (Lilja, 2015). "The concept is seen as a practical solution to the planet's emerging resources problem, which is linked to the increasing global population and rapid growth in developing countries and emerging economies" (Ragossnig & Jovovic, 2016).

In more developed countries, reuse, repair and recycling are becoming key activities in many sectors and are setting the principle of business. But companies have also shown an increasing interest in this economic model. Circular economy principle is highly recommended as an appropriate solution for meeting the goals of sustainable development (Saidani et al., 2019). "The most important benefit in a circular economy-based approach is the ability to retain added value in products for as long as possible" (Smol et al., 2017), extracting their maximum value as well as removing waste. The circular economy approach "implies resource efficiency, integrates cleaner production and industrial ecology into a wider system including industries, networks or chains of firms, eco-industrial parks and regional infrastructure to support resource optimization" (Li, 2012).

Also, French Environment and Energy Management Agency (ADEME) defines circular economy as "an economic system based on methods of exchange and production such that, at each stage of the product life cycle (goods and services), they aim to increase resource efficiency and reduce environmental impact while improving the well-being of individual citizens" (Magnier, 2017; Gallaud & Laperche, 2016).

It is important to say that the circular economy promotes systemic innovations intending to design waste, increase resource efficiency and achieve a better balance between the economy, the environment and society. From a practical point of view, "the circular economy approach is vital since, in the twenty-first century, science implies effective restorative approaches that will give future generations the opportunity for more sustainable development" (Smol et al., 2017).

In his work *The Environmental Emergency* (*L'urgence écologique*), Lévy (2009) highlights the main elements that make up the foundation of circular economy:

- Use of non-renewable resources efficiently and moderately;
- Respect during the exploitation of renewable sources, in a way that the conditions for the renewal of these resources are constantly met: ecological design and clean production; ecological consumption; recycling the waste as a resource and non-polluting waste treatment.

The circular economy is an economic system that replaces the "end of life" concept with the reuse, reduction, recycling and recovery of materials, both in production and in the process of consumption. In the twentieth century, the "preventive approach" was replaced by the "restorative approach" both in Europe and around the world (Smol et al., 2017).

The model of disturbing the natural balance of the environment, by depleting natural resources, uncontrolled production, consumption and disposal of products after use, by generating increasing amounts of waste is a linear model of the economy (Milanović et al., 2019).

Simply put, having the linear model of the economy, with depletion, production, use and discarding of resources, which leads to the generation of a great amount of waste and environmental degradation, and, on the other hand, having the circular economy model, with the use of every possible resource, returning it into the production process, while creating minimal waste shows the basic difference between these two concepts.

The circular economy represents the closing of the circle on the relation resource: final product–waste–resource. It fundamentally differs from the linear economy, that is, in the plan of steps to be followed, the perspective on what sustainability is and the quality of reuse practices. The first big step in changing the mind-set is related to recycling and waste treatment. Circular economy model is shown in Fig. 6.1.

The circular economy operates on several levels. It can be applied and can act at the micro level, meso level and macro level, to achieve sustainable development, while creating environmental quality, economic prosperity and social equality, for the benefit of present and future generations (Kirchherr et al., 2017). The micro level generally refers to the product level, the company level, as well as the consumer level. The meso level includes eco-industrial parks. And the level of the city, region and state falls under the macro level.

However, implementing circular economy principles is not an easy task at all. Decision-makers need tools to support the setting of adequate goals and to monitor the effects of actions taken to get moving from a linear to a circular model. However, in order to be able to apply and deal with the circular economy in general, it is also necessary to develop a wide range of measuring instruments, related to it, which will ensure a successful transition to the circular economy and will also measure and report on its progress (Milanović et al., 2020a).

In the last few years, there have been increasing attempts to develop indicators, i.e. circularity indicators (Saidani et al., 2019). Measurement methods are necessary

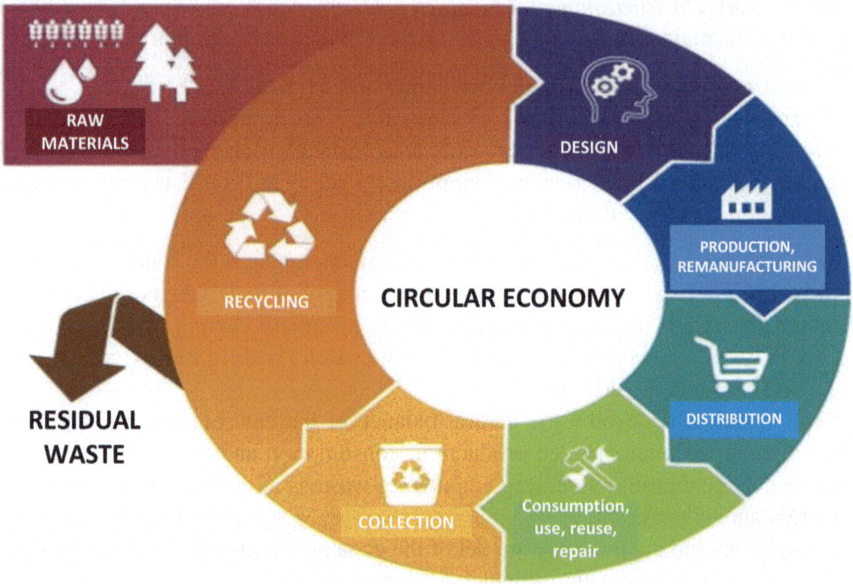

Fig. 6.1 Circular economy model. (EC, 2014a)

in a large number of applications, such as product design, material selection, progress monitoring, supporting internal decision-making, for example, regarding investment selection. These different uses will require different types of metrics based on different datasets (MacArthur Foundation and Granta Design, 2015b).

6.3 Circular Economy Indicators and Measuring Methods

Numerous scientists, researchers, organisations, as well as many others, have been examining and researching the methods to measure the circular economy alongside some basic indicators that can help in its application. Research on this topic has been done and set up by Potting et al. (2017) in their paper on "how to measure the progress of the transition to the circular economy?"; then how is the circular economy measured in enterprises and economies? (Saidani et al., 2019); how is circular economy measured at the product level? (Linder et al., 2017). The transition to the circular economy is a serious task for all, and what companies lack, according to EASAC (2015), are precisely indicators and targets, as one of the three key shortcomings of the transition to circular economy, which further leads to lack of information (Saidani et al., 2019).

Furthermore, according to the EEA (2016), upon analysing indicators on monitoring progress towards the circular economy, it was noticed that the knowledge base is quite incomplete and that information on decision-making is necessary to improve the investment in the circular economy business.

The review of different literature leads to a conclusion that in the last few years, a wide range of indicators has been developed, which helps with measuring the circularity. Likewise, to assess whether the circular economy principles lead to significant changes, it is necessary to develop an adequate measurement system (Cayzer et al., 2017; Geng et al., 2012). However, as there are several definitions of the term circular economy, there are numerous indicators for different purposes and needs, developed by various scientists, countries, government agencies, various companies and many others (Milanović et al., 2020a; Vercalsteren et al., 2018).

In general, indicators are variables that provide relevant information for decision making (Gallopin, 1996). The circular economy indicators can work on promoting the circular economy theme; enabling new quality standards, as well as comparing companies in terms of investing in sustainability and markets. However, Beratan et al. (2004) warn that indicators must be linked to decision-making and implementation. Therefore, indicators alone do not achieve a successful transition to the circular economy but are a very important tool to help progress towards this goal (Cayzer et al., 2017).

An indicator is a variable (parameter) or function of variables to provide information about circularity (technological cycles) or effects (cause-and-effect modelling). Besides, the indicator may be the result of complex information on quantitative and qualitative data (Moraga et al., 2019).

Quantitative indicators are essential for assessing the performance of an organisation or product system regarding the circular economy principles and "to broader national and international sustainability goals" (Su et al., 2013). "Indicators should be supported by rigorous scientific accounting and method assessment" (Saidani et al., 2019).

In addition to quantitatively estimating indicator values, there are several linking options. Hence, the value of the indicator can be related to (Vercalsteren et al., 2018):

- *Economic product*: e.g. Gross Domestic Product (GDP), at the level of countries, regions and value-added, at the level of sectors, products. This refers to the provision of information on the productivity or intensity of an economy or sector of economic activity.
- Per capita *data*: linking the value of the indicator to the per capita or household. Per capita figures allow comparisons of cities, regions or countries, avoiding the problem of country size and population.
- *Input indicators*: e.g. the input of domestic material, the input of raw materials. They are used to describe the materials mobilized or used to maintain economic activities, including the production of products for export. They are closely related to the way production works in a particular country or region and are sensitive to changes in the level and patterns of foreign trade and other factors such as a country's giftedness with natural resources as well as the level of technological development.
- *Output indicators*: These indicators (e.g. domestic processed output) describe material outflows related to a country's production and consumption activities. These are the materials that have been used in the economy and that have remained, either in the form of emissions and waste or are meant for export.

- *Consumption indicators*: These indicators (e.g. consumption of raw materials, domestic consumption of materials) describe the material consumed by a particular economic activity. They are closely related to the way they are consumed, but they are quite stable over time. The difference between consumption and input indicators is an indication of the degree of economic integration (i.e. the larger the difference, the greater the global economic integration due to exports).

Indicators related to data on economic production and capital, which allow comparison between countries to be made, are most commonly used. Economic product-related indicators are often referred to as efficiency and productivity indicators (Vercalsteren et al., 2018). However, the lack of academic and scientific knowledge about the circular economy indicators is an obstacle for its further implementation (Akerman, 2016).

6.4 Overview of the Initiatives Applied to Develop the Circular Economy Indicators

A review of the circular economy initiatives within the European region indicates that most of the leading countries in this field are the EU member-states, especially from the western and northern Europe (WHO, 2018). A large number of European countries are involved in the initiatives of the circular economy and its implementation, as well as in the work on the development of measurement indicators, which would help the process of monitoring and further development in this field and area. Upon developing various national initiatives, through action plans, visions, the circular economy roadmaps, strategies, etc., many European countries, such as Denmark, Finland, Luxembourg and the Netherlands, have joined in terms of the political initiative for the circular economy implementation (WHO, 2018).

In addition, topics such as waste management, resource efficiency and enactment of new national waste laws help promote the circular economy principles across European countries and beyond.

For example, Sweden has accepted the promotion of the circular economy as a business model and seeks to be the leader in innovative and sustainable industrial manufacturing through its "smart industry" vision, although it is still without its roadmap or the circular economy vision (Government Offices of Sweden, 2016). Furthermore, through its Resource Efficiency Program, Germany has included the development and expansion "of the circular economy as a guiding principle" (WHO, 2018).

On the other hand and looking at the countries outside of Europe, upon developing excellent national strategies for the initiatives and their implementations, China and Canada have become good examples for this. Also, Japan is a pioneer in the recycling industry, and although it does not have the circular economy strategy, through its regulation on waste management, it has long applied the principles of the circular economy.

Regarding the development of the circular economy indicators, circular economy metrics can be divided into two major groups: national measurement tools at the national level and business tools based on company activities or company products (EC, 2018b). This section will present an overview of existing initiatives and everything that has been done so far and is relevant for the circular economy indicators.

Measurement tools at the national level. Regarding the initiatives taken regarding the development of indicators, there are three levels: national, private and the level of the European Union. The ability to measure the circular economy at the national level is highly important since its application is a key driver of sustainable growth and a country's development (Garcia-Bernabeu et al., 2020).

At the national level, some of the examples of initiatives and development of major strategies are China, the Netherlands, France, Germany, as well as various circular economy indicators used in Japan. The initiative of each of these countries is briefly explained further below. For instance, regarding China, a national system of indicators has been developed and is based on the analysis of material flow and the review of indicators. Furthermore, in 2017, France has also developed key indicators for the circular economy, i.e. 10 indicators that monitor the French economy, covering seven pillars of the economy. In 2018, the Netherlands developed a system of indicators for circular economy monitoring. This report can only be found in Dutch and provides an overview of 21 indicators for measuring circularity in the Netherlands. Indicators used in Japan were presented in 2013 within the document "The Third Fundamental Plan for Establishing a Sound Material-Cycle Society" (2013). The indicators are based on the dimensions of material flow in the economy (input, turnover and output) with a focus on material stocks. Moreover, in 2012, Germany launched its initiative, through the Resource Efficiency Program (ProgRess II) for the development of the circular economy indicators, to support the sustainable use and conservation of natural resources. And, since 2016, it has used the total productivity of raw materials as the main indicator.

An example of developing national tools at the macro level, an initiative related to indicators developed at the private level is the 2017 "Cotec Evaluation of Circular Economy in Spain." The evaluation contains 20 indicators for the circular economy assessment. This framework also refers to Spain and its comparison with other countries (EC, 2018b).

Another example of the national level of private level indicator development can be found in the 2018 "Circularity Gap Report." In this report, the Global Circularity Metric is proposed as the only CE measure. As for previous EU initiatives related to the circular economy, an overview will be given in a separate chapter, below.

The second type of initiative involves tools and activities at the company level or products focused on the company's activity. For this level, the best overview of indicators was given by the Ellen MacArthur Foundation in collaboration with a software engineering company – Granta Design – and the indicators are focused on product and company assessment (Akerman, 2016).

They developed tools for products and business activities to measure circularity, and as the main indicator, the focus was on the circularity of materials. Some examples are the Circular Economy Toolkit, for self-assessment and focus on enterprise products. Also, another one, developed in 2017, is the Circle Economy Assessment, an internet tool for businesses. This tool focuses on seven elements to improve organisational activities and support the implementation of circular economy strategies at the company level (MacArthur Foundation and Granta Design, 2015a).

6.5 Overview of Groups of Indicators

By researching various literature, the following section will provide an overview of a couple of classifications of circular economy indicators that aim to monitor, improve, assess circular economy performance, circular economy application and circular economy principles.

The circular economy-related indicators can be classified according to different criteria (Vercalsteren et al., 2018):

1. Micro, meso, macro level
2. Circular economy strategies
3. Technology versus socio-institutional characteristics

Further classification of the circular economy indicators may be in terms of criteria such as levels of circular economy implementation (e.g. micro, meso, macro); circular economy loops (maintenance, reuse, recycling, recycling), performance, circularity perspective (actual, potential), they take into account or their degree of transferability (generic, sectoral, specific) (Saidani et al., 2019).

Therefore, Smol et al. (2017) proposed three groups of indicators, which are directly related to innovation and are based on "environmental innovation." These indicators "take into account the principles of circular economy: circular economy – input data for eco-innovation, circular economy – eco-innovation activities and circular economy – eco-innovation results." In addition to these indicators, they have also developed indicators related to the effects of the circular economy, as well as indicators of ecological innovation of the circular economy, which are resource efficiency results and socio-economic results. Observing at the regional level "this method of measurement proved to be a systematic and integrated approach to the concept of circular economy, and the developed indicators can be used in the transition phase of the circular economy, but also as a basis for further creation and development of final circular economy indicators."

Further classification according to Saidani et al. (2019) is done upon identifying 55 circular economy indicators and developing a classification of indicators following 10 steps of criteria:

1. Level of circular economy implementation (i.e. level of products and consumers, level of industrial parks and level of city, region or country)
2. Circular economy loops (i.e. maintenance/extension, reuse/renewal, recycling)

3. Performance (i.e. inner circle, consequent circle)
4. Circular perspective (actual, potential)
5. Possibility of use (purpose of information, decision making, communication, learning)
6. Degree of transversality (generic, sector-specific)
7. Dimensionality (one number, several indicators)
8. Measurability (quantitative, semi-quantitative, qualitative)
9. Estimation framework format (formulas for manual calculation, calculation tool)
10. Background and origin of development (academics, companies, agencies)

6.6 Overview of Indicators According to the Level of Implementation

Different levels of circular economy implementation and different characteristics of companies, industries or regions require different assessment indicators. One of the indicators presented in Banaitė (2016) stems from the fact that circular economy can be applied at three different levels of circular economy measurement indicators:

1. Macro (global, national, regional, city)
2. Meso (industrial symbiosis, eco-industrial parks)
3. Micro (individual company, product)

This way the system of circular economy assessment should show results that could be useful to decision-makers at all levels of decision making (Banaitė, 2016). All grading systems can be divided into three groups with different evaluation systems. "The first step of good assessment systems is to set appropriate indicators at each level of implementation" (Banaitė, 2016).

6.7 Circular Economy Assessment at the Micro Level

The company level, the consumer level and the product level can be considered as micro level. Each company sets specific circular economy indicators according to its characteristics, such as existing business strategies, the current state of the company, level of business, ways of entering and placing itself on the market, product range, method of production and other characteristics. Therefore, depending on the characteristics of the company, indicators should be developed to correspond to those characteristics. However, it should be noted that: "the circular economy indicators set for one company do not have to be in line with the indicators of another company. Setting a very unique and only one set of indicator standards may fail to capture the full development of circular economy in different companies" (Su et al.,

2013). "Companies should increasingly adapt their production and business as a whole in a more sustainable way and to the transfer of consumer concerns to a specific business, i.e. the purchase of environmentally friendly products," and in that sense the development of circular economy indicators and eco marketing will play a crucial role (Milanović et al., 2020b).

Adoption of the circular economy principle implies that the company implements various strategies to improve the circularity of its production system (Banaitė, 2016). Micro-level indicators provide detailed information on a particular substance or individual products for specific local decision-making processes, as well as to support the implementation of policies and decisions in areas such as product policy, energy efficiency and integrated waste management (Vercalsteren et al., 2018).

According to Akerman (2016), it is necessary to mention the importance and significance of studying circular economy indicators related to companies, because many countries have identified (as one of the two main barriers that can hinder the development and efficient use of circular economy indicator systems) the lack of performance indicators of circular economy for applications in companies (Milanović et al., 2020a).

The usefulness of circular economy indicators for practical application in companies is determined with special emphasis on supporting the decision-making process depending on the needs and capacities of the company. A way to increase economic viability is to use already created data for the analysis of circular economy indicators, although the economic aspect should be studied in more detail and should not rely on only one factor (Akerman, 2016).

The circular economy indicators have been developed to assess how well a company or product adapts to the concept of circularity. Indicators are specifically made to be a "decision-making tool" for designers but can also be used for internal reporting and evaluation of companies. The focus of these circular economy indicators is technical cycles and materials from non-renewable sources, and they primarily consist of material circularity indicators. Material circularity indicators include four different indicators (Table 6.1) that contain: material input, utility, waste management and recycling efficiency (MacArthur Foundation and Granta Design, 2015b). This indicators are relevant for the product evaluation, and they include measuring the amount of intake coming from raw materials and recycled materials and reused components; the duration and intensity of the product; the amount of material that ends on landfills, and the quantity of it that is collected for recycling and efficiency of recycling processes used to produce the recycled material introduced into the production and recycling process of the material after its use.

The indicators shown in Table 6.1 were developed specifically for product evaluation, but the Ellen MacArthur Foundation and Granta Design also developed a framework of evaluation for companies (Design, 2015).

The product-level methodology focuses in particular on the following possible uses of cases (MacArthur Foundation and Granta Design, 2015a):

- Indicators can be used in designing new products that will consider circularity as a criterion and present the basis for design decisions. The indicators make it pos-

Table 6.1 Material circularity indicators for the product

Indicators	Description
Access to the production process	To measure the amount of intake coming from raw materials and recycled materials and reused components
Benefits during the use phase	To measure the duration and intensity of the product use in comparison to an average industrial product of a similar type
Destination after use	To see the amount of material that ends on landfills (or energy recovery) and the quantity of it that is collected for recycling and what components are assembled for reuse
Recycling efficiency	To see the efficiency of recycling processes used to produce the recycled material introduced into the production and recycling process of the material after its use

Source: MacArthur Foundation and Granta Design (2015b)

sible to compare different versions of the product, given its roundness at the design level.

- Indicators can be used for internal reporting purposes. Companies can compare different products by their circularity. It also allows stakeholders from different departments to learn from each other about product circularity.
- Companies can also make their product indicators available to the public or selected organisations. This would allow the aforementioned organisations to use the indicator as part of their procurement decisions, for example, by defining a minimum threshold for the products they purchase.

6.8 Circular Economy Assessment at the Meso Level

The circular economy indicators at the meso level, in the simplest terms, represent economic, environmental and social impacts, as well as mutual symbiosis, i.e. connections at the level of regions, product groups, and industries. Meso-level indicators enable more differentiated monitoring of information and more detailed analysis of material flows in the economy, distinguishing not only categories of materials but also industries or branches of production and categories of consumption. These meso-level indicators focus on industry, consumption activity or a particular material level (Vercalsteren et al., 2018).

The best example of the application of circular economy indicators at the meso level is applied by China. This country, as one of the world's largest producers, has a large number of production facilities, industrial networks and industrial parks. It is also "the second-largest energy producer in the world, and also the second-largest energy consumer. China's energy consumption, i.e., its GDP (gross domestic product) per unit is twice as high as the world average" (Vercalsteren et al., 2018; Banaitė, 2016). Hence, high-energy consumption in the manufacturing industry creates a lot of serious environmental problems. For this reason, production plants, industrial parks and industrial networks in China function in a specific way which makes the production derivatives to be valued at the meso level.

By applying the concept of symbiosis, i.e. connecting companies at the meso level, it contributes to continuity in work and as well as the possibility of eliminating any unnecessary type of downtime. This way of connecting implies the joint use and utilisation of infrastructure and services of all production plants, industrial parks and industrial networks. Furthermore, all this enables the possibility of cooperative management of resource flows, "trading in industrial products that reduce environmental problems while reducing the dependence of companies on resources." And it follows that "reducing production costs increases industrial productivity and competitiveness" (Heshmati, 2015). Applying the concept of circular economy and developing measuring instruments and indicators for the application of circular economy will help control and interconnect these parks, and at the same time help company executives to make appropriate decisions (Banaitė, 2016). An overview of China's indicators will be presented below.

6.9 Circular Economy Assessment at the Macro Level

The macro level is much more complex and comprehensive, so it in itself requires the development and application of a wide range of sustainability and circular economy indicators. Indicators developed for the macro level are necessary and are used to assess, monitor and improve various programmes and policies. Manufacturers at this level choose certain indicators that will fully cover the strategic goal of development and sustainability of the circular economy and, therefore, must have access to a wide range of information (Banaitė, 2016).

Based on various data, some countries have developed specific approaches to monitoring progress towards circular economy at the national (macro) level. Interestingly, most of these initiatives are based on material flow calculations and waste management data. In several tools for monitoring data on resources and waste, broader macroeconomic data are added as a supplement. For example, the Japanese system includes the size of the lease market, the French system includes employment data in circular economy, and the Chinese and European Union systems include the added value of recycling industries, and the European Union system also includes data on patents (EC, 2018b).

Table 6.2 shows the presentation, i.e. the overview of micro-, meso- and macro-level indicators.

Table 6.2 shows the distribution of indicators according to Vercalsteren et al. (2018), i.e. according to the level of implementation at the micro, meso and macro level. Of the 17 indicators analysed, 4 can be applied only at the macro level, 3 can be applied at the meso level, while 6 can be applied at the micro level.

Table 6.2 Indicators according to micro, meso and macro level

Indicators	Macro level	Meso level	Micro level	Is the indicator also available on another level? Does it have anything to do with other indicators?
Raw material consumption	X			Macro-level indicator
Material system analysis	X	X		An indicator of macro levels, potentially at the meso level
Material flow monitor	X	X		Micro-level potential
Leakages from material cycles		X		Meso levels with links at the macro level
Resource footprint indicator			X	Micro-level indicator, with potential for meso and macro level
Cyclical material use rate	X	X		Macro-level indicator, with meso-level potential
Material circularity indicator, or MCI.			X	Budget methodology is available, GABI software is used
End-of-life recycling input rate	X			Macro-level indicator
Recycling rates	X	X		Macro-level indicator can also be applied at the meso level
Recyclability benefit			X	Micro-level indicator
Trade-in secondary raw materials			X	Micro-level indicator
Waste generation	X	X		Macro/meso-level indicator, the potential for micro level
Waste of electrical and electronic equipment, or WEEE management		X		Applicable to other flows and product categories
Basket of products	X		X	Macro and micro-level indicator
Product environmental footprint			X	Micro-level indicator
Private investments, jobs		X		Meso-level indicator
Others: recycled content, recyclability or repairability).			X	Floor development

Source: Vercalsteren et al. (2018)

All in all, five indicators can be applied at two levels, at the macro and meso levels: (1) *material system analysis indicator*; (2) *material flow monitor indicator*; (3) *cyclical material use rate indicator*; (4) *recycling rates indicator* and (5) *waste generation indicator*. Interestingly, basket *of products indicator* can be applied to both the macro and micro level. Also, there is the fact that *Resource Footprint indicator* is a micro-level indicator, with the potential of reaching both meso and macro level. *Material flow monitor*, with a purpose to provide insight into material flows in the economy, is an indicator of the macro and meso level, with the potential of reaching micro level.

6.10 Overview of Indicators Developed by Different Countries

This chapter will present indicators of the circular economy of the following countries: China and the countries of the European Union. Each of these countries is taken as an example due to its specificity. In this chapter, the emphasis will be on China, for the reason that for further development of indicators, CE indicators of China and CE indicators of EU countries could be compared with each other.

China is the first country to publish national indicators on the status of the circular economy, which are important indicators for policymakers and therefore provide assistance in achieving circular economy goals and results (Milanović et al., 2020a; Akerman, 2016; Geng et al., 2012). In 2009, China passed the Circular Economy Promotion Law that came into force in January 2009, which created the legal and political basis and started activities aimed at promoting the concept of the circular economy (National People's Congress, 2008).

The circular economy indicators review, developed by China and reviewed by Akerman (2016), starts from the fact that the Chinese government implements the model of sustainable development based on the circular economy concept at three levels; micro, meso and macro. Micro level is an individual company and includes measures such as eco-design and cleaner production. The meso level consists of industrial parks in which companies cooperate within an industrial symbiosis. Finally, the macro level is a temporary and national level that aims to create through production and consumption recycling companies (Akerman, 2016).

The circular economy indicators at the macro level consist of 22 indicators divided into four categories. The table of indicators can be seen in the review of the paper (Akerman, 2016; Geng et al., 2012) alongside with the category of resource utilisation rates that contains the largest number of indicators, a total of nine indicators, where five of them fall under the material. There is also a category of resource consumption, where indicators are mainly focused on water and energy consumption (Akerman, 2016; Geng et al., 2012).

The meso-level indicators, developed by China, have quite similar categories as the macro-level indicators. "The meso-level has 12 circular economy indicators, much less in comparison to the macro-level indicators of China's circular economy" (Akerman, 2016; Geng et al., 2012). However, the main difference is that the meso level has more indicators in the category of resource exit rates, while the other categories contain a smaller number of indicators, compared to the macro level. Meso-level indicators are more focused on the flow of different resources, such as energy, water and minerals (Akerman, 2016; Geng et al., 2012).

Looking at the Chinese government's circular economy indicators at the meso and macro level, "there are many shortcomings as circular economy is defined as a concept of sustainability, encompassing all three pillars, such as social, environmental and economic" (Akerman, 2016; Geng et al., 2012), so it can be concluded that circular economy indicators developed by China are one-way and the few, i.e. most resource-oriented categories (resource outflow, resource consumption, and use, i.e. resource use) and both levels include the waste disposal category and pollution emissions.

The shortcomings of China's circular economy indicators, according to Geng et al. (2012), are related to five specific ones: lack of social indicators, lack of industrial symbiosis indicators, insufficient number of micro-level indicators, an absolute reduction of materials and energy and, finally, lack of indicator-oriented according to prevention. Since an important part of the circular economy concept "is the inclusion of all aspects, economic, environmental and social," the main drawback is the lack of indicators referring to social aspects. Furthermore, the Chinese circular economy concept of all levels, including the micro level, has been applied, which includes enterprises and facilities, but indicators of this level are not shown.

France is interesting because it has developed a set of indicators for the circular economy, based around three areas of action and seven pillars of the economy (Magnier, 2017). Finally, the focus of the chapter is on the indicators of the circular economy of the European Union, which will be explained in detail later in the chapter.

6.11 Overview of the Circular Economy Indicators in France

In France, the spreading of the circular economy concept has begun within various environmental organisations in 2007. Several definitions include a range of relatively broad activities with the common goal of optimising resource use and removing or reducing waste (Magnier, 2017; Gallaud & Laperche, 2016).

France stated in its Law on Energy Transition for Green Growth (LTECV) that its goal was to move to circular economy, meaning to separate from the linear model of the economy, which is based on the "take it, do it, put it off system." This would also mean focusing on more moderate and responsible consumption of natural resources and raw materials, alongside with, as a priority, prevention of waste production, especially reuse of products and, following the established hierarchy of waste treatment, reuse, recycling or fails, by replacing waste materials (Magnier, 2017).

Interestingly, in France, circular economy indicators have been developed around three areas of action and seven pillars of the economy (Magnier, 2017). Areas of action are *Waste Management*, *Supply from Economic Stakeholders* and *Consumer Demand and Behaviour*. Hence, these are the indicators developed by France to monitor the implementation of circular economy.

Based on these areas and the seven pillars of the economy, the French Environment and Energy Management Agency (Ademe) has adopted the following indicators, shown in Table 6.3.

As seen in Table 6.3 France, unlike China, covers both economic and social and environmental pillars. Also, there is a variety of indicators, which can cover many areas in which the principles of circular economy can be applied, and thus can provide a greater possibility of applying circular economy in general. By covering all

Table 6.3 Areas of action, pillars and 10 key circular economy indicators

Action areas	Pillars of economy	Indicators
Waste management	Recycling	1. Quantities of waste sent to landfill 2. Use of recycled raw materials in the production process
Supply from economic stakeholders	Responsible consumption: Shopping Collective consumption Use	3. Food waste 4. Household spending on product repair and maintenance
	Product life extension: Reuse Repair Recycling	5. Employment in circular economy (although this indicator can be applied for each pillar)
Demand and consumer behaviour	Extraction/Production and sustainable supply chain	6. Domestic consumption of materials per capita 7. Resource productivity
	Eco-product design	8. Eco-label holders
	Industrial and territorial ecology	9. Number of industrial ecology projects
	Functional ecology	10. Car sharing

Source: Milanović according to Magnier (2017)

seven pillars of circular economy, each indicator provides information and details about a specific goal, current trends, analysis, as well as points of international comparison (Magnier, 2017).

6.12 Overview of the European Union Circular Economy Indicators

The attitude towards the environment today represents the most important policy of the European Union. European environmental policy is based "on the principles of precaution, prevention of pollution at source and the principle of polluter pays" (Ohliger, 2015). The European Union waste and circular economy policy have evolved over the last 30 years through a series of environmental action plans and circular economy-related policies, intending to reduce unwanted environmental impacts, improve human health and create an efficient economy.

In Europe, the application of the circular economy, and what is the main difference in relation to China, is that "the perspective of the development of the circular economy has appeared in response to huge amounts of waste and is mainly focused on companies." The application of China's circular economy is focused and developed as a "response to rapid industrialization and increased waste contamination caused by overuse of resources" (Kyriakopoulos et al., 2019). Central role in the European Union circular economy is about waste management, and it determines the way of implementing the waste hierarchy in practice.

Hence, the European Union has developed three important documents named: *seventh European Environmental Action Plan* (EEAP7), *Resource Efficiency Roadmap* and *Raw Materials Initiative*. Each document was very important for the design of the circular economy package issued by the European Commission in December 2015.

Furthermore, EEAP7 shares the vision of Europe 2050 and opens space for long-term thinking and real systemic transitions of society, economy and a more thoughtful relationship with the environment (Mitrović & Veselinov, 2018). EEAP7 documents contain three main objectives of the EC (2014c): *(1) To protect, preserve and improve natural capital; (2) to make the Union resource-efficient, green and competitive; (3) to protect the citizens of the Union from environmental pressures and risks to human health and well-being*.

The Resource Efficiency Roadmap (EC, 2011) is fully dedicated to the material aspect of development and focuses on critical raw materials for Europe. The importance of this document is reflected in the fact that the dimension of material flows is built-in, because it offers a long-term way to separate resource consumption from economic growth in the European Union (Mitrović & Veselinov, 2018; Vasileios Rizos et al., 2017).

The European Union Action Plan (EC, 2015a), as the third document, through which the European Union recognised the necessity to develop CE indicators, states that: "to assess the progress of the transition to circular economy and the effectiveness of action at European Union and national level, it is important to have set reliable indicators." The European Union Action Plan for circular economy also promotes "a green employment initiative, a new skills agenda and a green action plan for SMEs" (EC, 2014b).

The European Commission's circular economy Action Plan (COM (2015) 614/2 Closing the loop: an European Union action plan for the circular economy) defines circular economy "as an economy in which the value of products, materials and resources is maintained in the economy for as long as possible, and waste generation is minimized. The procedures outlined in the action plan will encourage the European Union transition to the circular economy, which is a priority of the European Union Commission to support the development and job creation." The circular economy model has also been supported by European business and civil society, reflecting the growing appreciation of the economic, environmental and social opportunities that exist in reducing the material use of our economy, as well as dependence on resources and energy. For example, one estimate suggests that circular economy initiatives by 2030 could bring Europe an economic benefit of 1.8 trillion euros and reduce carbon emissions by 450 million tonnes. The growth of initiatives outside Europe also suggests a global reach and policy relevance to this concept (EC, 2018b).

> In the European Union Action Plan for the Circular Economy for 2015 (COM (2015) 614/2 Closing the loop: the European Union Action Plan for the Circular Economy), the Commission of the European Union published a development framework for monitoring the circular economy, with the aim of measuring progress and assessment of the efficiency of the circular economy in the European Union and the Member States.

To this end, and based on existing data, a set of significant indicators has been identified that will cover the various phases of circular economy. The framework is "the key tool that shows whether existing policy initiatives are successful in achieving the expected results and allows the identification of those areas where the action is needed" (EC, 2018a).

6.13 Overview of the Circular Economy Monitoring Framework

"The framework for monitoring the circular economy is the European Commission's proposal to measure the progress of the circular economy in the Member States of the European Union and the European Union" (EC, 2018a). The framework contains four areas related to: production and consumption, waste management, secondary raw materials and competitiveness and innovation. Furthermore, based on these four areas, 10 indicators were developed, and further their sub-indicators that can be used to monitor the circular economy (EC, 2015a).

These 10 indicators are shown in Fig. 6.2 and are divided according to the following four thematic areas (EC, 2018a):

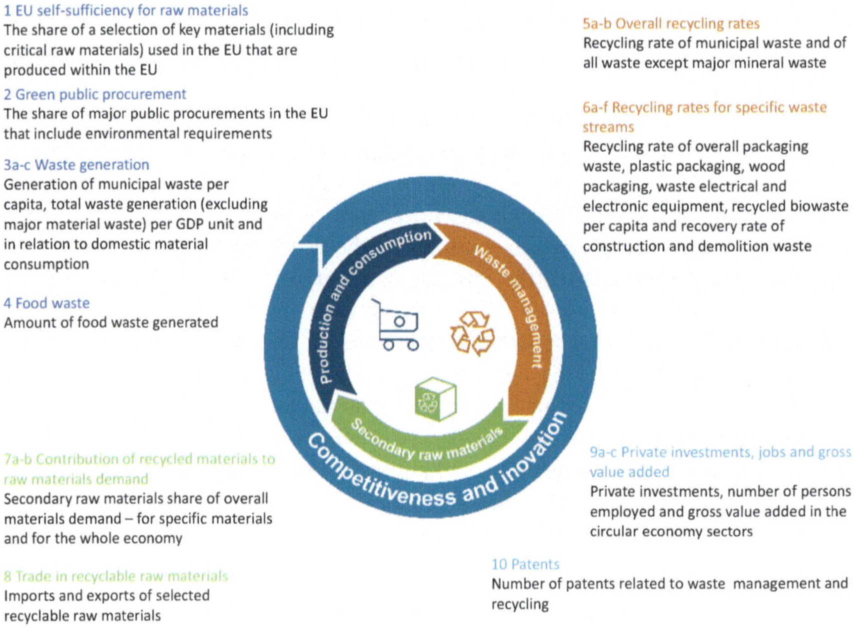

Fig. 6.2 Circular economy indicators in the European Union. (EC, 2018a)

1. *Production and consumption area* includes four indicators:
   (a) Self-sufficiency of raw materials for production in the European Union
   (b) Green public procurement (as an indicator for aspects of financing)
   (c) Waste production (as an indicator of consumption aspects)
   (d) Food waste.

The indicator "Monitoring the production and consumption phase is essential for understanding progress towards a circular economy" (EC, 2018a). Here, it is important to strive as much as possible to reduce the generation of waste within households and the economic sector, because "such behavior can contribute to increasing the self-sufficiency of selected raw materials for production in the European Union" (EC, 2018a). The indicator of the share of green public procurement in the economy is also a useful indicator of "how much public funds contribute to the circular economy" (EC, 2018a).

2. *Waste management area* includes two indicators:
   (a) Recycling rate (share of recyclable waste)
   (b) Specific waste streams (packaging waste, biowaste, e-waste, etc.).

One of the most important segments of the transition to a circular economy is recycling and its increasing application. Recycling means the part of waste that is recycled and returned to production and use. Two main indicators are related to this area: overall progress in waste recycling and recycling in specific waste streams (EC, 2018b).

3. *Secondary raw materials area* includes two indicators:
   (a) The contribution of recycled materials to the demand for raw materials;
   (b) Trade of recyclable raw materials between the European Union Member States and the rest of the world.

The essence of the circular economy is that everything that is used and produced, to be returned to the economy after use and recycling, where the basic principle of the circular economy is achieved, and that is to close the loop. Returning to the economy can be in the form of finished and processed products, but also in the form of new products. One of the most important indicators for circular economy is how materials that replace the extraction of natural resources are recycled, i.e. how much waste materials are re-incorporated into the economy (EC, 2018b). To have an increasing share of secondary raw materials in the economy, it is important to establish stable markets for them, so another relevant indicator is related to trade in recyclable raw materials between the European Union member-states and the rest of the world.

4. *Competitiveness and innovation area* includes two indicators:
   (a) Private investment, business and gross value added
   (b) Patents relating to recycling and secondary raw materials as a breakthrough for innovation.

The development of innovative technologies, in addition to promoting innovative industrial processes, also improves and enables product design for ease of use. All this leads to the fact that with the development of such technologies, the circular economy contributes to growth and development, through the creation of new jobs.

Two indicators are included in this area to monitor developments: one on the economics of the circular economy sector (recycling, repair and reuse) in terms of "jobs, investment and gross value added" and the other on patents related to recycling and secondary raw materials, as an intermediary for innovation" (EC, 2018b).

Table 6.4 shows the European Union areas/topics, indicators and sub-indicators.

The evaluation of the selected indicators was done according to their work in terms of relevance, acceptability, credibility, lightness and robustness. Indicators for green public procurement and food waste are included, even if statistical work is still on-going, and data will only be available in the coming years (EC, 2018b). As it can be seen from the table, most of the indicators relate to the area of waste management, while to a much lesser extent it covers other phases of the loop, such as production, consumption, reuse, repair and some other activities.

The table also shows and has developed sub-indicators for each of these indicators. The largest number of sub-indicators was done under the indicator *Recycling of specific waste streams* which includes six sub-indicators; the next one is the indicator *Trade of recyclable raw materials between the European Union member-states and with the rest of the world*, which contains four developed sub-indicators and the third one is the indicator of *Waste production*, with three developed sub-indicators.

Regarding other European Union initiatives related to sustainability monitoring, the European Union has also developed sustainability indicators (EC, 2015c). The European Commission's sustainability indicators are focused on the national level presented in Table 6.5.

As shown in Table 6.5, 10 different categories have been singled out, which include economic, social and environmental aspects that are important in the context of sustainable development. The category of good governance does not include the indicator. Furthermore, the "common bird index" indicator in the category of natural resources includes operational indicators of biodiversity, freshwater resources, marine ecosystems and land use. The general objective of these natural resources is to bypass over-exploitation, improve management and identify the value of the ecosystem service (EC, 2015b). Energy-related to climate change is included with a focus on greenhouse gas emissions although there are many other environmental problems.

Table 6.4 European Union areas/topics, indicators and sub-indicators

Area/topic	Indicators	Sub-indicators
I. Production and consumption	1. Self-sufficiency of raw materials for production in the European Union	No
	2. Green procurement*	*Indicator in development
	3. Waste production	(a) Production of municipal waste per capita (b) Waste production without major mineral waste per unit of GDP (c) Generation of waste without major mineral waste per unit of consumption of domestic material
	4. Food waste*	*Indicator in development
II. Waste management	5. Recycling rates	(a) Municipal waste recycling rate (b) Recycling rate of total waste excluding mineral waste
	6. Recycling for specific waste streams	(a) Recycling rate of total packaging waste (b) Recycling rate of plastic packaging waste (c) Wood packaging recycling rate (d) Recycling rate of electrical and electronic waste (e-waste) (e) Recycling of biowaste per capita (f) Recovery rate of construction and demolition waste
III. Secondary raw materials	7. Contribution of recycled materials to raw material demand	(a) End-of-life recycling input rates (b) Circular use rate of materials
	8. Trade in recyclable raw materials between European Union member states and the rest of the world	(a) Imports from non-European Union countries (b) Exports to non- European Union countries (c) Imports from European Union countries (d) Exports to European Union countries
IV. Competitiveness and innovation	9. Private investment, jobs and gross value added related to the circular economy sectors	(a) Gross investment in tangible goods (b) Number of employees (c) Value added at factor cost
	10. Number of patents related to recycling and secondary raw materials	No

Source: Milanović according to EC (2018b)

Table 6.5 European Commissions sustainability indicators

Category	Name of Indicator
Socio-economic development	Real GDP per capita
Sustainable consumption and production	Resource productivity
Social inclusion	People at risk of poverty or social exclusion
Demographic change	The employment rate of older workers
Public health	Years of healthy living Life expectancy
Climate change and energy	Gas emissions Primary energy consumption
Sustainable transport	Energy consumption in transport regarding GDP
Natural resources	"Common bird index" indicator
Global partnership	Official development assistance as a share of GDP
Good management	There is no indicator

Source: EC (2015c)

6.14 Discussion and Conclusions

The circular economy model represents a new business concept whose implementation has started in all areas, starting from the production, the product usage, to the iterative process of waste inclusion in the production process and returning the product into the usage once again. The implementation of circular economy and its principles affects the entire economy of a country, and its purpose is to emphasise effective use of the resources.

The circular economy indicators represent the measuring tools for the circular economy implementation achievements and mostly are the clear representation of a country's success within a certain area when implementing the model of circular economy. This chapter shows the theoretical review of the circular economy indicators and their measuring method, as well as the review of the initiatives taken due to the development of the circular economy indicators. Besides, the review of the indicator groups, aligned with the implementation levels and with different countries, is also given stressing the necessity and the importance of the circular economy implementation. Simultaneously, this chapter represents a significant review bibliography of the circular economy indicators in a broader sense, pointing out those circular economy indicators developed in the European Union member-states to gain insight into the level of implementation and development. The European Union member-states have developed their indicators in four areas and many of them have already achieved significant results (in the first place countries such as Germany, the Netherlands, Austria, Belgium) completing the model of circular economy principles.

Most certainly, the further development of the circular economy indicators is highly significant for the development of the model for calculating the composite circular economy index that enables comparative analysis of the states in the circular economy implementation. There are numerous research studies where the DEA

Method for calculating the circular economy composite index and country ranking has been applied, and the review can be found in Milanović et al. (2022) (to be published).

Regarding the achieved projects in this area within the European Union member-states, the European Union Commission fully supports any kind of projects related to research and innovation within the CE area. Furthermore, through the programme *Horizon 2020,* the initiative *Industry 2020 in the Circular economy* has been included, setting aside 650 million euro funds for innovative projects that support the circular economy goals, industrial competition within the European Union including the process industry, production and new business models (Horizon, 2013). Moreover, *Horizon 2020* supports innovative circular economy projects in other areas, such as prevention and waste management, food waste, processing, sustainable processing industry, industrial symbiosis and bioeconomy (Horizon, 2013).

Information and the circular economy indicators review in this chapter represent the "relevant tool for managers to measure the efficiency of circular economy in tasks inherent to the planning process, such as the monitoring and control of action implemented, decision – making, and comparative analysis in the time and space" (Sánchez-Ortiz et al., 2020), but these could also be a good starting point and direction for project managers when it comes to establishing and developing ideas for their projects within this area.

Finally, the obtained review of circular economy indicators provides a precise starting point for developing some further research and analyses. Recommendations for further studies will be that developing indicators is a quite good basis for recognising those areas where it is necessary for countries to further develop their renewable capacities and potentials, to reorient their strategies and policies, for them to meet the established goals of efficient circular economy principles, as well as for those areas where clearly defined indicators are missing, further development is being done.

References

Akerman, E. (2016). *Development of circular economy core indicators for natural resources: Analysis of existing sustainability indicators as a baseline for developing circular economy indicators*.

Banaitė, D. (2016). Towards circular economy: Analysis of indicators in the context of sustainable development. *Social Transformation in Contemporary Society, 4*(9), 142–150.

Beratan, K. K., Kabala, S. J., Loveless, S. M., Martin, P. J., & Spyke, N. P. (2004). Sustainability indicators as a communicative tool: Building bridges in Pennsylvania. *Environmental Monitoring and Assessment, 94*(1–3), 179–191.

Biggs, D., Biggs, R. O., Dakos, V., Scholes, R., & Schoon, M. (2011). Are we entering an era of concatenated global crises? *Ecological Society, 16*(2), 27.

Borojević, T., Maletič, M., Petrović, N., Radaković, J. A., Senegačnik, M., & Maletič, D. (2017). Youth attitudes towards goals of a new sustainable development agenda. *Problemy Ekorozwoju/ Problems of Sustainable Development journal, 12*(2), 161–172.

Cayzer, S., Griffiths, P., & Beghetto, V. (2017). Design of indicators for measuring product performance in the circular economy. *International Journal of Sustainable Engineering, 10*(4–5), 289–298.

Design, G. (2015). *Circularity indicators: An approach to measuring circularity.*

European Academies Science Advisory Council – EASAC. (2015). *Circular economy: A commentary from the perspectives of the 1047 natural and social sciences.*

European Commission – EC. (2011). *Roadmap to a resource efficient Europe*. European Comission. (COM/2011/571). Retrieved at: http://www.europarl.europa.eu/meetdocs/2009_2014/documents/com/com_com(2011)0571_/com_com(2011)0571_en.pdf

European Commission – EC. (2014a). *Communication from the Commission to the European Parliament, The Council, The European Economic and Social Committee and the Committee of the Regions – Towards a circular economy: A zero waste programme for Europe*. Retrieved at: https://ec.europa.eu/environment/circular-economy/pdf/circular-economy-communication.pdf

European Commission – EC. (2014b). *Green action plan for SMEs Enabling SMEs to turn environmental challenges into business opportunities*. European Commission. (COM/2014/0440).

European Commission – EC. (2014c). *General Union environment action programme to 2020.* Retrieved at: https://ec.europa.eu/environment/action-programme/

European Commission – EC. (2015a). *Closing the loop – An EU action plan for the circular economy*. COM (2015) 614 Final. Retrieved at: https://eur-lex.europa.eu/legal-content/EN/TXT/?uri=CELEX%3A52015DC0614

European Commission – EC. (2015b). *Natural resources*. Retrieved at: https://ec.europa.eu/eurostat/web/sdi/indicators/natural-resources

European Commission – EC. (2015c). *Sustainable development indicators*. Retrieved at: https://ec.europa.eu/eurostat/web/sdi/indicators

European Commission – EC. (2018a). *Communication from the Commission to the European Parliament, the Council, the European Economic and Social Committee and the Committee of the Regions on a monitoring framework for the circular economy*. SWD (2018) 17 final. January 16, 2018. Retrieved at: https://eur-lex.europa.eu/legal-content/EN/TXT/?uri=COM%3A2018%3A29%3AFIN

European Commission – EC. (2018b). *Commission Staff Working Document measuring progress towards circular economy in the EU – Key indicators for a monitoring framework*. SWD (2018) 17 final. January 16, 2018. Retrieved at: https://ec.europa.eu/environment/circular-economy/pdf/monitoring-framework_staff-working-document.pdf

European Environment Agency – EEA. (2016). *Circular economy in Europe – Developing the knowledge base*. EEA Report No 1065 2/2016.

Gallaud, D., & Laperche, B. (2016). *Circular economy, industrial ecology and short supply chain*. Wiley.

Gallopin, G. C. (1996). Environmental and sustainability indicators and the concept of situational indicators. A systems approach. *Environmental Modeling & Assessment, 1*(3), 101–117.

Garcia-Bernabeu, A., Hilario-Caballero, A., Pla-Santamaria, D., & Salas-Molina, F. (2020). A process oriented MCDM approach to construct a circular economy composite index. *Sustainability, 12*(2), 618.

Geng, Y., Fu, J., Sarkis, J., & Xue, B. (2012). Towards a national circular economy indicator system in China: An evaluation and critical analysis. *Journal of Cleaner Production, 23*(1), 216–224.

Government Offices of Sweden. (2016). *Smart industry: A strategy for new industrialization for Sweden*. Ministry of Enterprise and Innovation. Retrieved at: http://www.government.se/498615/contentassets/3be3b6421c034b038dae4a7ad75f2f54/nist_statsformat_160420_eng_webb.pdf

Heshmati, A. (2015). *A review of the circular economy and its implementation*. IZA Discussion Papers, No. 9611. Institute for the Study of Labor (IZA), Bonn. Retrieved at: https://www.econstor.eu/bitstream/10419/130297/1/dp9611.pdf

Horizon. (2013). *Horizon 2020 work programme for 2014–2015; Climate action, environment, resource efficiency and raw materials*. European Commission. Retrieved at: http://ec.europa.eu/research/participants/data/ref/h2020/wp/2014_2015/main/h2020-wp1415-climate_en.pdf

Houshfar, E., Skreiberg, Ø., Todorović, D., Skreiberg, A., Løvås, T., Jovović, A., & Sørum, L. (2012). NOx emission reduction by staged combustion in grate combustion of biomass fuels and fuel mixtures. *Fuel, 98*, 29–40.

Hugé, J., Block, T., Waas, T., Wright, T., & Dahdouh-Gueba, F. (2016). How to walk the talk? Developing actions for sustainability in academic research. *Journal of Cleaner Production, 137*, 83–92. https://doi.org/10.1016/j.jclepro.2016.07.010

Iannuzzi, A. (2017). *Greener products: The making and marketing of sustainable brands*. CRC Press.

Kirchherr, J., Reike, D., & Hekkert, M. (2017). Conceptualizing the circular economy: An analysis of 114 definitions. *Resources, Conservation and Recycling, 127*, 221–232.

Kyriakopoulos, G. L., Kapsalis, V. C., Aravossis, K. G., Zamparas, M., & Mitsikas, A. (2019). Evaluating circular economy under a multi-parametric approach: A technological review. *Sustainability, 11*(21), 6139.

Lévy, J. C. (2009). *Économie circulaire: l'urgence écologique: monde en transe. Chine en transit.* Presses de l'École Nationale des Ponts et Chaussées.

Li, S. (2012). The research on quantitative evaluation of circular economy based on waste input-output analysis. *Procedia Environmental Sciences, 12*, 65–71.

Lilja, R. K. (2015). Policy instruments for promoting material efficiency: Case of Finland. *Clean Technologies and Environmental Policy, 17*(7), 2029–2040.

Linder, M., Sarasini, S., & van Loon, P. (2017). A metric for quantifying product-level circularity. *Journal of Industrial Ecology, 21*(3), 545–558.

MacArthur, E., Foundation, and Granta Design. (2015a). *Circularity indicators – An approach to measure circularity*. Methodology. Retrieved at: http://www.ellenmacarthurfoundation.org/assets/downloads/insight/CircularityIndicators_Methodology_May2015.pdf.

MacArthur, E., Foundation, and Granta Design. (2015b). *Project overview: An approach to measuring circularity*. Retrieved at: http://www.ellenmacarthurfoundation.org/assets/downloads/insight/CircularityIndicators_Project-Overview_May2015.pdf

Magnier, C. (2017). *10 key indicators for monitoring the circular economy*. March 2017. The Monitoring and Statistics Directorate (SOeS). Retrieved at: http://www.epsilon.insee.fr/jspui/bitstream/1/86506/1/22978_2017_ENG.pdf

Mihajlov, A., Mladenović, A., & Jovanović, A. (2019). *Circular economy in Serbia: The process started.* Environmental Ambassadors for Sustainable Development.

Milanović, T., Petrović, N., & Ćirović, M. (2019). "Upravljanje otpadom i cirkularna ekonomija", Zbornik radova XII skup privrednika i naučnika SPIN'19 "Lin transformacija i digitalizacija privrede Srbije". Fakultet organizacionih nauka, Centar za operacioni menadžment, Beograd, 2019, pp. 328–334. Fakultet organizacionih nauka. [In Serbian].

Milanović, T., Popović, V. M., Vučković, S., Rakaščan, N., Popović, S., & Petković, Z. (2020b). Analysis of soybean production and biogas yield to improve eco-marketing and circular economy. *Economics of Agriculture, 67*(1), 141–156. https://doi.org/10.5937/ekoPolj2001141M

Milanović, T., Jovović, A., Petrović, N., & Martić, M. (2020a). Pregled indikatora cirkularne ekonomije razvijenih zemalja. Zbornik radova SYM-OP-IS 2020 XLVII Simozijum o operacionim istraživanjima, 20–23.09.2020, Beograd (online). [In Serbian].

Milanović, T., Savić, G., Martić, M., Milanović, M., & Petrović, N. (2022). Development of the waste management composite index using DEA method as circular economy indicator: The case of European Union countries. *Polish Journal of Environmental Studies, 31*(1), 771–784. https://doi.org/10.15244/pjoes/139896

Mitrović, Đ., & Veselinov, M. (2018). Measuring countries competitiveness in circular economy-composite index approach. In J. Kočović, J. Selimović, B. Boričić, V. Kašćelan, & V. Rajić (Eds.), *Quantitative models in economics*. University of Belgrade, Faculty of Economics Publishing Centre.

Moraga, G., Huysveld, S., Mathieux, F., Blengini, G. A., Alaerts, L., Van Acker, K., et al. (2019). Circular economy indicators: What do they measure? *Resources, Conservation and Recycling, 146*, 452–461.

Murray, A., Skene, K., & Haynes, K. (2017). The circular economy: An interdisciplinary exploration of the concept and application in a global context. *Journal of Business Ethics, 140*(3), 369–380.

National People's Congress. (2008). *People's Republic of China circular economy promotion law*. Standing Committee of the Fourth Session of the Elevent National People's Congress 29 August 2008, Beijing.

Ohliger, T. (2015). *Environment policy: General principles and basic framework*. European Parliament.

Petrovic, N., Savic, G., Andrijasevic, D., Stanojevic, M., Cirovic, M., Slovic, D., & Radakovic, J. A. (2016). Evaluating eco-efficiency of beverage packaging materials: A data envelopment analysis approach. *Fresenius Environmental Bulletin, 25*(8), 2958–2963.

Potting, J., Hekkert, M. P., Worrell, E., & Hanemaaijer, A. (2017). *Circular economy: Measuring innovation in the product chain*. PBL Netherlands Environmental Assessment Agency. Retrieved at: https://www.researchgate.net/profile/MP_Hekkert/publication/319314335_Circular_Economy_Measuring_innovation_in_the_product_chain/links/5a83e8baaca272d6501efa7b/Circular-Economy-Measuring-innovation-in-the-product-chain.pdf

Radaković, J. A., Petrović, N., Milenković, N., Stanojević, K., & Đoković, A. (2017). The improvement of students' higher environmental and climate change knowledge: A case study. *Polish Journal of Environmental Studies, 26*(6), 2711–2719. https://doi.org/10.15244/pjoes/69645

Ragossnig, A. M., & Jovovic, A. (2016). Waste management life cycle: Sensitisation–implementation/integration–transition–optimisation. *Waste Management & Research, 34*(9), 813–815.

Saidani, M., Yannou, B., Leroy, Y., Cluzel, F., & Kendall, A. (2019). A taxonomy of circular economy indicators. *Journal of Cleaner Production, 207*, 542–559.

Sánchez-Ortiz, J., Rodríguez-Cornejo, V., Río-Sánchez, D., & García-Valderrama, T. (2020). Indicators to measure efficiency in circular economies. *Sustainability, 12*(11), 4483.

Smol, M., Kulczycka, J., & Avdiushchenko, A. (2017). Circular economy indicators in relation to eco-innovation in European regions. *Clean Technologies and Environmental Policy, 19*(3), 669–678.

Smyth, J. C. (2006). Environment and education: A view of a changing scene. Reprinted from Environmental Education on Research (2005) 1(1), 3–20. *Environmental Education Research, 12*(3–4), 247–264.

Staniškis, J. K. (2012). Sustainable consumption and production: How to make it possible. *Clean Technologies and Environmental Policy, 14*(6), 1015–1022.

Su, B., Heshmati, A., Geng, Y., & Yu, X. (2013). A review of the circular economy in China: Moving from rhetoric to implementation. *Journal of Cleaner Production, 42*, 215–227.

Vasileios Rizos, V. R., Katja Tuokko, K. T., & Arno Behrens, A. B. (2017). *The circular economy: A review of definitions, processes and impacts*. CEPS Research Report No 2017/8, April 2017. Retrieved at: http://aei.pitt.edu/85892/1/RR2017-08_CircularEconomy_0.pdf

Vercalsteren, A., Christis, M., & Van Hoof, V. (2018). *Short-term assignment indicators for a circular economy*. SUMA Circular Economy Policy Research Centre. Retrieved at: https://circulareconomy.europa.eu/platform/sites/default/files/summa_-_indicators_for_a_circular_economy.pdf

Vukadinović, B., Popović, I., Dunjić, B., Jovović, A., Vlajić, M., Stanković, D., et al. (2016). Correlation between eco-efficiency measures and resource and impact decoupling for thermal power plants in Serbia. *Journal of Cleaner Production, 138*, 264–274.

World Health Organization – WHO. (2018). *Circular economy and health: Opportunities and risks*. Retrieved at: https://www.euro.who.int/__data/assets/pdf_file/0004/374917/Circular-Economy_EN_WHO_web_august-2018.pdf

Chapter 7
Financing Start-Up Projects in Circular Economy: Does Crowdfunding Fit?

Isidora Ljumović and Aida Hanić

Abstract Financing a green and circular economy is a multi-level problem for entrepreneurs, businesses, local governments, and nations worldwide. Alternative methods of finance have become increasingly popular as a means of obtaining necessary funds due to the advancement of modern technology. Crowdfunding is one example of such a capital network. This chapter emphasizes the role of crowdfunding in financing start-ups oriented towards sustainable, green, or circular projects, exploring their likelihood of success. We hypothesized that start-up projects that use circular economy principles have a better chance of raising the desired amount of money from the crowd. We collected data for the study from the "Kaggle.com" open-source repository. Our findings show that campaigns oriented to the concept of circularity differ from others in several parameters. Campaigns with circularity elements target higher amounts of funds and raise more money. They are also more often chosen as a staff pick. Along with this, the results of econometric estimates support the conclusion that campaigns with circular orientation are more likely to be successful.

Keywords Circular economy · Crowdfunding · Start-ups · Kickstarter campaigns

7.1 Introduction

Economic activity and a society based on consumption have led to the emergence of a large amount of waste, which is no longer merely an environmental issue. In that aspect, many researchers criticized growth models employed thus far from social and environmental perspectives, requiring the repair of the current socio-economic paradigm. One of the approaches that is often viewed as a solution to this issue is circularity. A circular economy (CE) enables economic prosperity by creating new sectors and jobs.

I. Ljumović (✉) · A. Hanić
Institute of Economic Sciences, Belgrade, Republic of Serbia
e-mail: isidora.ljumovic@ien.bg.ac.rs; aida.hanic@ien.bg.ac.rs

V. Obradović (ed.), *Sustainable Business Change*,
https://doi.org/10.1007/978-3-031-23543-6_7

The transition to CE necessitates the reuse of materials or product designs and the implementation of new economic models. A circular business model (CBM) explains how an organization creates, delivers, and captures value with and within closed material loops (Mentink, 2014). Start-ups can become circular pioneers using this principle. Their environmentally friendly products or services provide a positive environmental impact, contribute to a greener economy (Bergset & Fitcher, 2015), and on the other hand, are extremely interesting for consumers.

Financing in a green and circular economy is a challenge on multi-levels for entrepreneurs, companies, local self-government, and states worldwide. The typical perception of investing in environmental activities is that it reduces profitability and that environmental investments have a negative impact on the stock market (Hamilton, 1995; Halme & Niskanan, 2001). These facts are unsustainable in the twenty-first century (Ljumović & Pavlović, 2016; Lukić et al., 2018). Companies that implement environmental principles can reduce costs and increase earnings because they have: better access to specific markets, easier product differentiation, sell pollution control technology, better relationships with stakeholders, lower costs of inputs (Porter, 1991; Porter & van der Linde, 1995), and various national and international support programs (Ambec & Paul, 2008). Although CE creates a new economic paradigm, access to finance and limited financial opportunities are significant barriers for start-ups to realize new business ideas and take growth opportunities.

We organized the remainder of the chapter as follows. In Sect. 7.2, we provide a literature review and explore the phenomena of the circular economy, circular start-up, and the role of crowdfunding. Furthermore, we elaborate on the research hypotheses. Section 7.3 presents data, methodology, and descriptive statistics. Finally, we report our findings and summarize the results.

7.2 Literature Review

7.2.1 Circular Economy

Research with CE elements has developed intensely over the last few decades. For instance, in the 1960s, Boulding (1966) used the idea of closed systems in terms that the outputs of all parts of the system are linked to the inputs of other components. In the 1970s, Stahel and Reday-Mulvey (1976) pointed out the possibility of the life extensions of products concerning the ecological aspect of the process. Their focus was on the dematerialization of the industrial economy by observing it as a loop to prevent waste, create jobs, and resource efficiency. Stahel (1982) proposed a spiral-loop system based on reuse (loop 1), repair (loop 2), reconditioning (loop 3), and recycling (loop 4), emphasizing the role of the private sector in this new self-replenishing economy. Equally, Pearce and Turner (1990) investigated the linear and open-ended characteristics of modern economic systems, while Geissdoerfer et al. (2017) referred to Boulding's (1966) work in terms that natural resources can provide inputs for production and consumption. Although it happened for a relatively long period, various research contributed to creating the circular economy concept, or simply the circularity.

Even now, a circular economy is not an easy term to define. It gained attention among scholars and practitioners worldwide, which has led to the emergence of many definitions and understandings of the concept (Kirchherr et al., 2017; Merli et al., 2018). Kirchherr et al. (2017) identified 114 definitions of CE, classified into 17 dimensions, pointing out that CE is a combination of reducing, reusing, and recycling activities. This can be summarized in the definition by the European Commission (2015) that highlights the importance of extended use of products, materials, and resources with minimum of waste. CE is an economy trying to redefine growth by overcoming the take-make-dispose linear pattern (Merli et al., 2018) on micro, mezzo, and macro levels (Bauwens et al., 2020). According to Blomsma and Brennan (2017), CE's main principles are reuse, recycling, and remanufacturing. However, the list of R's spreads out the literature, and we found several R's as a basis for a circular business model. For instance, Brennan et al. (2015) or Kirchherr et al. (2017) classify it as Regenerate, Reduce, Reuse, Recycle, and Recover.

Although several terms can be associated with the CE, in their literature review, Beaulieu et al. (2016) note that the following concepts provided the frame for circularity: (a) sustainable development, (b) ecological transition, (c) green economy, (d) functional economy, (e) life cycle thinking, (f) cradle to cradle thinking, (g) shared value, (h) industrial ecology, (i) extended producer responsibility, and (j) eco-design. If we observe these elements together, CE is a sustainable economic system based on R's with value creation throughout the supply chain, which requires fundamental changes in legislation, innovations, and socio-economic model (Schenkel et al., 2015; Reichel et al., 2016; Corona et al., 2019).

World Economic Forum (Global Risk Report, 2020) presented the circular economy as a win-win option instead of a trade-off that is currently in use, especially in terms of GHG emissions and habitat loss. Besides the ecological aspect, one of the main questions of the circular economy is how to make a profit while reducing dependence on natural resources and how circularity can be a driver for business competitiveness (Bocken et al., 2016). In a study done by McKinsey (2016), the analysis showed that six circular-economy activities could improve performance and reduce costs: Regenerate (shift to renewable energy); Share (prolonging product life); Optimize (better product efficiency); Loop (remanufacturing and recycling); Deliver and use (utilize virtually); Exchange (the use of new technologies).

It is necessary to involve all stakeholders to implement these six elements because innovation circulates in the circular economy. According to Millette et al. (2020), stakeholders interact to provide needed information for the circular economy's implementation and development. This approach is a circularly focused incubator where stakeholders include entrepreneurs, companies on both sides of seeking and making it an added value, government, academia, and NGOs. In that aspect, how can a circular economy boost circular projects, especially start-ups?

7.2.2 Circular Economy and Start-Ups

For a company to be the pioneer of circularity, it is not enough to change current materials or product design. Implementing new business models or so-called circular business models (CBM) is vital. According to Geissdoerfer et al. (2020), this term was first introduced by Schwager and Moser (2006). Geissdoerfer et al. (2018), observes CBM as a sustainable business model with the aim to create additional monetary and nonmonetary value in a long-term perspective. At the same time, Osterwalder and Pigneur (2010) highlight the rationality in creating, delivering, and capturing values within the closed material loop.

Urbinati et al. (2017) note that circular business models can be focused on improving the circularity of the value creation systems downstream or combining both. Also, different CBM can be used at various stages of circularity transition. This can include innovating the current business model. At the same time, start-ups can adopt a circular business model from the start, based on the principle design to last (Henry et al., 2019). In that aspect, circular start-ups overlap with other different environmental models. Regarding the approach to new markets, start-ups can use different circular business models such as sustaining innovations, low-end disruptions, new-market disruptions (Vuorio, 2020), or their combination.

On the example of 147 circular start-ups in the Netherlands, Bauwens et al. (2019) found that circular start-ups develop higher circularity strategies regarding waste management and are more open to innovations. Henry et al. (2019) did similar research on a sample of 128 circular start-ups in the Randstad region in the Netherlands, Berlin, and London. The authors concluded that there are five circular start-up archetypes: design-based (circular innovations are adopted in the pre-market phase); waste-based (exploring external waste streams); platform-based (use of share/trade business model in B2B, B2C, or C2C); service-based (increase efficiency in service-systems) and nature-based start-ups (use the nature-based systemic solution in products and services).

But even if there is a defined circular start-up archetype and CBM, one of the principal issues is how limited resources influence the start-ups and what type of financing can they use.

7.2.3 Circular Economy and Crowdfunding

Access to finance for companies is a central issue for enterprises worldwide that can strongly influence the success or failure of a start-up (Carter & Van Auken, 1990; Gimeno et al., 1997; Ljumović et al., 2015a, b; Ljumović & Jakšić, 2015; Kee et al., 2019; Irwin et al., 2019). Challenges in access to finance can arise due to a wide specter of reasons, but at the same time, it is an essential condition for the innovation ecosystem (Wyman, 2017). Obstacles can arise due to the lack of collateral or profit (Cosh et al., 2009, Ljumović et al., 2015a), low reputation and small size

(Cassar, 2004), lack of valid documentation – balance sheet figures, or proof of success (Bernstein et al., 2017; Jones & Jayawarna, 2010), information asymmetries and moral hazard problems between start-ups and investors (Lee et al., 2015; Nofsinger & Wang, 2011). While start-ups find it difficult to gain a foothold in using financial services, companies with already established credit histories are offered ease of access due to their prior inclusion into the financial market. The valuation of start-ups can be challenging because of their characteristic (Ljumović et al., 2012). Ortas et al. (2013) note that ecological investments vary across countries and regions in terms of the level of development of the financial system. This includes capital availability, degree of development of the banking sector, the existence of financial regulators, and technology risks. In practice, start-ups are pragmatic and use a range of financial instruments, not all targeting new companies (Bergset, 2015, 2018). Traditional sources of finance include all internal sources (founders' funds and returns from business activities, such as retained earnings, sale of inventories, fixed assets or other assets, and debt collection), financing from family and friends, banks, microfinance institutions, leasing company, and capital market.

With the development of modern technologies, alternative sources of financing are an increasingly popular source of financing the company's operations, and among them are social capital networks – crowdfunding. The scope of alternative products ranges from financing based on future income, online loans, peer-to-peer loans, cryptocurrencies (Bitcoin, Ethereum, XRP, Tether, and others), social bonds, and similar mechanisms beyond formal financial systems. Although it is a matter of raising relatively small amounts of money, in this way, it is possible to acquire significant amounts of funds (Ljumović & Pejović, 2020). Recently, crowdfunding has become increasingly relevant as a source of funding for start-ups (Bocken, 2015; Angerer et al., 2017; Cumming & Hornuf, 2018; Bergset, 2018; Brown et al., 2019).

According to Mollick (2014), crowdfunding represents efforts made by entrepreneurs to fund their venture ideas based on small contributions made by a large number or group of individuals only by using the internet, excluding traditional financial intermediaries. In other words, financial resources are raised directly from a large audience or the crowd (Belleflamme et al., 2014), based on stranger's willingness to support other strangers (Testa et al., 2019), which is a much more democratic way to access the capital (Mollick & Robb, 2016) and accelerate the innovation process.

Crowdfunding can take several forms: reward-based, donation-based, lending-based, and equity-based (Stanko & Henard, 2016; Vismara, 2019, Table 7.1). In their literature review, Böckel et al. (2020) discovered that donation-based crowdfunding was the most explored type of crowdfunding. However, this type constitutes only 8% of the global crowdfunding market (Massolution, 2015). According to Petruzzelli et al. (2019), every crowdfunding project needs a project creator, the backers, the crowdfunding platform, the campaign itself, and the crowdfunding outcomes. With these five elements, crowdfunding serves as a novel socio-technical practice (Testa et al., 2019) that has the potential to transform financial structures, overcome geographical barriers (Agrawal et al., 2015), be more flexible than traditional sources of financing, and become an effective marketing tool (Efrat & Gilboa, 2019).

Table 7.1 Forms of crowdfunding

Reward based	Proponents seek financial contributions from a crowd of backers in exchange for rewards or customized products or services. There are two types of reward-based crowdfunding: keep-it-all (KIA) and all-or-nothing (AON)
Donation based	Charitable giving with no material delivery to donors
Lending based	Peer-to-peer lending – Fixed interest rates for landers
Equity based	Entrepreneurs make an open call to sell a specific amount of equity in their company

Source: Forbes and Schaefer (2017); Wang et al. (2018); Vismara (2019)

Because it is gaining more attention, researchers analyze different aspects and the factors that influence the role and success of crowdfunding. For instance, Ordanini et al. (2011) examined how the crowd's behavior affects crowdfunding, while Zhou et al. (2016) focused on project description: length, readability, tone, experience, and past expertise. On the other hand, Gerber et al. (2012) identified that financing, forming relationships and networks, self-affirmation, replication of success stories, and increased awareness of the product influence the decision to use crowdfunding.

Stanko and Henard (2017) emphasize that backers generate word-of-mouth awareness. Their research concluded that the amount of funding raised during a crowdfunding campaign does not significantly impact the later market performance of the crowdfunded product. At the same time, the number of backers attracted to the campaign does. In similar research done on a sample of 959 projects in China, Wang et al. (2018) note that comment quantity, comment score, reply length, and reply speed by backers are positively associated with fundraising success.

At the same time, Block et al. (2018) emphasize the importance of crowdfunding as a tool to foster sustainability. In that aspect, we will focus more on this relationship because the connection between crowdfunding projects and environmental issues is increasingly analyzed, and authors are putting this phenomenon at the center of their research. In this context, Thompson et al. (2011) state that no distinction should be made between environmental and sustainable entrepreneurship but observed as a link between entrepreneurship and sustainable development. According to Böckel et al. (2020), the first article addressing the relationship between sustainability and crowdfunding was published in 2011. Jovanovic (2019) analyzed 90 scientific papers published between 2011 and 2016 and found that 8% of all research on crowdfunding relates to sustainability. This can be explained by the high expectations that crowdfunding will help to accelerate sustainability (Böckel et al., 2020).

In that sense, Bocken et al. (2014) identify crowdfunding as an example of a business model that can help develop and scale-up sustainable innovations by bringing together like-minded individuals, firms, and investors. In doing so, Petruzzelli et al. (2019) identified five aspects to comprehend the sustainability implications of crowdfunding properly. For instance, in the case of creators, they need to set up effective communication with potential backers because these types of projects often provide a public good, and if the crowd is focused on social issues and doing social good, then crowdfunding is an ideal tool to fund sustainable entrepreneurs or green start-ups (Calic & Mosakowski, 2016). Authors note that crowdfunding can support social entrepreneurship that emphasizes sustainability. They found that projects with social or sustainable components will be more successful than commercial-only projects, which is closely related to their conclusion that sustainability impacts creativity, which increases the success of crowdfunding. Confirmation can be found in Böckel et al. (2020), concluding that 74% of analyzed articles have a social component in the sustainability dimension of crowdfunding.

The communication process (Petruzzelli et al., 2019), public discourse about crowdfunding and sustainability, social media (Mollick, 2014), and other factors can all be important, mainly because they can reach geographically dispersed people (Saxton & Wang, 2013) and play a vital role in the success of crowdfunding campaigns (Lu et al., 2014; Beier & Wagner, 2015). In that aspect, it is essential to assess the interpolation between crowdfunding and sustainability in social media. Using Social Media Analytics (SMA) to track public discussions regarding crowdfunding showed that social media debate on sustainability and sustainability-oriented campaigns receives limited attention (Laurell et al., 2019). In other words, social and sustainable entrepreneurs should focus on specific user segments. Dos-Santos et al. (2020) did similar research and used the same approach as Laurell et al. (2019) but included the Google Trend in the analysis. Their results show that crowdfunding has been increasing since 2014, and its sustainable dimension is considered a proxy of marketing strategy.

Because crowdfunding encourages innovation, it can help close the funding gap for sensitive projects whose primary goal is to benefit the public good. One such example is cleantech because a cleantech crowdfunding campaign delivers more than just a product and accelerates the transition to a low-carbon economy (Bento et al., 2019). But the issue of new technologies, especially cleantech, is differently treated by countries. For instance, Cumming et al. (2017) analyzed 20,000 different cleantech projects on the Indiegogo platform in 81 countries worldwide. Results show that cleantech crowdfunding is more common in countries with low levels of individualism and is more common when oil prices are rising. This is consistent with Adhami et al. (2017).

On the other hand, Bento et al. (2019) discovered that after the country's risk is considered, the returns are not consistent with the risks associated with the technology employed in the projects, based on a sample of 365 European cleantech projects. Regarding the effect that institutional settings in a country can have on green crowdfunding campaigns, Butticè et al. (2019) notice that green campaigns are more diffused in countries with a limited environmental sustainability orientation. They based their sample on the population of 48,598 campaigns launched on Kickstarter between July 1, 2009, and July 1, 2012. Adhami et al. (2017) analyzed the determinants of the funding success of a sample of 423 green projects published in 27 specialized crowdfunding platforms in Europe from 2011 to 2017 using two different indexes: the Environmental Performance Index and the Social Sustainability

Index. The result shows a significant positive effect of green crowdfunding activity on these two indexes. Finally, Ljumović et al., (2021a, b) found that sustainable campaigns in the agri-food industry are more successful in countries with relatively lower importance of agriculture in the country's economy.

We expect significant differences between the project campaign oriented to the circular economy concept and others following the analyzed literature. Thus, we derive our first hypothesis:

H1. There is a statistically significant difference between the campaigns of the projects oriented to the circular economy and others.

Crowdfunders are often driven by normative or altruistic motives (Lindenberg & Steg, 2007), usually focus on the entrepreneurs' core values and ideas (such as sustainability, social agenda, and similar) instead of focusing on business plans (Lehner, 2013), and are initiated by intrinsic motives (Allison et al., 2015). Although the number of papers on this topic is growing, there is still no conclusive evidence about whether the environmental orientation of crowdfunding projects can influence their likelihood of successful funding. Several types of research tried to find the relationship between sustainability orientation and crowdfunding success, and while Lehner (2013); Belleflamme et al. (2014); Calic and Mosakowski (2016) note that the social aspect of crowdfunding is the reason to have a positive likelihood of success of crowdfunding campaigns, opposite to this, Hörisch (2015) found no positive effect of environmental orientation in terms of its likelihood of success. Their results show that sustainability-oriented projects do not present a significant advantage in terms of crowdfunding success. This is consistent with Moss et al. (2015) that crowd-investors often focus on profit-seeking opportunities.

Motylska-Kuzma (2018) found similar results in the case of Polish crowdfunding sites, where the project's long-term sustainability was less significant. There are also studies with mixed results, such as Cumming et al. (2017) in the case of clean-tech projects. According to the authors, on average, these projects are not significantly more successful, but at the same time, they attract significantly higher total pledges and more backers. Vismara (2019) made the same conclusion on a sample of 345 crowdfunding projects in the UK. In other words, although projects with a sustainability orientation don't have better chances of success, they attract a higher number of restricted investors. Butticè et al. (2019) notice that in countries with higher Environmental Performance Index, sustainability or green crowdfunding campaigns do not positively impact the likelihood of success.

Following this, we have set our second hypothesis:

H2. Projects oriented to the circular economy are more likely to succeed at crowdfunding than projects without the orientation to the circular economy, ceteris paribus.

7.3 Methods

Kickstarter is one of the oldest and largest crowdfunding platforms, and it has been used in several studies (e.g., Pitschner & Pitschner-Finn, 2014; Mollick, 2014; Colombo et al., 2015; Butticè et al., 2017; Courtney et al., 2017; Butticè et al., 2019; Böckel et al., 2020). According to the Kickstarter website (https://www.kickstarter.com/about?ref=global-footer), its mission is to help bring creative projects to life and make ideas into reality. Creators share new visions for creative work with the communities that will come together to fund them. Kickstarter is a reward-based crowdfunding platform with an "all-or-nothing" funding model, meaning that if a campaign fails, the project creators do not get funding, nor do the bakers get a reward. Creators can cash in the money pledged only if the campaign reaches the funding goal. However, there is no upper limit to the amount of money creators can attract during the campaign. Rewards are products, services, or gadgets, while financial rewards, equity shares, and interest for a loan are now allowed. For a small contribution, creators can offer a symbolic gift, such as thank you note or a small reward (promo material and similar).

In contrast, rewards can include the pre-purchase of the product for a full contribution. The platform hosts 15 categories: art, comics, crafts, dance, design, fashion, film, food, games, journalism, music, photo, publishing, technology, and theatre. Statistics on crowdfunding change daily, and currently, there is no official, unified statistics. As of December 2020, the success rate of fully funding a project on the Kickstarted was 38.28%, with 507,318 launched projects and 4.93 billion U.S. dollars pledged. Identifying what leads to the success of a crowdfunding campaign can be very helpful for project creators (Greenberg et al., 2013; Xu et al., 2014) because, according to Zhou et al. (2016), literature shows that less than 50% of projects were successfully funded on Kickstarter.

7.3.1 *Sample*

We collected data for the study from the "Kaggle.com" open-source repository. The initial full dataset provided on the repository contained data on 430,938 Kickstarter campaigns from 2009 to 2019. The dataset provided detailed information on crowdfunding campaigns, including attributes such as the title of the project (campaign); project goal; funding goal as the amount of money a creator needs to complete the project; blurb; short description displayed under the name of the project and on the browse part of the platform page; pledged funds, as the amount of money the project raised; backers, as the number of people that have supported the project; state of the project as successful, failed, canceled, live or suspended; country of origin of the campaign creator; currency; category, and similar.

On the Kickstarter platform, projects can have five statuses: active, successfully funded, failed, canceled, and suspended. In the case of active projects, fundraising is ongoing. When projects achieve funding goals, they are defined as successfully funded. On the contrary, they are marked as failed projects if they do not meet the funding goal. Canceled projects are void by the creator before the end of the duration. Kickstarter bans suspended projects for reasons such as violating the rules, misrepresentation, or others (Liang et al., 2020). Following Pitschner and Pitschner-Finn (2014), a project is coded as "successful" if the target amount defined by the initiators is reached and as "unsuccessful" if the project fails to fund the targeted amount.

Kickstarter is an international platform where entrepreneurs may post amounts in different currencies. All currencies other than USD were converted into USD using a yearly average exchange rate.

Before the analysis, we modified the dataset. First, we excluded double entries and data for the project that were live (ongoing) at the time of data collecting since we could not know their outcome (whether they succeeded or failed). In line with Liang et al., 2020, we removed all suspended projects and canceled where pledged value did not reach the target. We dropped off all campaigns with a goal below USD 5000 since they often target friends and family members (Cumming et al., 2017; Mollick, 2014; Liang et al., 2020). A final filter was to remove extreme values, so-called outliers, or those beyond the 99-percentile distribution in our case with a value of over $500,000 (Butticè et al., 2019). This leads to a final dataset of 130,528 project campaigns as presented in Table 7.2.

7.3.2 *Variables*

We founded hypotheses on the idea that a start-up project incorporating circular economy principles has a higher chance of launching a crowdfunding campaign. In line with this and the literature analyzed, our primary concern was to identify campaigns in the sample that have the element of the circular economy. To test our hypotheses, first, we had to identify projects that integrate the circular economy principles. We further applied econometric estimates, where the dependent variable is a dummy, indicating whether a crowdfunding campaign integrates the principles of the circular economy. A variable is a dummy equal to one if we identified such elements and zero elsewise. In other words, if the campaign contained

Table 7.2 Process of the database modification

All projects	430,938
Double entries, ongoing, canceled, and suspended projects	133,801
Projects with a goal of less than 5000 and over 600.000 dollars	166,609
Total sample projects	130,528

circular-economy activities, it was classified as a circular economy project, elsewise as a non-circular economy project.

We followed the work of Cumming et al. (2017) in testing the hypotheses and identifying projects that integrate the principles of the circular economy. For this purpose, we performed a text analysis technique by searching predefined keywords related to the circular economy concept. After a detailed literature review (presented in the previous part of this research), we have identified several words. Then, we have been searching for them in the project description: "circular economy", "reuse", "renewable", "recycle", "renewable", "remanufacture", "regenerate", "fuel consumption", "waste", "cleantech", "Greentech", "GHG", "low-carbon", "environmental", "sustainable", "ecology", "eco-", "solar", "biomass", "hydro", and "wind". This way, we got the 2871 campaigns identified as circular. Here are some examples of the campaigns included in the sample as circular-oriented. *Food waste to energy converter – re-invent the food cycle.* A new way to convert food waste within your community into fertilizer & renewable fuel. Portable and affordable machine that converts food scraps into energy and plant food using microbes with zero waste. *Bluecup – Refillable capsule for Nespresso.* A reusable and refillable capsule for use in Nespresso® machines. The system provides a choice of espresso capsule for the customer and is 99 times more environmentally friendly than pre-filled capsules and cost-efficient (data for projects: Kickstarter.com website).

We expect, for all variables, except for target capital (where we expect negative correlation), to be positively correlated with the campaign's success.

7.3.3 Descriptive Statistic

Overall, among the crowdfunding 430,938 campaigns posted on Kickstarter during the considered time window, after the modification, we have received 130,528 campaigns. We classified 2871 as circular projects economy (2.2%). In Table 7.3, we have reported the descriptive statistic of the related sample used in this study. The number of successful projects is in line with the general statistic on the Kickstarter platform. It amounts to a bit above 40% (41% for the whole sample and 43.3% for the campaigns related to the circular economy). Comparing the two groups, the success rate is slightly higher at 3.3% for campaigns related to the circular economy concept.

Two categories stand out in absolute and relative terms, namely technology and food. These two are the project categories, including most crowdfunding campaigns related to the circular economy (in absolute 591 and 540 respectively, or 20.6% and 18.8% in relative terms). Fashion and design are the following categories with 438 and 376 campaigns or 15.3% and 13.1% share (the results are consistent with Butticè et al., 2019). All other categories contain a limited number of campaigns related to circular economy projects that are below 10%, or 160 campaigns. It is not surprising that there is a relatively small number of campaigns in other categories, considering the nature of projects related to the concept of circularity. The majority

Table 7.3 Descriptive statistic

Characteristic	Sample	Oriented towards the circular economy
No. of projects	130,528	2871
Successful projects (%)	41.0%	43.3%
Average no of investors mean (median)	206.61 (27.00)	211.35 (38.00)
Average funding target (in 000) mean (median)	26.77 (11.30)	31.79 (15.00)
The average amount of pledged (in 000) mean (median)	18.95 (2.37)	20.82 (4.82)
Duration of the campaign	35,11 (35.00)	35.05 (30.00)
Staff pick	19,550 (15.0%)	513 (17.9%)
Spotlight	53,579 (41%)	1244 (43.3%)
Year	*% of campaigns*	*% of campaigns*
2009	0.2	0.1
2010	1.9	2.0
2011	4.9	3.9
2012	9.5	7.2
2013	10.6	8.5
2014	17.0	17.1
2015	20.8	20.0
2016	13.6	13.2
2017	11.1	13.4
2018	9.0	12.7
2019	1.4	1.8
Category	*% of campaigns*	*% of campaigns*
Art	5.6	5.5
Comics	2.3	0.6
Crafts	1.4	3.3
Dance	0.9	0.5
Design	5.4	13.1
Fashion	6.7	15.3
Film & video	15.8	6.5
Food	8.7	18.8
Games	10.2	5.4
Journalism	1.7	1.5
Music	11.8	1.6
Photography	2.2	1.0
Publishing	11.1	5.9
Technology	14.6	20.6
Theatre	1.8	0.5

of the campaigns in the sample were published in 2015, followed by 2014, 2016, and finally 2017 and beyond. The campaigns related to the circular economy are distributed similarly to the whole sample.

Several studies identified that particular types of campaigns could differ in different dimensions. Differences in "green" and "clean" campaigns exist along several dimensions, such as the number of backers, average, target goals, number of visuals, external links and networks, comments, and education (Butticè et al., 2019). Cumming et al. (2017) confirmed these findings for target goals. They expanded the list of the following features: funding model, digital output, teams, soft information, total pledge, and periods when there has been an increase in oil prices. However, they did find duration, comments, and social networks to be statistically insignificant features.

In our study, we found that campaigns with the element of circularity have greater success (0.433 opposite to 0.410, p-value <0.05), and target higher amounts of funds, namely $31,785.32 against $26,653.59 compared to non-circular campaigns (p-value <0.05). In addition, they are more often chosen as a staff pick (0.18, against 0.15, p-value <0.05) and more frequently use the spotlight function (0.43 vs 0.41, p-value <0.05). These elements favor our first hypothesis that projects related to the circular economy concept differ. On average, circular campaigns raise more money than non-circular campaigns ($20,814.47 vs. $18,906.32), but this difference was not statistically significant. There is no significant difference in the number of backers or duration (Table 7.4).

Table 7.4 Comparison between circular and non-circular campaigns

	Circular	*Non-circular*
Observation	2871	127,657
Success	0.433*	0.410*
	(0.496)	(0.492)
Backers	211.35	206.50
	(810.86)	(1181.39)
Goal	31,785.32*	26,653.59*
	(50,344.29)	(43,716.58)
Duration	35.05	35.11
	(11.48)	(11.97)
StaffPick	0.18*	0.15*
	(0.383)	(0.356)*
Spotlight	0.43*	0.41*
	(0.496)	(0.492)
Amount pledged	20,814.47	18,906.32
	(83,742.93)	(118,101.62)

* Significance level: 0.05

7.3.4 Results

In the second step of the analysis, we used the dummy variable on funding success to test the second hypothesis. We have run the following binary logistic regression model (as in Hörisch, 2015; Calic & Mosakowski, 2016; Cumming et al., 2017; Butticè et al., 2019; Vismara, 2019):

$$P\left(Y_i = 1\right) = \frac{1}{1 + e^{-\left(\alpha + \beta_1 X_{1i} + \beta_2 X_{2i} + \beta_3 X_{3i} + \beta_4 X_{4i} + \beta_5 X_{5i} + \beta_6 X_{6i} + \beta_7 X_{7i}\right)}}$$

$X_{1=}$ No_Backers
$X_{2=}$ Circular_Economy
$X_{3=}$ Target_Capital
$X_{4=}$ Campaign_duration
$X_{5=}$ Staff_Pick

For the analysis, we used variables as described in Table 7.5. The dependent variable is the dummy differentiating projects that have reached the funding goal (success = 1) from those that did not (success = 0). This way, we estimated which factors, including the orientation to a circular economy, foster the likelihood of achieving the self-set targets.

The proposed model is significant ($p < 0.01$), and overall model fit information shows that the model explains the relevant share of the variation of the dependent variable (Cox & Snell R Square 0.552). Table 7.6 reports the results of our estimates. The effects of analyzed parameters on the funding success are statistically significant ($p < 0.01$). We found positive, statistically relevant relation between circularly oriented projects, the number of backers, and staff picks. Campaigns with the element of circularity are positively related to their likelihood of success (0.293). The odds of being successfully funded increase by 34% for circular projects compared to non-circular ones. Also, the number of backers is positively related to the

Table 7.5 Variable definition

Variable	Definition
Success	Dummy = 1 to one if the funding amount is higher than the target amount; 0 elsewise
No_backers	Numbers of backers that have invested in the campaign
Circular_economy	Dummy =1 if the identified activities are related to the concept of the circular economy, 0 elsewise
Target_capital	The logarithm of the target capital. For campaigns based on a currency other than USD, we converted the amount into USD at an annual average exchange rate
Campaign_duration	Length of the campaign duration
Staff_pick	Indicating whether Kickstarter team members designated a campaign as a "favorite" while it was active

Table 7.6 Binary logistic results

Model summary	
Dependent variable	Funding Success
Pseudo R^2 (Cox & Snell R Square)	0.552
Significance of the model	0.000
Parametric rating	
CircBool	0.293*** (0.065)
Duration	−0.003*** (0.001)
Backers	0.033*** (0.000)
StaffPick	0.499*** (0.032)
LnGoal	−1.945*** (0.018)
Constant	15,529 (0,162)

Standard errors are in parentheses and *** Significance level: 0.01

campaign's success (0.033), with a 3.3% chance of greater success with each increase in the number of backers. If Kickstarter team members designated the campaign as "favourite" while it was active, the odds for success increased by 64.7%. We recorded a highly significant, negative regression coefficient for the ln of the funding target. This reveals, as expected, that projects with higher targets are less likely to be successful (−1.945). Surprisingly, we found a negative regression coefficient for the duration too. According to our results, the longer the campaign, the odds of success are lower (−0.003). Each increase in duration lowers the odds of success by 0.3%.

7.4 Discussion

Acquiring resources is critical for starting a business that influences all future ventures. Without external funding, companies and individuals have a limited capacity to develop new ideas and projects. Traditional financial institutions, such as banks, are generally unwilling to invest in new ventures. Without diverse sources of finance, good ideas can fail, where the growth potential for the economy is lost.

Crowdfunding was triggered by the development and increased number of social networks users, that are potential small investors. It can provide a critical link in start-up financing, filling the financial gap for start-ups. To address the question whether crowding platforms are suitable source of finance for a circular economy, we sought to understand the underlying factors that influence the decision to invest in projects related to the circular economy. The analysis results are somewhat mixed but provide enough evidence to support our first hypothesis. Circular-oriented and non-circular-oriented campaigns on Kickstarter differ in several features. We can conclude that projects related to the circular economy concept, on average, have a higher chance for success, target and receive higher amounts of money, attract more backers, and are more frequently chosen as a favorite by the Kickstarter team members. However, we did not find pledged funds and numbers of backers to be statistically significant. Our results are consistent with Cumming et al. (2017) and partially with Butticè et al. (2019), except for the number of backers.

Concerning the success of campaigns with circularity, results of econometric estimates support the findings that circular-oriented projects are likely to belong to the group of successful projects. This is consistent with Belleflamme et al. (2014) and Calic and Mosakowski (2016), who found that projects with the elements of sustainability and social aspects have more success in funding. However, these results differ from Hörisch (2015), who found no positive effect of environmental orientation in terms of its likelihood of success, or Moss et al. (2015), claiming that crowd-investors are often focused on profit-seeking opportunities. Furthermore, results related to the duration of this research surprisingly point out the negative relationship between these two concepts. Although the variations are minimal, they are statistically significant, as in Mollick (2014), and are in line with Butticè et al. (2019), Cumming et al. (2017), but are not consistent with Hörisch (2015). Regarding the number of backers, we found week statistically significant association as in Vismara (2019). Nevertheless, because we only analyzed data from one reward-based platform, caution should be applied when interpreting our findings. Backers may be lenders (lending-based), owners (equity-based), philanthropists (donation-based) or consumers (reward-based). The type of the backer considerably impacts on the factors that determine a crowdfunding campaign's success. So far, no prior studies have taken into account nor addressed the connection between the influence of Kickstarter team members and success. Our results show a positive relationship between these two concepts, as we supposed when defining the variables in the model. This connection is relatively strong and increases the odds by 64.7%. Finally, in line with Mollick (2014), Hörisch (2015), Butticè et al. (2019), we found a negative connection between the goal of the campaign and success, referring to the fact that projects with higher funding targets are less likely to reach their funding targets.

7.5 Conclusion

This chapter provided an empirical analysis of the crowdfunding campaigns related to the circular economy that contributed to the academic literature. We examined over 130,000 campaigns from the Kickstarter platform around the world, where 2.2% were projects related to the concept of the circular economy to understand what factors correlate with the success of crowdfunding campaigns, taking into account the idea of circularity. We chose to look at the whole picture without considering specific categories, as some authors prefer.

Overall, our results suggest that we can consider crowdfunding as a possible way to finance projects related to the concept of circularity. However, several limitations exist. In discussing the results, we interpret them as associations and not as causal relationships, so the odds for every specific variable depend on the simultaneous inclusion of other variables in the model. Next, we restricted the dataset to rewards crowdfunding campaigns collected from a single platform, so the results cannot be replicated on other crowdfunding platforms or different crowding types. Likewise, most project campaigns are US-based 75.8 percent, opposite to only 68 percent of circular campaigns based in the US. In interpreting results, we must consider this, considering that different economic, cultural, and political surroundings can affect crowdfunding success. Finally, our research includes only a limiting number of general success factors, whereas crowdfunding success depends on many more. To plan a successful crowdfunding campaign, you'll need to come up with a promotional concept, create promotional materials, and identify appropriate media channels for campaign promotion.

Acknowledgment This paper is a result of the research financed by the Ministry of Education, Science, and Technological Development of the Republic of Serbia.

References

Adhami, S., Giudici, G., & Anh, H. P. N. (2017). *Crowdfunding for green projects in Europe: Success factors and effects on the local environmental performance and wellbeing*. Available at: http://www.crowdfundres.eu/wp-content/uploads/2017/11/Crowdfunding-for-green-projects-in-Europe-2017.pdf

Agrawal, A., Catalni, C., & Goldfarb, A. (2015). Crowdfunding: Geography, social networks, and the timing of investment decisions. *Journal of Economics and Management Strategy, 24*(2), 253–274.

Allison, T. H., Davis, B. C., Short, J. C., & Webb, J. W. (2015). Crowdfunding in a prosocial microlending environment: Examining the role of intrinsic versus extrinsic cues. *Entrepreneurship Theory and Practice, 39*(1), 53–73.

Ambec, S., & Paul, L. (2008). Does it pay to be green? A systematic overview. *Academy of Management Perspectives, 22*(4), 45–62.

Angerer, M., Brem, A., Kraus, S., & Peter, A. (2017). Start-up funding via equity crowdfunding in Germany: A qualitative analysis of success factors. *The Journal of Entrepreneurial Finance, 19*(1), 1–34.

Bauwens, T., Mees, R., Gerardts, M., Friedl, J. V. D. H., & Von Daniels, C. et al. (2019). *Disruptors: How circular start-ups can accelerate the circular economy transition*. Available at: https://www.uu.nl/sites/default/files/disruptors_circular_start-ups_uu_e-version.pdf

Bauwens, T., Hekkert, M., & Kirchherr, J. (2020). Circular futures: What will they look like? *Ecological Economics, 175*, 1–14.

Beaulieu, L., van Durme, G., & Arpin, M-L. (2016). Circular economy: A critical review of concepts. In *International reference Centre for the Life Cycle of Products, Processes, and Services*.

Beier, M., & Wagner, K. (2015). Crowdfunding success: A perspective from social media and E-commerce. In *Thirty sixth international conference on information systems*. Fort Worth.

Belleflamme, P., Lambert, T., & Schwienbacher, A. (2014). Crowdfunding: Tapping the right crowd. *Journal of Business Venturing, 29*(5), 585–609.

Bento, N., Gianfrate, G., & Groppo, S. V. (2019). Do crowdfunding returns reward risk? Evidences from cleantech projects. *Technological Forecasting and Social Change*, Elsevier, *141*, 107–116.

Bergset, L. (2015). The rationality and irrationality of financing green start-ups. *Administrative Sciences, 5*(4), 260–285.

Bergset, L. (2018). Green start-up finance–where do particular challenges lie? *International Journal of Entrepreneurial Behavior & Research, 24*(2), 451–475.

Bergset, L., & Fitcher, K. (2015). Green start-ups – A new typology for sustainable entrepreneurship and innovation research. *Journal of Innovation Management, 3*(3), 118–144.

Bernstein, S., Korteweg, A., & Laws, K. (2017). Attracting early-stage investors: Evidence from a randomized field experiment. *Journal of Finance, 72*(2), 509–538.

Block, J., Hornuf, L., & Moritz, A. (2018). Which updates during an equity crowdfunding campaign increase crowd participation? *Small Business Economics, 50*(1), 3–27.

Blomsma, F., & Brennan, G. (2017). The emergence of circular economy: A new framing around prolonging resource productivity. *Journal of Industrial Ecology, 21*(3), 603–614.

Böckel, A., Hörisch, J., & Tenner, I. (2020). A systematic literature review of crowdfunding and sustainability: Highlighting what really matters. *Management Review Quarterly, 71*, 433. https://doi.org/10.1007/s11301-020-00189-3

Bocken, N. M. (2015). Sustainable venture capital–catalyst for sustainable start-up success? *Journal of Cleaner Production, 108*, 647–658.

Bocken, N. M. P., Samuel, W. S., Padmakshi, R., & Steve, E. (2014). A literature and practice review to develop sustainable business model archetypes. *Journal of Cleaner Production, 65*, 42–56.

Bocken, N. M. P., de Pauw, I., Bakker, C., & van der Grinten, B. (2016). Product design and business model strategies for a circular economy. *Journal of Industrial and Production Engineering, 33*(5), 308–320.

Boulding, K. E. (1966). *The economics of the coming spaceship earth*. Avaliable at: http://www.ub.edu/prometheus21/articulos/obsprometheus/BOULDING.pdf

Brennan, G., Tennant, G., & Blomsma, F. (2015). Business and production solutions: Closing loops & the circular economy. In H. Kopnina & Shoreman-Ouimet (Eds.), *R "sustainability-key issues"*. Routledge.

Brown, R., Mawson, S., & Rowe, A. (2019). Start-ups, entrepreneurial networks and equity crowdfunding: A processual perspective. *Industrial Marketing Management, 80*, 115–125.

Butticè, V., Colombo, M. G., & Wright, M. (2017). Serial crowdfunding, social capital, and project success. *Entrepreneurship: Theory and Practice, 41*(2), 183.

Butticè, V., Colombo, M. G., Fumagalli, E., & Orsenigo, C. (2019). Green oriented crowdfunding campaigns: Their characteristics and diffusion in different institutional settings. *Technological Forecasting and Social Change, 141*, 85–97.

Calic, G., & Mosakowski, E. (2016). Kicking off social entrepreneurship: How a sustainability orientation influences crowdfunding success. *Journal of Management Studies, 53*(5), 738–767.

Carter, R. B., & Van Auken, H. E. (1990). Personal equity investment and small business financial difficulties. *Entrepreneurship Theory and Practice, 15*(2), 51–60.

Cassar, G. (2004). The financing of business start-ups. *Journal of Business Venturing, 19*(2), 261–283.

Colombo, M. G., Franzoni, C., & Rossi-Lamastra, C. (2015). Internal social capital and the attraction of early contributions in crowdfunding. *Entrepreneurship Theory and Practice, 39*(1), 75.

Corona, B., Shen, L., Reike, D., Rosales Carreón, J., & Worrell, E. (2019). Towards sustainable development through the circular economy-a review and critical assessment on current circularity metrics. *Resources, Conservation & Recycling, 151*, 1–15.

Cosh, A., Cumming, D., & Hughes, A. (2009). Outside entrepreneurial capital. *The Economic Journal, 119*(540), 1494–1533.

Courtney, C., Dutta, S., & Li, Y. (2017). Resolving information asymmetry: Signaling, endorsement, and crowdfunding success. *Entrepreneurship Theory and Practice, 41*(2), 265.

Cumming, D., & Hornuf, L. (2018). *The economics of crowdfunding*. Palgrave Macmillan.

Cumming, D. J., Leboeuf, G., & Schwienbacher, A. (2017). Crowdfunding cleantech. *Energy Economics, 65*, 292–303.

Dos-Santos, M. J. P. L., Mota, M., Ferreiro, M. F., Ahmad, N., Ahmed, R. R., Machado-Santos, C., & Silva, E. (2020). Assessing the sustainability of crowdfunding in social media and Google trends. In *Advances in human factors, business management and leadership*. Springer.

Efrat, K., & Gilboa, S. (2019). Relationship approach to crowdfunding: How creators and supporters interaction enhances projects' success. *Electron Markets, 30*, 899. https://doi.org/10.1007/s12525-019-00391-6

European Commission. (2015). *Closing the loop – An EU action plan for the circular economy*. European Commission.

Forbes, H., & Schaefer, D. (2017). Guidelines for successful crowdfunding. *Procedia CIRP, 60*, 398–403.

Geissdoerfer, F., Savaget, P., & Bocken, N. (2017). The circular economy – A new sustainability paradiagm? *Journal of Cleaner Production, 143*, 757–768.

Geissdoerfer, M., Morioka, S. N., de Carvalho, M. M., & Evans, S. (2018). Business models and supply chains for the circular economy. *Journal of Cleaner Production, 190*, 712e721. http://linkinghub.elsevier.com/retrieve/pii/S0959652618311867

Geissdoerfer, M., Pieroni, M. P. P., Pigosso, D. C. A., & Soufani, K. (2020). Circular business models: A review. *Journal of Cleaner Production, 277*, 1–17.

Gerber, E., Hui, J., & Kuo, P-Y (2012). Crowdfunding: Why people are motivated to post and fund projects on crowdfunding platforms. In *CSCW workshop*.

Gimeno, J., Folta, T. B., Cooper, A. C., & Woo, C. Y. (1997). Survival of the fittest? Entrepreneurial human capital and the persistence of underperforming firms. *Administrative Science Quarterly, 42*, 750–783.

Greenberg, M. D., Hui, J., & Gerber, E. (2013). Crowdfunding: A resource exchange perspective. In *Conference: CHI'13 extended abstracts on human factors in computing systems*.

Halme, M., & Niskanan, J. (2001). Does corporate environmental protection increase or decrease shareholder value? The case of environmental investments. *Business Strategy and the Environment, 10*(4), 200–214.

Hamilton, T. (1995). Pollution as news: Media and stock market reaction to the toxics release inventory data. *Journal of Environmental Economics and Management, 28*, 98–113.

Henry, M., Bauwensa, T., Hekkerta, M., & Kirchherr, J. (2019). A typology of circular start-ups – Analysis of 128 circular business models. *Journal of Cleaner Production, 245*, 118528.

Hörisch, J. (2015). Crowdfunding for environmental ventures: An empirical analysis of the influence of environmental orientation on the success of crowdfunding initiatives. *Journal of Cleaner Production, 107*, 636–645.

Irwin, K. C., Gilstrap, C. M., Drnevich, P. L., & Tudor, C. M. (2019). From start-up to acquisition: Implications of financial investment trends for small-to medium-sized high-tech enterprises. *Journal of Small Business Strategy, 29*(2), 22–43.

Jones, O., & Jayawarna, D. (2010). Resourcing new businesses: Social networks, boot- strapping and firm performance. *Venture Capital, 12*(2), 127–152.

Jovanovic, T. (2019). Crowdfunding: What do we know so far? *International Journal of Innovation and Technology Management, 16*(01), 1950009.

Kee, D. M. H., Yusoff, Y. M., & Khin, S. (2019). The role of support on start-up success: A pls-sem approach. *Asian Academy of Management Journal, 24*, 43–59.

Kirchherr, J., Reike, D., & Hekkert, M. (2017). Conceptualizing the circular economy: An analysis of 114 definitions. *Resources, Conservation and Recycling, 127*, 221–232.

Laurell, C., Sandström, B., & Suseno, Y. (2019). Assessing the interplay between crowdfunding and sustainability in social media. *Technological Forecasting and Social Change, 141*, 117–127.

Lee, N., Sameen, H., & Cowling, M. (2015). Access to finance for innovative SMEs since the financial crisis. *Research Policy, 44*(2), 370–380.

Lehner, O. M. (2013). Crowdfunding social ventures: A model and research agenda. *Venture Capital, 15*(4), 289–311.

Liang, X., Hu, X., & Jiang, J. (2020). Research on the effects of information description on crowdfunding success within a sustainable economy-the perspective of information communication. *Sustainability, 12*(2), 650.

Lindenberg, S., & Steg, L. (2007). Normative, gain and hedonic goal frames guiding environmental behaviour. *Journal of Social Issues, 63*(1), 117–137.

Ljumović, I., & Jakšić, K. (2015). Dostupnost finansisjkih resursa preduzećima na severu Kosova. *Zbornik Matice Srpske za društvene nauke, 152*(3), 557–568.

Ljumović, I., & Pavlović, D. (2016). Izvori finansiranja aktivnosti u oblasti zaštite životne sredine. *Ecologica: nauka, privreda, iskustva, 23*(84), 727–730.

Ljumović, I., & Pejović. (2020, November 16–17). Financing of innovation: Are crowdfunding and venture capital complements or substitutes? In *Book of proceedings, PaKom conference. Proceedings in print.*

Ljumović, I., Cvijanović, J., & Lazić, J. (2012). Valuation of biotechnology companies: Real options approach under uncertainty. *Economics of Agriculture, 59*(1), 51–62.

Ljumović, I., Lazić, J., & Vesić, A. (2015a). *Specifičnosti finansiranja u preduzetništvu*. Ekonomski institut.

Ljumović, I., Viduka, B., & Cvijanović, J. M. (2015b). Access to finance for organic producers in Serbia: Demand side. *Economics of Agriculture, 62*(4), 989–1002.

Ljumović, I., Hanić, A., & Kovačević, V. (2021a). The role of reward-based crowdfunding in farm financing: What characterizes successful campaign? *Economics of Agriculture, 68*(3), 773–788.

Ljumović, I., Pavlović, V., & Knežević, G. (2021b). Financing agri-food business in the Mediterranean area through crowdfunding: Do environmental issues matter? *New Medit, 3*, 101–117.

Lu, C.-T., Xie, S., Kong, X., & Yu, P. S. (2014, February 24–28). Inferring the impacts of social media on crowdfunding. In: *ACM WSDM conference*. New York, NY.

Lukić, R., Lalić, S., Sućeska, A., Hanić, A., & Bugarčić, M. (2018). Carbon dioxide emissions in retail food. *Economics of Agriculture, 65*(2), 859–874.

Massolution. (2015). *2015CF, The crowdfunding report*. Available at: https://reports.crowdsourcing.org/index.php?route=product/product&product_id=54

McKinsey Center for Business and Environment. (2016). *The circular economy: Moving from theory to practice*. McKinsey & Company.

Mentink, B. (2014). *Circular business model innovation: A process framework and a tool for business model innovation in a circular economy* (Master thesis). Industrial Design Engineering.

Merli, R., Preziosi, M., & Acampora, A. (2018). How do scholars approach the circular economy? A systematic literature review. *Journal of Cleaner Production, 178*, 703–722.

Millette, S., Hull, C. E., & Williams, E. (2020). Business incubators as effective tools for driving circular economy. *Journal of Cleaner Production, 266*, 121999.

Mollick, E. R. (2014). The dynamics of crowdfunding: An exploratory study. *Journal of Business Venturing, 29*, 1–16.

Mollick, E., & Robb, A. (2016). Democratizing innovation and capital access: The role of crowdfunding. *California Management Review, 58*(2), 72–87.

Moss, T. W., Neubaum, D. O., & Meyskens, M. (2015). The effect of virtuous and entrepreneurial orientations on microfinance lending and repayment: A signaling theory perspective. *Entrepreneurship Theory and Practice, 39*(1), 27–52.

Motylska-Kuzma, A. (2018). Crowdfunding and sustainable development. *Sustainability, 10*(12), 4650.

Nofsinger, J. R., & Wang, W. (2011). Determinants of start-up firm external financing worldwide. *Journal of Banking and Finance, 35*(9), 2282–2294.

Ordanini, A., Miceli, L., Pizzetti, M., & Parasura, A. (2011). CrowdFunding: Transforming customers into investors through innovative service platforms. *Journal of Service Management, 22*(4), 443–470.

Ortas, E., Burritt, R. L., & Moneva, J. M. (2013). Socially responsible investment and cleaner production in the Asia Pacific: Does it pay to be good? *Journal of Cleaner Production, 52*, 272–280.

Osterwalder, A., & Pigneur, Y. (2010). *Business model generation: A handbook for visionaries, game changers, and challengers*. Wiley.

Pearce, D. W., & Turner, R. K. (1990). Economics of natural resources and the environment. *American Journal of Agricultural Economics, 73*(1), 227–228.

Petruzzelli, A. M., Natalicchio, A., Panniello, U., & Roma, P. (2019). Understanding the crowdfunding phenomenon and its implications for sustainability. *Technological Forecasting and Social Change, 141*, 138–148.

Pitschner, S., & Pitschner-Finn, S. (2014). Non-profit differentials in crowd-based financing: Evidence from 50,000 campaigns. *Economics Letters, 123*(3), 391–394.

Porter, M. (1991). America's green strategy. *Scientific American, 264*(4), 168.

Porter, M., & van der Linde, C. (1995). Toward a new conception of the environment-competitiveness relationship. *Journal of Economic Perspective, 9*(4), 97–118.

Prieto-Sandoval, V., Jaca, C., & Ormazabal, M. (2016). Towards a consensus on the circular economy. *Journal of Cleaner Production, 179*, 605–615.

Reichel, A., De Schoenmakere, M., & Gillabel, J. (2016). *Circular economy in Europe: Developing the knowledge base*. European Environment Agency. Available at: http://www.eea.europa.eu/publications/circular-economy-in-europe

Saxton, G. D., & Wang, L. (2013). The social network effect. *Nonprofit and Voluntary Sector Quarterly, 43*(5), 850–868.

Schenkel, M., Krikke, H., Caniëls, M. C. J., & van der Laan, E. (2015). Creating integral value for stakeholders in closed loop supply chains. *Journal of Purchasing and Supply Management, 21*(3), 155–166.

Schwager, P., & Moser, F. (2006). The application of chemical leasing business models in Mexico. *Environmental Science and Pollution Research, 13*, 131–137.

Stahel, W. R. (1982). "The product life factor. An inquiry into the nature of sustainable societies: The role of the private sector", (series: 1982 Mitchell prize papers), NARC.

Stahel, W. R., & Reday-Mulvey, G. (1976). *The potential for substituting manpower for energy: A report to the European Commission. Subsequently published as jobs for tomorrow*. Vantage Press (1981).

Stanko, M. A., & Henard, D. H. (2016). How crowdfunding influences innovation. *MIT Sloan Management Review, 57*, 15.

Stanko, M. A., & Henard, D. H. (2017). Toward a better understanding of crowdfunding, openness and the consequences for innovation. *Research Policy, 46*(4), 1–45.

Testa, S., Nielsen, K. R., Bogers, M., & Cincotti, S. (2019). The role of crowdfunding in moving towards a sustainable society. *Technological Forecasting and Social Change, 141*, 66–73.

Thompson, N., Kiefer, K., & York, J. G. (2011). Distinctions not dichotomies: Exploring social, sustainable, and environmental entrepreneurship. In *Social and sustainable entrepreneurship* (Advances in entrepreneurship, firm emergence and growth) (Vol. 13). Emerald Group Publishing Limited.

Urbinati, A., Chiaroni, D., & Chiesa, V. (2017). Towards a new taxonomy of circular economy business models. *Journal of Cleaner Production, 168*, 487–498.

Vismara, S. (2019). Sustainability in equity crowdfunding. *Technological Forecasting and Social Change, 141*, 98–106.

Vuorio, J. (2020). *Business model innovation for circular economy start-ups* (Master Thesis).

Wang, N., Li, Q., Liang, H., Ye, T., & Ge, S. (2018). Understanding the importance of interaction between creators andbackers in crowdfunding success. *Electronic Commerce Research and Applications, 27*, 106–117.

World Economic Forum. (2020). *The global risks report 2020*. World Economic Forum.
Wyman, O. (2017). *Supporting the circular economy transition: The role of the financial sector in the Netherlands*. Available at: https://www.oliverwyman.com/our-expertise/insights/2017/sep/the-circular-economy.html
Xu, A., Yang, X., Rao, H., Fu, W.-T., Huang, S.-H., & Bailey, P. B. (2014). Show me the money! An analysis of project updates during crowdfunding campaigns. In *CHI 2014*. One of a CHInd, Toronto, ON, Canada.
Zhou, M. J., Lu, B., Fann, W. G., & Wang, G. A. (2016). Project description and crowdfunding success: An exploratory study. *Information Systems Frontiers, 20*, 259–274.

Part III
Integrating Sustainability into PM Methodology

Chapter 8
Methodology Hybridization for Sustainable Strategic Management of Circular Projects and Programs

Sergiy Bushuyev, Natalia Bushuyeva, and Victoria Bushuieva

Abstract Models and methods of hybridization of project management methodologies and sustainable development programs in the transition to a circular economy are considered. A conceptual model of a sustainable strategy for the implementation of projects and programs of the circular economy based on the entropy model is defined. A strategy is a long-term action plan aimed at achieving a specific goal that is important to the organization. The strategy in different areas of activity is defined differently. In the implementation of projects and programs, the role of strategy is to implement an organizational and analytical structure that justifies and prepares solutions for the implementation of rational actions. Developed sustainable strategies should be implemented taking into account the real state of affairs, but cannot always be implemented as planned.

Purpose: To solve the scientific problem of creating a hybrid methodology for sustainable strategic management of infrastructure programs in the condition of the circular economy.

Keywords Program · Sustainable strategies · Methodology · Entropy · Circular economy projects

8.1 Introduction

The fast development of project management as a scientific discipline, and its active practical implementation, is essential today to ensure and maintain the competitiveness of domestic enterprises and organizations at the global level. At the same time, a huge range of knowledge and methodologies has already been formed in world practice, many of which contain unique models, methods and management

S. Bushuyev (✉) · N. Bushuyeva · V. Bushuieva
Kyiv National University of Construction and Architecture, Kyiv, Ukraine
e-mail: bushuiev.sd@knuba.edu.ua; bushuieva.ns@knuba.edu.ua; bushuieva.vb@knuba.edu.ua

V. Obradović (ed.), *Sustainable Business Change*,
https://doi.org/10.1007/978-3-031-23543-6_8

mechanisms borrowed from other subject areas and fields of knowledge. This proves the possibility and expediency of using the convergence principles of methods, models and tools of hybridization of project management methodologies for further development of mechanisms for the successful implementation of projects and programs. Most companies in global markets view project management methodology, programs and project portfolios as key elements in ensuring and maintaining competitiveness.

Modern technologies, means of communication, increased rates of accumulation and availability of information allow quick learning and applying the achievements of leaders in the transition to the circular economy (Bocken et al., 2016).

In this process, benchmarking technologies are widely used, the transfer of good practices, which does not lead to a heuristic increase in knowledge, models and methods. Most often new technologies are reduced to simple copying of existing methodologies, with small variations concerning the source, without an in-depth analysis of their applicability in certain conditions, incorrect application by different entities for inadequate project, program or portfolio management (Bondar et al., 2021). This situation leads to an increase in methodological uncertainty, confusion, systemic chaos and illusory pictures of the diversity of the diverse project management world. Project management practices, looking into such a kaleidoscope of methodology, are lost in a variety of technical and managerial tools, often forced to randomly choose methods for their needs. Hopes for benchmarking are not justified, as it is a local tool that cannot provide effective convergence schemes (consolidation of knowledge) of different subject areas, which form new, more effective knowledge systems, approaches and methodologies for project management, programs and project portfolios management (Bushuyev et al., 2022).

On the other hand, globalization and global trends also affect project management methodologies. The trends have positive and negative sides. If the global spread of standardization of project management methodology is carried out systematically, taking into account national characteristics and traditions that do not reduce the accumulated national experience and do not introduce methodological confusion, it is possible to improve productivity and quality of project activities, enriched by advanced achievements in this field.

Thus, the construction of a paradigm of hybridization based on a convergent approach shows that an effective mechanism for this can be a combination of the best elements of methodologies, which provides a good level of requirements for the quality of the management process (e.g., PRINCE-2; standards family ISO 21500 2012, ISO 21502, ISO 21503, ISO 21504 and ISO 21505, PMBoK PMI 2017) with methods that meet the requirements for the quality of competence of project participants. This is, for example, on the one hand, competence standards IPMA (ICB4, 2015) and (IPMA OCB, 2013), on the other hand, (PMAJ P2M, 2015), a methodology that requires high competence in goal setting, management values and expectations. Today Agile methodology becomes to the project management world from information technology and keeps a leading position on the lower level of managing projects (ICB4 in Agile world, 2018). The uncertainty of the internal environment and the strong influence of Agile transformation of the circular projects are characterized by shortcomings that are inherent in our research problem in the existing conditions of the organization activity.

8.2 Research Methodology, Approach and Design

In the practice of strategic management, several definitions of strategic sustainability are used. Let's consider some of them.

Strategic sustainability is an assessment of the likelihood of success in implementing an organization's strategy, project or program (Azarov et al., 2011).

Strategic sustainability is considered an assessment of the achievement of goals based on the use of an accepted project and/or program management methodology (Todorović et al., 2015). In the process of assessing strategic sustainability based on the results of strategic analysis, a strategy is selected, its implementation and monitoring of success in achieving the goals determined by the strategy (Bushuyev et al., 2020). Quantitative indicators of goal setting and goal achievement become dominant in performance evaluation. To ensure strategic sustainability, the development of hybrid methodologies, models and methods is required (Azarov et al., 2011).

Let's present a functional-competence model for the analysis of strategic sustainability. The conceptual scheme of the proposed model is presented in Fig. 8.1.

The functional-competence model contains three levels of strategic sustainability analysis (Fig. 8.1).

The first level is the management of the organization. At the same time, the mission, vision and strategy of the organization are declared (Bushuyev & Verenych, 2018).

The second level ensures the functioning of the organization within the operational and project activities. At this level, many functions are formed that are implemented by the organization within the framework of the management system (Bushuyev et al., 2019).

The third level is a role model of production and management processes. At the same time, many roles are formed that perform the functions of production and management. For each, a set of competencies necessary for the successful

Fig. 8.1 The functional-competence model for analysis of strategic sustainability

implementation of the selected functions is determined. At the same time, a benchmark score is formed that determines the success of the implementation of the function for each competency. At this level, the competence of the personnel performing roles at the third level is assessed. For this, the authors used the IPMA OCB model (Bushuyev & Verenych, 2017).

The conceptual model for the implementation of sustainable strategy based on a hybrid methodology of management circular projects is presented in Fig. 8.2.

The formation of projects that implement the strategy takes place taking into account their priority (Cockburn, 2000).

We propose a two-stage method for the formation of projects for the implementation of the sustainable development strategy:

- The first stage is connected with determining the degree of importance of the project to achieve the mission or transfer the system to an ideal state.
- The second stage ensures the formation of the content of projects in the areas of strategy implementation.

Let $S_i, i = \overline{1,n}$ be the set of strategic goals that ensure the achievement of the ideal state of the system. At the same time $p_j, j = \overline{1,m}$ – the priority of strategic goals. Priority ratings can be formed by any of the known methods (e.g., in the form of a rank or the form of a score on a 100- or 10-point scale, the main thing is that the principle is observed – the higher the value, the more important).

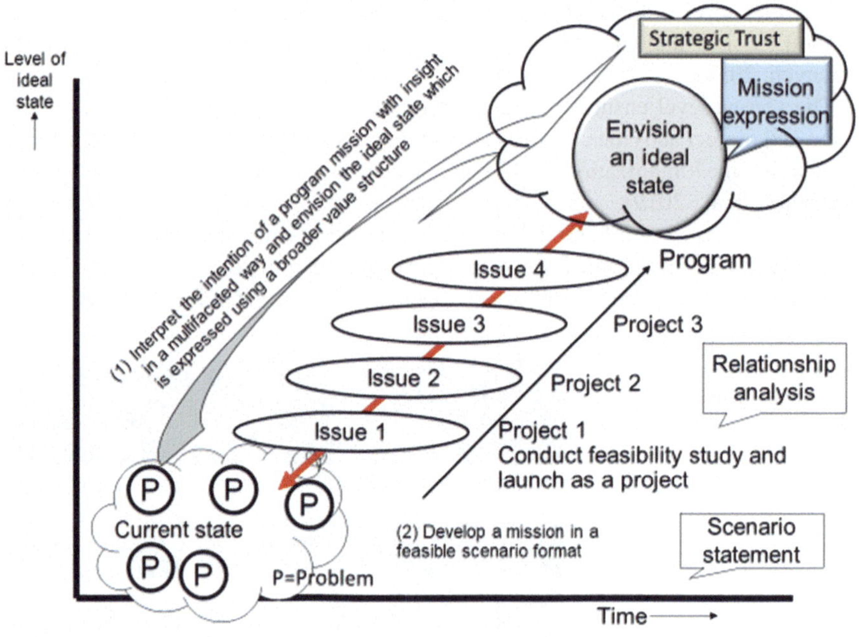

Fig. 8.2 The conceptual model for implementation of the sustainable strategy based on a hybrid methodology

Based on these assessments r_{ij}, which we define as the "degree of importance of the goal" for the implementation of a sustainable strategy, $_i, i = \overline{1,n}$ can be formed:

$$A_i = S_i \sum_{j=1}^{m} P_j \cdot r_{ij}, i = \overline{1,n}. \tag{8.1}$$

By setting a certain threshold value A^*, those goals that at this stage are most important for the sustainability of the strategy can be selected:

$$Z = \left\{ A_i \geq A^*, i = \overline{1,n} \right\}, \tag{8.2}$$

where Z is the set of the most significant goals of the strategy. After setting the most significant goals, we implement the stage of selecting alternatives for the content of projects based on a system of criteria. Let L be the number of established criteria. Among the many criteria may be finances, resources, implementation period, financial risks, environmental risks, etc.

Thus, for each alternative project $v = \overline{1, V_i}$ in each most important area, it is possible to establish $i \in Z$ the degree of compliance with the criterion:

$$0 \leq K_{lvi} \leq 1, l = \overline{1, L}, v = \overline{1, V}, i \in Z. \tag{8.3}$$

To find the rating of alternative projects, we use the following estimates:

$$I_v^i = \sum_{l=1}^{L} K_{lvi}, v = \overline{1, V}, i \in Z, \tag{8.4}$$

or

$$I_v^i = \sum_{l=1}^{L} d_l \cdot K_{lvi}, v = \overline{1, V}, i \in Z, \tag{8.5}$$

where $d_l, l = \overline{1, L}$ is the weight of the criterion based on the traditional approach and $\sum_{l=1}^{L} d_l \cdot K_{lvi} = 1$. These weighting factors can be differentiated for each specific direction, that is $\sum_{l=1}^{L} d_l^i \cdot K_{lvi} = 1$.

Thus, the condition

$$\max \left\{ I_v^i, v = \overline{1, V} \right\}, i \in Z \tag{8.6}$$

allows choosing the most appropriate project for each identified priority area of the sustainable development strategy.

Entropy as a measure of the uncertainty of the state of a system has long gone beyond the scope of thermodynamics and information theory. Today, entropy is a universal characteristic of the state of a system, including projects and programs. The papers (Bondar et al., 2021) justified the use of entropy for organizations and society as a whole and established the relationship between information (Shannon entropy) and energy entropy (entropy in the context of energy dissipation). Since almost all sustainable development strategies can be considered either from the standpoint of benefits or values or from the standpoint of information that reduces uncertainty, the proposed relationship fully reflects these properties:

$$S = \lambda \cdot H, \tag{8.7}$$

where S "entropy" is an indicator of the state of the strategic development program in the context of its energy losses (dissipation) in the process of energy turnover, H information entropy, λ is the coefficient, $H = -\sum_{k=1}^{K} p(A_k) \cdot \ln(p(A_k))$ the expression for which is presented in, A_k is the state of the sustainable development strategy, $p(A_k)$ are the probabilities of these states. At the same time, it determines the efficiency of resource use, taking into account their current level.

The universality of the fundamental laws of physics, substantiated in the works of many scientists, makes it possible to use the law of conservation of energy or the first law of thermodynamics for systems that are diverse, including organizations and society as a whole (Collier, 2011). In this case, the stability of systems of different classes can be interpreted in terms of the system's energy conservation law.

The developed entropy concept of control (Bondar et al., 2021), which formulated the main patterns of the structure and dynamics of entropy for the analysis of the uncertainty of the state of organizations, made it possible to assess their stability based on modelling the behaviour of entropy. At the same time, energy exchange, dissipation and interaction with the external environment through management and entrepreneurial potential indicators of the effectiveness of this interaction taking into account the reduction of uncertainty – all this is the result of applying the second law of thermodynamics within projects, organizations and society as systems. In the context of the entropy concept of control, entropy reflects the result of the interaction between the system and the circular environment (Geissdoerfer et al., 2017). It evaluates the state of the system from an external point of view.

The function of system states also includes enthalpy, which reflects the amount of entrepreneurial energy available for conversion into the results of the system. Enthalpy in an organizational context reflects the "available resource or energy" of a system for energy exchange with the external environment and maintaining a stable system structure (Hart et al., 2019).

Consider a specific type of energy as "entrepreneurial energy" (Verenych & Bushuyev, 2018), such a resource is defined as the energy of labour resources. This is the knowledge, experience, skills and competence of staff and managers. This is a special type of energy, which is also valued in terms of money (salary, etc.), but plays the most important role in the efficiency of the organization's energy exchange with the external environment (Verenych & Bushuieva, 2018).

With the same costs for the resources of two organizations, the one with a "special type of energy" – the energy of labour resources or entrepreneurial energy will be more successful. Naturally, a significant role is played by motivation, which allows you to mobilize and realize the potential of employees, but all these issues are separate topics of study.

Entrepreneurial energy provides a certain level of interaction between the organization and the external environment, forming the flows of incoming and outgoing energy. In addition, the result of the interaction between the organization and the external environment is not only a certain ratio but also the level of information entropy H, reflecting the degree of uncertainty of the results of the "energy exchange" of the organization (Fig. 8.3).

The lower the level of H, the higher the qualitative level of the entrepreneurial energy of the organization's activities. The state of the organization in terms of its energy exchange with the external environment, the degree of control over it, as well as energy costs for the organization itself (maintaining structural stability) is estimated using the entropy S.

Consider the hypothesis that the state of systems can be estimated by entropy – uncertainty and enthalpy – entrepreneurial energy. In this case, the Gibbs-Helmholtz equation is the most appropriate model of organizational behaviour.

It should be noted that when assessing the states of the system, it is not the value of the estimated indicator that is important, but its change (Lewandowski, 2016). According to the Gibbs-Helmholtz equation, energy, enthalpy and entropy are related as follows:

$$\Delta G = \Delta I - T\Delta S, \tag{8.8}$$

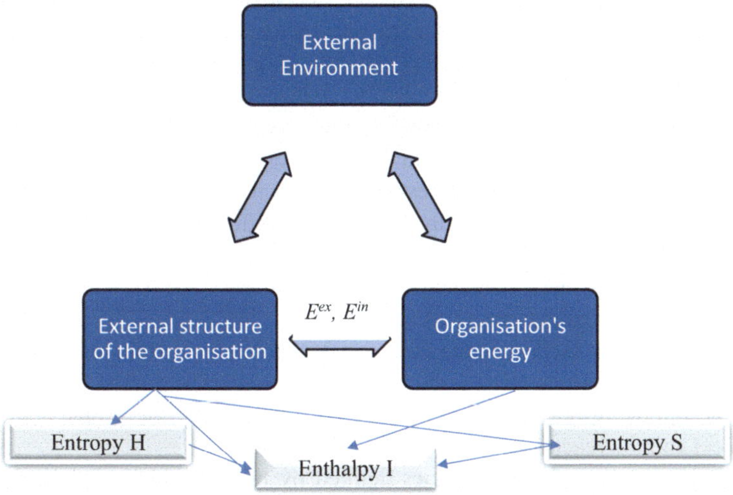

Fig. 8.3 Organizational enthalpy formation scheme

where ΔI is the change in enthalpy, ΔG is the change in energy, T is the temperature, ΔS is the change in entropy.

From (8.8) follows:

$$\Delta I = \Delta G + T\Delta S. \tag{8.9}$$

Note that in this case, the energy of the organization is an analogue of the Gibbs energy (free energy, Gibbs potential or thermodynamic potential).

From (8.2) it follows that a certain amount of heat $T\Delta S$ is spent to increase entropy, this part of the energy is lost to perform useful work, it is sometimes called "bound energy". Another part of the heat ΔG can be used to do "useful" work, so the Gibbs energy is often also called free energy.

Enthalpy, therefore, is an assessment of the energy potential of an organization, taking into account both informational entropy and entropy in a thermodynamic context.

In terms of the entropy control concept:

$$\Delta G = E^{\text{in}} - E^{\text{ex}}, \tag{8.10}$$

where E^{in}, E^{ex}, respectively, incoming and outgoing energy – the inflow of resources and their costs in the course of the organization's activities.

The Gibbs energy indicates how much of the total internal energy of the system is used for "work", so in terms of the organization and its energy, the Gibbs energy can be interpreted as the "realized energy potential" of the organization, while the enthalpy is the "full energy potential" of the organization.

The equations in classical thermodynamics are usually established empirically, rather than "derived" from certain laws. The use of such laws for organizations based on generalized experience, and logical relationships of various categories of organizational management make it possible to justify the adequacy of the application of known models existing in the framework of thermodynamics or statistical physics.

Typically, a project management team in a stakeholder management activity usually determines for interested parties only their basic functions and the degree of impact on the project activity. This makes it necessary to abandon the use of the term "governance" concerning interested parties in its traditional sense. Therefore, in the further application of this term, it will have a different meaning, which will be disclosed in the proposed definitions.

In this case, the project management team's activities will be related to the performance of two key functions.

The first function facilitates the common broadcast understanding of messages between interested parties related to the project situation. The goal of facilitating the project design of further activities. In this case, we consider the actual situation in the project and the reaction of interested parties.

The key category of competencies defined by the "attitude" of the interested party to the project situation.

The second function concerns the measurement, evaluation and interpretation of strategic sustainability according to the criterion of the uncertainty of the current state of the interaction environment in a specific project situation.

The role of the key categories of competencies within the implementation of these functions is played by the basic characteristics of the connection with the environment – the project context of the corporate culture of the interested parties and the cultural context of the project.

Analysis of the definitions of "mechanism" shows that in the social sciences, it is interpreted ambiguously as a methodology, procedure, interaction of forces, form of communication, as well as the system that determines the order of a particular activity, set of procedures and rules and more. In this case, the mechanism differs from the algorithm, which is traditionally understood as certain actions performed in a certain order. Unlike the algorithm, the mechanism necessarily reveals the essence of the actions (blocks of actions) that are associated with the transformation of information into one that can be used in the next step for further transformation. These actions determine the features of the movement of information at the stages of the mechanism and the choice of tools for its transformation.

8.3 Methodology for Implementation of Projects in Circular Environments

Convergence of knowledge between different domains arises as a result of ensuring the innovation of processes in various fields of science, technology, education, etc. It is an intertwining, business game, interaction and interpenetration of the elements of the existing system (Kalmykova et al., 2018).

The convergence hypothesis is a statement that does not matter how different systems are in the process of development, they reject ineffective tools while maintaining effective ones. Because they become relatively similar after such a step-by-step selection (Korhonen et al., 2018). This is happening due to the creation of knowledge and methodologies, methods, models and technologies that are slowly reducing the gap between systems in different industries. This is the process of creating a single methodological framework, taking into account the characteristics domain of management.

Also, if the object of control at the initial stage is further from the position of stable equilibrium, the rate of its development will be higher than the system that is closer to equilibrium according to the convergence hypothesis. In this case, in the long run, differentiation can be smoothed out (European Commission, 2015).

The proposed conceptual model is based on the organization's life cycle, uncertainty, problems and challenges related to specific situations during the life cycles of the project, early warning indicators and proactive impact model applied during program implementation (Bushuyev et al., 2019).

Let's present the example of the Hybrid Project Management Methodology developed by the authors. It is based on the PM^2 methodology developed by the European Commission and applies to the program level and the Agile methodology applies to the low level for managing projects. Its purpose is to enable Project Managers (PMs) to deliver solutions and benefits to their organizations by effectively managing the entire life-cycle of their projects. Project management methodology has been created with the requirement of European Union Institutions and projects in mind but is transferable to projects in any organization (Ethical Corporation, 2019).

Waterline between application of waterfall and Agile methodology (Larman, 2004) defined by the program manager and management teams. The model of the Hybrid Project Management Methodology structure is presented in Fig. 8.4.

The proposed hybrid project management methodology incorporates the elements from globally accepted project management good practices methodologies (Planing, 2015). Its development has also been influenced by proven experience on various projects both within European Union Institutions and global users (Schroeder et al., 2019).

Can we say that project management methodologies use hybrid technologies? Yes, it is possible when the project one methodology (a fragment of a methodology) is replaced/supplemented by another methodology or methodological improvisation takes place (Bushuyev et al., 2020).

In many cases of applying Agile methodologies, hybridization does not resemble systematically thought-out and ahead-of-time planned processes of the synergistic combination of certain methodologies.

Correspondingly, in Agile project management methodologies, the convergence of methodologies can be present only fragmentarily and more often generally not (Lüdeke-Freund et al., 2019). Today Agile is actively used in specific conditions, in software development and extreme programming projects (Collier, 2011). From a system point of view, the creation of a specific situation is the result of uncertainty and an insufficiently mature management process (IPMA ICB4, 2018).

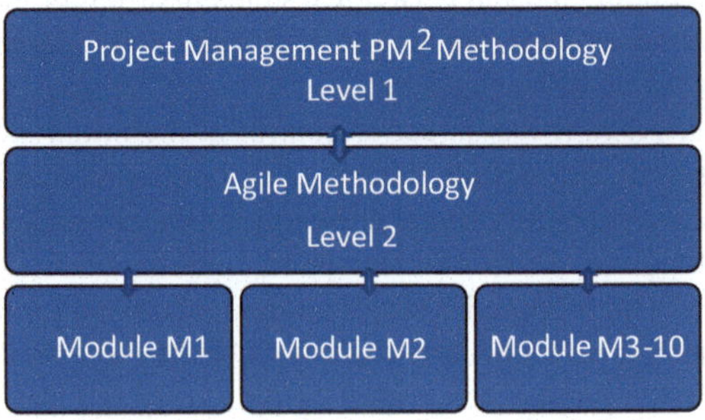

Fig. 8.4 The model of hybrid project management methodology structure

8.4 Conclusion

The hybridization of project management methodologies in the strategic management of infrastructure programs is a tool for the convergence and harmonization of project management systems that are different in nature and life cycle. Hybrid methodologies allowed the authors to build an adaptive system model of hybrid dual project management.

The proposed conceptual scheme for managing programs based on a hybrid methodology allows for organically combining projects with waterfall and Agile project life cycle models.

The genome model of hybrid program management methodologies makes it possible to compactly store the DNA of each methodology and, through the processes of harmonization, integration, convergence and actualization, create the hybrid model of project and program management systems.

8.5 Research Findings

1. The model allows the building of trajectories for the sustainable strategic development of circular projects and programs.
2. The implementation of the proposed methodology for profiling a sustainable strategy has proven the practical significance and effectiveness of models and methods of strategic management of projects and programs in Ukraine.

8.6 Research Limitations

The experience of recent global crises has shown that in conditions of instability, even small fluctuations can trigger a process with unpredictable consequences for the global economic system.

Latent value is manifested in the conflicts that determine the direction of the future development of circular projects and programs. Differences in the values of the subjects of the program do not necessarily cause tensions in their interactions. System modelling allows displaying the functioning and development of the organization with non-linear changes in the internal and external environment.

The functioning of the system is usually understood as its transition from one state to another in the direction of achieving the goal with its characterizing properties in the dynamics, and an uncertain environment is not accompanied by a change of purpose. The development of the same system is understood as a change in its quality (i.e. its composition, structure or both together).

References

Azarov, N. Y., Yaroshenko, F. A., & Bushuyev, S. D. (2011). *Innovative mechanisms for managing development programs* (564 p). Summit Book.

Bocken, N. M., De Pauw, I., Bakker, C., & Van Der Grinten, B. (2016). Product design and business model strategies for a circular economy. *Journal of Industrial and Production Engineering, 33*(5), 308–320.

Bondar A., Bushuyev S., Bushuieva V., & Onyshchenko S. (2021). Complementary strategic model for managing entropy of the organization. In *CEUR workshop proceedings* (pp. 2851–302). http://ceur-ws.org/Vol-2851/paper27.pdf

Bushuyev, S., & Verenych, O. (2017). Competencies as an instrument for the blended mental space management of a project. In *The 11th international conference on project management (ProMac2017)* (pp. 850–859). Munich, Germany.

Bushuyev, S., & Verenych, O. (2018). Chapter 6: Organizational maturity and project: Program and portfolio success. In G. Silvius & G. Karayaz (Eds.), *Developing organizational maturity for effective project management under the head* (p. 349). IGI Global. (chapter 6 P. 104–127).

Bushuyev, S., Kozyr, B., & Zapryvoda, A. (2019). Nonlinear strategic management of infrastructure programs. *Innovative Technologies and Scientific Solutions for Industries, 4*(10), 14–23.

Bushuyev, S., Babayev, J., Bushuiev, D., & Kozyr, B. (2020). Emotional infection of management innovation SMART government projects. In *IEEE European technology & engineering management summit* (pp. 53–58).

Bushuyev S., Bushuyeva N., Bushuiev D., Bushuieva, V. (2022). Inspirational emotions as a driver of managing information-communication projects. In 17th IEEE International Scientific and Technical Conference on Computer Sciences and Information Technologies (pp. 438–441).

Cockburn, A. (2000). Selecting a project's methodology. *IEEE Software, 17*(4), 64–71.

Collier, K. W. (2011). *Agile analytics: A value-driven approach to business intelligence and data warehousing* (p. 121). Pearson Education. ff. ISBN 9780321669544. What is a self-organizing team?

Ethical Coorporation. (2019). Circular economy briefing. In *Responsible business summit Europe 2020*. Reuters Events.

European Commission. (2015). *Closing the loop – An EU action plan for the circular economy*. Brussels. https://eur-lex.europa.eu/legal-content/EN/TXT/?uri=CELEX:52015DC0614

Geissdoerfer, M., Savaget, P., Bocken, N. M. P., & Jan, H. E. (2017). The circular economy – A new sustainability paradigm? *Journal of Cleaner Production, 143*, 757–768. https://doi.org/10.1016/j.jclepro.2016.12.048

Hart, J., Adams, K., Giesekam, J., Tingley, D. D., & Pomponi, F. (2019). Barriers and drivers in a circular economy: The case of the built environment. *Procedia Cirp, 80*, 619–624.

International Project Management Association. (2015). *Individual competence baseline for project, programme & portfolio management IPMA ICB 4* (Version 4, 415 p). International Project Management Association.

IPMA. (2018). *IPMA reference guide ICB4 in an agile world* (Version 2.3, 71 p). IPMA.

IPMA OCB. (2013). *IPMA organisational competence baseline (IPMA OCB)* (67 p). IPMA.

ISO 21500:2012. (2012). *Guidance on project management* (36 p). Project Committee ISO/PC 236.

Kalmykova, Y., Sadagopan, M., & Rosado, L. (2018). Circular economy–from review of theories and practices to development of implementation tools. *Resources, Conservation and Recycling, 135*, 190–201.

Korhonen, J., Honkasalo, A., & Seppälä, J. (2018). Circular economy: The concept and its limitations. *Ecological Economics, 143*, 37–46.

Larman, C. (2004). *Agile and iterative development: A manager's guide* (p. 27). Addison-Wesley. ISBN 978-0-13-111155-4.

Lewandowski, M. (2016). Designing the business models for circular economy—Towards the conceptual framework. *Sustainability, 8*(1), 43.

Lüdeke-Freund, F., Gold, S., & Bocken, N. M. (2019). A review and typology of circular economy business model patterns. *Journal of Industrial Ecology, 23*(1), 36–61.

Planing, P. (2015). Business model innovation in a circular economy reasons for non-acceptance of circular business models. *Open Journal of Business Model Innovation, 1*(11), 1–11.

PMAJ P2M. (2015). *Guide to managing innovative projects and programs* (Version 3, 366 p). PMAJ

Project Management Institute. (2017). *A guide to the project management body of knowledge (PMBOK Guide)* (6th ed.). Project Management Institute.

Schroeder, P., Anggraeni, K., & Weber, U. (2019). The relevance of circular economy practices to the sustainable development goals. *Journal of Industrial Ecology, 23*(1), 77–95.

Todorović, M., Petrović, D., Mihić, M., Obradović, V., & Bushuyev, S. (2015). Project success analysis framework: A knowledge-based approach in project management. *International Journal of Project Management, 33*(4), 772–783.

Verenych, O., & Bushuieva, V. (2017). The blended mental space as one of the factors for sustainability and success in the project implementation process. In *International research conference* (pp. 118–125). Dortmund, Germany.

Verenych, O., & Bushuyev, S. (2018). Interaction researching mental spaces of movable context, stakeholder and project manager. *Organization, Technology and Management in Construction: An International Journal, 10*(1), 1684–1695.

Chapter 9
The System Dynamics Model for the Impact Assessment of Project Management on Circular Economic Processes

Ekaterina Andreevna Khalimon, Irina Stanislavovna Brikoshina, and Mariya Nikolaevna Guseva

Abstract The study tested the scientific hypothesis about the possibility of creating a neural recurrent model of the circular economy for an adequate representation of its key processes. To develop such a model, various conceptual models, mathematical descriptions of certain circular economic processes, and approaches were taken for analysis. Next, the parameters and indicators used by some groups of researchers to measure indicators of the circular economy were reviewed. Simulation modeling based on system dynamics and deep recurrent neural network modeling based on the multilayer feed-forward neural network configuration were used to build a model containing many deep multi-level feedbacks. The simulation and neuronet tools not only allowed to work with quantitative data obtained during the simulation model run, but also to evaluate the information suitability of the constructed model. The results confirmed the scientific hypothesis and permitted to make conclusions that should form the basis for further research of circular economy processes, apply project management tools at the level of public administration and corporate governance to address the identified threats when planning strategic development programs for countries, regions, cities, industries, and companies.

Keywords Circular economy · System dynamics · Simulation modeling · Recurrent neural network · Project management

E. A. Khalimon (✉) · I. S. Brikoshina · M. N. Guseva
State University of Management, Project Management Department, Moscow, Russian Federation
e-mail: guu.konf@yandex.ru; is_brikoshina@guu.ru; boxgusevoy@yandex.ru

V. Obradović (ed.), *Sustainable Business Change*,
https://doi.org/10.1007/978-3-031-23543-6_9

9.1 Introduction

The linear economy paradigm, based on the use of fossil raw materials, has provided significant global socio-economic and technological development, but this has come at the cost of escalating resource use, global environmental degradation, and an unprecedented scale of anthropogenic impact on the climate. The industrial age brought about global economic convergence, but at the risk of sacrificing the safe working space of our planet.

As a result of research conducted in 2015 that identified nine planetary boundaries to ensure a safe working space for humanity, it was concluded that four of them are crossed: climate change, loss of integrity of the biosphere, changes in the land use system, and changes in biogeochemical cycles (phosphorus and nitrogen) (Steffen et al., 2015). Two of them – climate change and the integrity of the biosphere – are "fundamental boundaries", meaning that changing them will bring the Earth's system into a new state that can no longer support our current economic system.

In the transition from a fossil-based economy to a low-carbon economy, the focus of politics and the media tends to be on the energy sector. However, it has been estimated that 60–65% of environmental damage is related to the production of materials, while only 35–40% is related to energy (UNEP, 2017). This fact highlights the need to develop a closed-loop economy (or even closed-loop Bioeconomy (Hetemäki et al., 2017).

No country in the world today can be described as a country with a circular economy, but most countries are ready for transformation and are already on the threshold of a different world order. In order for the implementation of the strategic plans to be successful and the expectations of the country's leaders to be met, it is necessary to develop a full-fledged model of the closed-loop economy, which can be detailed, scaled, and adapted to any country. Such a model should not be conceptual, but live, and the best means for this is neural networks with reverse cycles.

The use of neural networks will not only allow to build a model and to determine the relationships between variables and resulting indicators but also to evaluate their suitability for further practical use.

Because of the absence of actual retrospective data from countries with circular economies (as they do not exist nowadays), adaptive learning should be used by training neural network. In this regard, from 100 to 500 experiments should be performed on the constructed model and the results obtained could be used as the input data for training the neural network and checking its characteristics. Based on the neural network's powerful ability to learn and integrate a variety of data from various factors, more accurate values can be obtained at the output.

9.2 CE Models, Concepts, and Approaches

The scale of the problems (environmental, economic, social, etc.) facing humanity in recent time makes it increasingly necessary to refer to the concept of CE or "closed-loop economy". It involves moving from linear systems to cyclical (or closed) systems by promoting Resource Economics and increasing resource productivity at all stages of production, distribution, and consumption.

In most scientific sources that write about this topic, it is customary to point to the 60 s of the twentieth century, as the period when the first theories on the problems of conservation of natural resources and the imperfections of the existing linear economic model appeared. However, it is worth noting that in the Soviet Union, the problems of rational use of resources, across the entire economic system of its countries, were the subject of close study, and many of these developments (delivery of used containers for reuse, collection of waste paper and recyclables, a developed system of rental points, the rejection of disposable tableware, increasing the service life of products) have not lost their relevance to this day. Studying the experience of the USSR in creating partially closed chains, creating products with an extended service life, and so on, can be the subject of a separate study. Moreover, many of the principles implemented in the USSR formed the basis of the basic principles of the circular economy.

Today there are many models of the CE and approaches to the description of its main processes. Next, some basic ones will be considered that will form the basis for creating a generalized model for its construction in terms of system dynamics and neural modeling.

1. The basis of the CE approach are 10 principles also called "types of activities" (Allwood et al., 2011; DenHollander & Bakker, 2012) or "10R", and they have gone through stages of development and transformation during their existence from 3 to 10 "R". The letter "R" is not a random choice. This is the letter with which many words begin, indicating the repeated use of the subject. "10R" at the present stage includes the following elements, principles, and activities:
    1. "Rethink" – make the use of the product more intensive (e.g. by sharing the product)
    2. "Reduce" – increase the efficiency of production or use of products by reducing the consumption of natural resources and materials
    3. "Reuse" – reuse by another consumer a product that is not needed by the original consumer, but is still in good condition and performs its original function; active development of resale processes among consumers, especially using popular Internet sites, online auctions, and stores (eBay, Amazon, etc.)
    4. "Repair" – repair and maintenance of a broken product so that it can be used with its original functionality
    5. "Refurbish" is to restore the old product and bring it to a modern state, for example, repair of buildings and structures, heavy equipment, etc.

6. "Remanufacture" – use of the unwanted/broken product to a new product with the same features
7. "Repurpose" – to use the unwanted/broken product for a new product with other functions
8. "Recycle" – recycle materials to get the same high-quality or lower-quality product
9. "Recover" – the burning of material with energy recovery
10. "Refuse" – offer the same product functionality using a different approach/materials/tools.

    It was noted by Reike D., Vermeulen W.J.V., and Witjes S. after having reviewed 69 contributions that they all summarized imperatives for reuse as a certain number of Rs (Reike et al., 2018). It seems obvious why this is popular: the "re-" in Latin means "again," "back," but also "afresh," "anew," fairly well expressing the essence of CE (Sihvonen & Ritola, 2015). However, the simplicity that makes such terminology attractive may simultaneously have contributed to confusions in CE literature and its related literature strands (Reike et al., 2018). The authors proposed their vision of a CE model based on the 9 "R" approach (Fig. 9.1).

2. A closed-loop economy, as defined by the Ellen MacArthur Foundation (Fig. 9.2), shows how technological and biological nutrient-based products and materials cycle through the economic system, each with their own set of characteristics (EMF, 2013).

   In their study, the authors also developed a high-level value driver tree (Fig. 9.3) that contains the input values of the CE model and the likely levers for improving the resulting indicators based on the described assumptions about what a circular CE system will entail in terms of collection and reverse processing for each of the products under study. The tree represents the architecture of the CE model.
3. Another point of view on CE was given through the creation of Bioeconomy by Hetemäki L., Hanewinkel M., and others in 2017 (Fig. 9.4). The authors agree with the model and the definition of CE made by Ellen MacArthur Foundation, and they suggest that the Bioeconomy is restorative and closed in nature and aims to ensure that products, components, and materials always remain as useful and valuable as possible, taking into account the differences between technical and biological cycles (Hetemäki et al., 2017). It's important they note that the Bioeconomy and the CE do not per se imply sustainability, they have to be made sustainable (Hetemäki et al., 2017).

Summarizing all the concepts, the CE is a new stage, the next stage in the development of economic relations in society. The approaches and principles implemented within this paradigm, in contrast to the linear economy, seem to be more integral, and systematic, with a plan of action and value from the moment the idea arises to the end of the product's operating cycle.

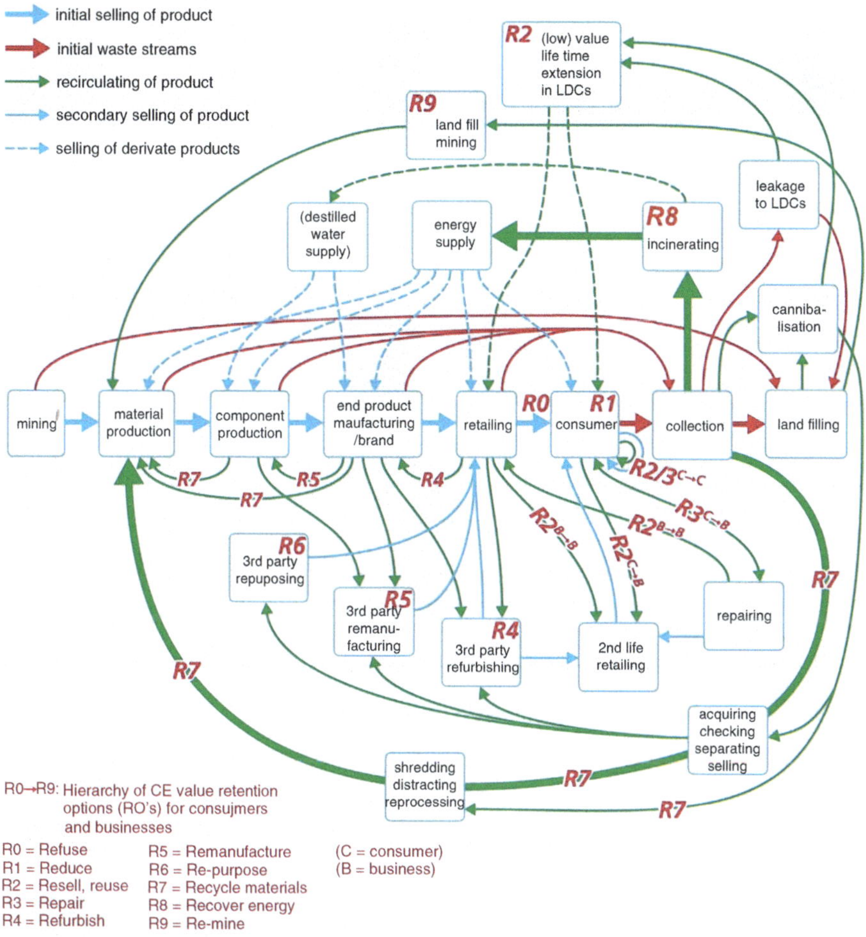

Fig. 9.1 Mapping circular economy retention options: The life cycle of production and use of the product. (Source: Reike et al., 2018)

9.3 Parameters and Indicators Measuring CE

In order to estimate the economic and environmental benefits of circularity interventions, it is important to assess their cost-effectiveness. Such an attempt was done through the application of analytical methods that assess the impact of particular policies (Elia et al., 2017). However, there is no recognized framework for measuring how effective a country is in making a transition to circularity. Such an approach needs to integrate indicators with a clear understanding of the circularity mechanism influencing multiple economic activities and their environmental performance (Aguilar-Hernandez et al., 2018).

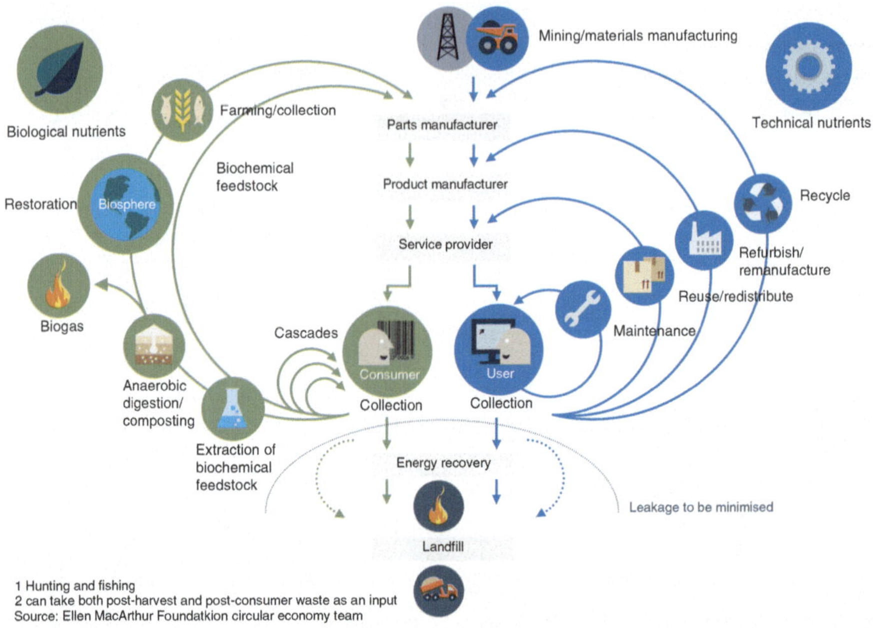

Fig. 9.2 The circular economy – an industrial system that is restorative by design. (Source: EMF, 2013)

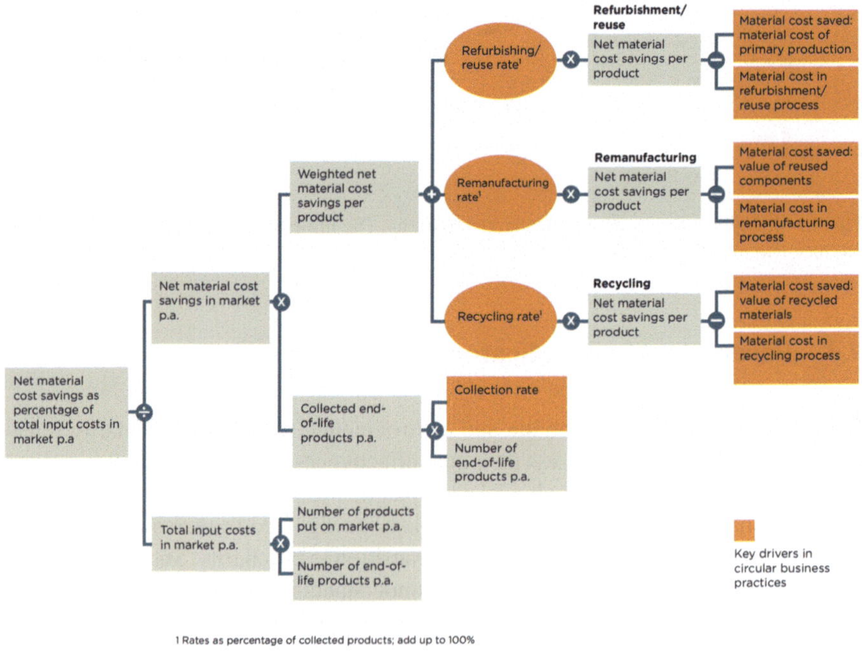

Fig. 9.3 Driver tree: factors affecting net material cost savings as a percentage of total input costs. (Source: EMF, 2013)

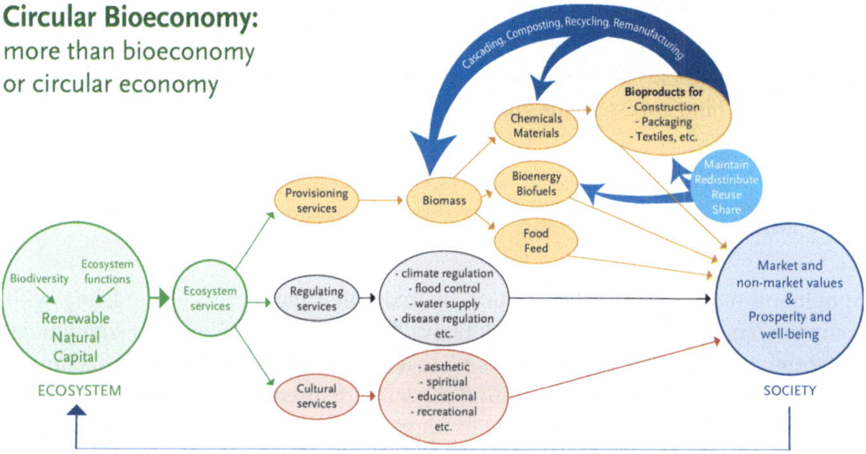

Fig. 9.4 Illustration of circular bioeconomy flows. (Source: Hetemäki et al., 2017)

It was pointed out by the group of researchers (EASAC, 2016) that GDP was not for measuring social well-being: on the basis of market transactions, it provides a monetary measure of the value of all final goods and services produced in a given period of time. It does not take into account social costs, environmental impacts and income inequality, as recognized by many environmental economists and politicians. Despite this, GDP remains the headline indicator against which economies' performance tends to be assessed (EASAC, 2016). Many alternative measures of progress have been devised. Some of them – the Index of Sustainable Economic Welfare (ISEW) and the Genuine Progress Indicator (GPI) – adjust GDP to incorporate social and environmental factors, for example the benefits from volunteer work, the costs of divorce, crime, and environmental pollution. Comparing these indicators with GDP the researchers (EASAC, 2016) revealed that none has yet succeeded in systematically repairing the shortcomings of GDP, and that an ideal indicator of social welfare has yet to be developed.

The recent research (Moraga et al., 2019) illustrates the classification framework with quantitative micro-scale indicators selected from literature and macro-scale indicators from the European Union's 'CE monitoring framework'. It was concluded that most of the indicators focused on the preservation of materials, with strategies such as recycling. However, micro-scale indicators focused on other CE strategies considering Life Cycle Thinking (LCT) approach, while the European indicators mostly accounted for materials often without taking LCT into account. Furthermore, none of the available indicators can assess the preservation of functions instead of products, with strategies such as sharing platforms, schemes for product redundancy, or multifunctionality (Moraga et al., 2019). Finally, the authors suggested that a set of indicators should be used to assess CE instead of a single indicator.

In another research, the authors (Qing et al., 2011) used the reference to the "Evaluation index system of circular economy (macro)" issued by China's National

Development and Reform Commission, which selects four factors: the resource consumption, resource recycling, environmental protection, and economic development as the main classified indexes to evaluate the circular economy development situation of China's western provinces.

9.4 Simulation Modeling

Simulation models are a combination of traditional mathematical modeling with computer technologies (Filyak & Zolotarev, 2015). The goal of building simulations is to bring the model as close as possible to a specific object and achieve maximum accuracy of its description. Simulation modeling is used when conducting experiments on a real system is impossible or impractical, most often due to their cost or duration. In various fields of Economics, business, and science, simulation helps to find optimal solutions and provides a clear understanding of complex systems. In addition, the simulation model can be analyzed in dynamics, as well as view animation in 2D or 3D, which allows to study the processes and make changes to the simulation model during its operation, better analyze the system, and quickly solve the problem.

In this research we'll follow the stages of simulation modeling presented below:

(a) Understanding the system: understanding what is happening in the system that is being analyzed, what its structure is, and what processes are taking place in it.
(b) Formulation of the system modeling goal: a list of tasks that are supposed to be solved using the future model. List of input and output parameters of the model, list of source data, criteria for completion of future research.
(c) Development of the conceptual structure of the model: the structure of the model, the composition of the essential processes to be displayed in the model, the fixed level of abstraction for each subsystem of the model (list of assumptions), the description of the control logic for the subsystems.
(d) Implementation of the model in the modeling environment: implemented subsystems, their parameters and variables, their behavior, implemented logic, and connections of subsystems.
(e) Implementation of the animation representation of the model: animated representation of the model, user interface.
(f) Checking the correctness of the model implementation: the belief that the model correctly reflects the processes of the real system that need to be analyzed.
(g) Model calibration: fixing the values of parameters, coefficients of equations, and distributions of random variables that reflect the situations that the model will be used to analyze.
(h) Planning and conducting a computer experiment: simulation results – graphs, tables, etc., giving answers to the questions.

There are four quite different systems of views: dynamic systems, system dynamics, discrete-event modeling, and multi-agent models (Karpov, 2005). These paradigms differ not so much in their areas of application, but in their concepts and views on the problem and approaches to solving the problem.

9.4.1 Dynamic Systems

Static models operate with characteristics and objects that do not change over time. In dynamic models, which are usually more complex, the change in parameters over time is significant. Static models usually deal with steady-state processes, balance-type equations, and limiting stationary characteristics. Dynamic system modeling consists of simulating the rules for a system's transition from one state to another over time (Karpov, 2006). The state of a system is defined as a set of values for its essential parameters and variables. A change in the state of a system over time in dynamic systems is a change in the values of system variables in accordance with the laws that determine the relationships of variables and their dependencies on each other over time.

Real physical objects function in continuous time, and to study many problems of physical systems, their models must be continuous. The state of such models changes continuously over time. These are models of motion in real coordinates, models of chemical production, and so on. At a higher level of abstraction, for many systems, models are adequate in which the transitions of the system from one state to another can be considered instantaneous, occurring at discrete moments of time (Karpov, 2006). Such systems are called discrete. An example of an instant transition is a change in the number of bank customers or the number of customers in a store. It is obvious that a discrete system is an abstraction, processes in nature do not happen instantly. At an even higher level of abstraction, continuous models are also used in system analysis, which is typical for system dynamics.

When modeling complex real-world systems, researchers often encounter situations in which random influences play a significant role. Stochastic models, in contrast to deterministic ones, take into account the probabilistic nature of the parameters of the simulated object. For example, in the oil port model, the exact time when tankers arrive at the port cannot be determined. These moments are random variables, because this model is stochastic: the values of the model variables that depend on the realizations of random variables themselves become random variables (Karpov, 2006). Analysis of such models is performed on a computer based on statistics collected during simulation experiments when the model is repeatedly run for different values of the initial random variables selected in accordance with their statistical characteristics.

Taking into account the complexity of the created model of the circular economy and the level of abstraction at which it is necessary to work, it can be stated that the developed CE model will be dynamic, continuous, and stochastic.

9.4.2 System Dynamics

The paradigm of computer modeling, in which graphical diagrams of causal relationships and global effects of some parameters on other parameters over time are constructed for the system under study, and then the model created on the basis of these diagrams is imitated on a computer, is called system dynamics (Karpov, 2006). Complex relationships and mutual influences of processes are often found in business, ecology, social systems, urbanism, and so on.

System dynamics is suitable for modeling complex systems at the highest level of aggregation. It is based on the idea of a system as a set of interdependent flows (money, products, human resources, etc.) that change over time. When constructing such models, a number of assumptions and simplifications are made. First, models of system dynamics abstract from individual characteristics and behaviors of system objects, and even from individual objects themselves (whether they are documents, personnel, animals, etc.). Second, these models usually depend on the physical characteristics of the environment in which the processes take place. Third, all variables, even if they characterize discrete quantities (e.g. population or number of customers), are considered continuous. Finally, individual events in the system are not highlighted here; all processes are considered to take place in continuous time (e.g. generational changes in the population or days for product sales are not highlighted).

Despite these simplifications, system dynamics models have proven to be very productive for investigating many complex problems. A small number of relatively simple structural elements, the repeated use of which allows to build models of business processes and city development, models of production and population dynamics, models of ecology and epidemic development. All this allows us to talk about system dynamics as a very effective, universal paradigm for studying complex systems using simulation.

From the mathematical point of view, models of system dynamics have variables and constants, differential equations, and functional dependencies. Therefore, formally, for the development of such models, there are enough tools that allow you to build models of dynamic systems, with the ability to represent formulas, algebraic and differential equations, as well as various functions (both built-in computer programs and user-defined). However, there is a significant difference in approaches to developing models of system dynamics and models of dynamic systems. This difference is that developers of system dynamics models do not think in terms of differential and algebraic equations, but in terms of graphical representations of flowcharts, in terms of functional dependencies of variables, the structure of variable relationships, and do not use the terminology of differential equations.

In addition, from the point of view of system dynamics, all systems, no matter how complex they may be, are built on closed feedback loops of dependencies of system variables. Therefore, graphical representation of flow diagrams, as well as relationships of variables that can be used to analyze cycles of parameter dependencies, is an essential requirement for system dynamics.

It should also be mentioned about the model that represents complex mutual dependencies of variables – the E. Lorentz model, which was originally created to simplify the description of processes that occur in the atmosphere and determine the weather.

From a formal point of view, the Lorentz model is not complicated: it is just three variables that are mutually connected by a system of differential equations. However, it has three different feedback loops, which makes the system difficult to understand and analyze. Despite the apparent simplicity of the system, it does not have an analytical solution. Complex relationships between variables make the model dynamics very interesting. First, under certain initial conditions, the model describes a so-called deterministic chaos – a deterministic model whose variables behave extremely chaotically (another name for the model is the Lorentz attractor). Second, this model demonstrates high sensitivity to initial conditions. For example, for certain parameter values, minor changes in the initial conditions significantly change the trajectories of variables. One of the experiments (Fig. 9.5) shows that sometime after the start, the values of x_i begin to differ significantly from the values of x. Lorenz interpreted this effect as follows: "even the flapping of a butterfly's wing somewhere in the forests of Brazil can cause a hurricane in Texas in a month" (Lorenz, 1963). Lorenz called it "the butterfly effect". Another conclusion from this model is that it is impossible to rely on the accuracy of long-term weather forecasts: all measurements are made with a certain error, within which processes in the atmosphere can develop over time in completely different directions. In other words, the accuracy of weather forecasts is objectively limited.

System dynamics models that are built to analyze the real economic and business processes, urbanization processes, population dynamics, and so on are usually much more complex than this Lorentz model. The cross-mutual influence of processes in them can sometimes cause completely unexpected effects, similar to the effect of deterministic chaos described by the Lorentz model. J. Forrester writes (Forrester, 1969) that a large corporation, a city, an economy, or a government are all examples of complex systems whose behavior is radically different from what we usually assume from the experience of observing simple systems. The Lorentz model demonstrates the possibility of chaotic behavior in such systems, in which there are cross-causal relationships of variables. One of the tasks of modeling is to introduce a managing mechanism that prevents the chaotic development of processes and directs them in the desired direction.

The most important practical value of system-dynamic models is that they provide a preliminary quantitative representation of the processes under study. The parameters used in them (e.g. the product usage index or production speed) have a certain meaning, and this allows you to check whether the model corresponds to the actual process that it describes. Based on the data obtained, the model can be calibrated — calculate the values of the model parameters and then use this model as a basis for further research. Moreover, feedback loops are the basic concept of system dynamics.

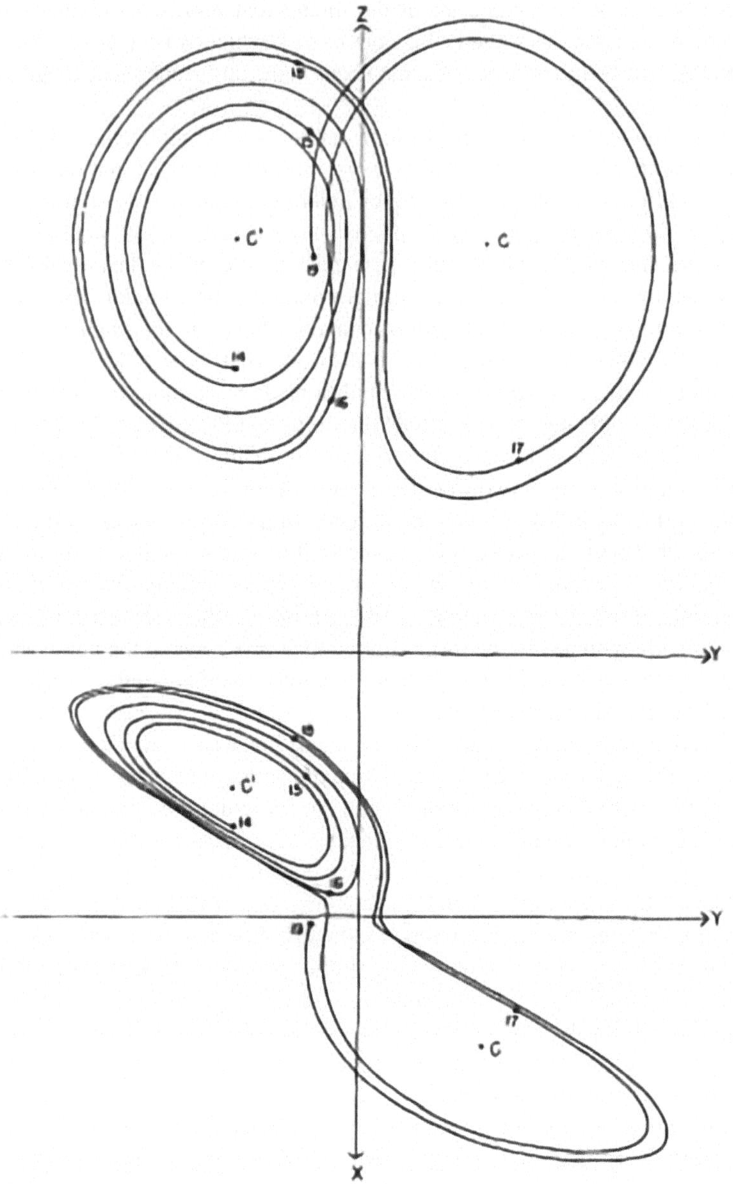

Fig. 9.5 The experiment demonstrates the influence of the weather, "the flap of a butterfly wing". (Source: Lorenz, 1963)

System dynamics is supported by several software products that are similar to each other, but AnyLogic (Source: AnyLogic) will be chosen to build a CE model, since:

- It supports development and modeling in terms of system dynamics.
- It presents all the advantages of an object-oriented approach in system dynamics. Complex models can be multi-level, created using objects connected by interface variables. System dynamics diagrams in this case are hidden inside objects. A user can create his own libraries from these objects, which are system-dynamic templates, and use them in other models. (www.anylogic.ru).
- It allows to export models, run them in the cloud, animate them, and integrate them with other software tools.
- It is the only tool that allows to combine the method of system dynamics with agent-based and discrete-event modeling (if needed).

In addition, in 2019. an attempt has already been made to prove the interdependence between people and their environment and how the development of a circular economy can affect energy use and the environment. The conceptual simplified model was developed by Clemens Dempers in AnyLogic simulation software (Source: Circular Economy Concept) and is a subject of discussion for further transdisciplinary research and development. The purpose of the conceptual model was to consider several factors that are minimally necessary and contribute to community prosperity in an urban environment: energy demand, jobs, tourism, transport, and the impact of waste on water and land (landfills). As the study continues, not only life-support factors should be added, but also the effects of carbon mitigation and climate change.

9.5 Neural Network Modeling

Artificial neural networks (ANN) are a family of statistical learning models based on how biological nervous systems, such as the brain, process information. By mimicking brain functions, they can detect patterns in data and then extrapolate predictions when they get new amounts of information. Neural networks can learn complex relationships in data. They process records and "learn" by comparing their developed system with a known actual classification record. Errors from the initial classification of the first record are fed back to the network and used to modify the network algorithm a second time, and so on for a large number of iterations in the training process to predict reliable results from complex or inaccurate data. The typical architecture of a neural network structure is shown in Fig. 9.6.

It consists of connected blocks called "nodes" or "neurons". There are three types of neurons in ANN: input nodes, hidden nodes, and output nodes. Neurons are arranged in layers. Input layer neurons receive input data for calculations. These values are passed to the neurons of the first hidden layer, which perform calculations on their inputs and pass their outputs to the next layer. This next layer can be another hidden layer, if there is one. The outputs from the neurons in the last hidden layer are passed to the neuron or neurons that generate the final outputs of the network.

Neural networks are used in a wide range of areas, including stock market forecasting, credit and loan risk allocation, credit fraud detection, sales forecasting,

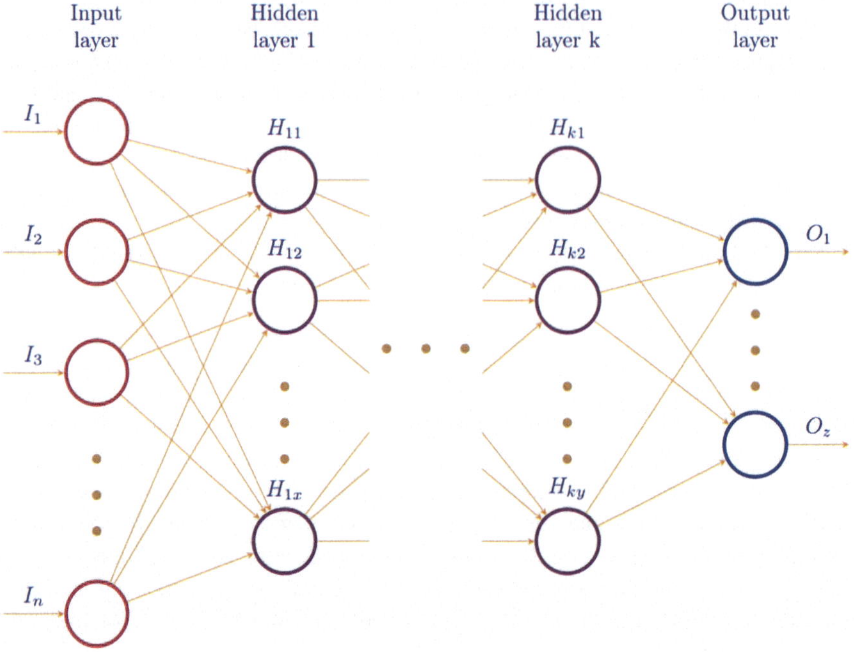

Fig. 9.6 The architecture of an ANN structure. (Source: Aqib et al., 2019)

general business forecasting, investment risk assessment, medical diagnostics, research in scientific fields, and management systems.

There are cases (Azadeh et al., 2014) of successful application of ANN to forecasting the completion dates of projects at oil refineries. There is an example (Cheng et al., 2011) of using a fuzzy hybrid neural network (HNN) to enhance project cash flow management and improve the performance of contractors in the construction industry, which is crucial in terms of providing the project manager with information about the progress of contractors, as well as for selecting appropriate contractors for a specific current or future project need. ANN in project management is used to identify factors and assess their impact on the effectiveness of project management (Wanyin, 2015). A study conducted by the American institute of project management PMI (Vargas, 2015) provides a detailed example of modeling the costs required for project management using ANN.

Recurrent neural networks (RNN) or feedback neural network is another kind of ANN model, in which the outputs from neurons are used as feedback to the neurons of the previous layer (Tarun & Khalid, 2019). In other words, the current output is considered as an input for the next output. In our research, RNN is introduced as an extension to feedforward networks, in order to allow the processing of variable-length sequences, and some of the most popular recurrent architectures in use, including long short-term memory (LSTM). The typical architecture of an RNN structure is shown and described in Tarun and Khalid (2019).

RNN architecture can be further extended to a deep recurrent neural network (DRNN) – another type of neural networks – with multiple layers between the input and output layers, presented and described in (Kong et al., 2019).

There are numerous examples of successful use of such deep multilayer recurrent networks for analyzing the processes of complex economic systems, presented in (Herbrich et al., 1999; Masloboev et al., 2014; Verstyuk, 2018).

Three stages are used to develop and test neural networks:

1. Preparing a basic data set to be used as a template for the neural network's "learning process". It is important to emphasize that usually, a correct data set is expensive and time-consuming to build. In most practical research works on the use of neural networks, the preprocessing technique is reduced to normalization, scaling, and initial initialization of weights. If the factor space is small, the specifics of the source data distribution should be taken into account for effective training of the neural network. With a large number of factors, this can be difficult to do. In this case, it is advisable to use clustering to form a training set of examples of features that are most unique in the aggregate. One of the most effective clustering methods is the use of self–organizing Kohonen maps (Pastukhov & Prokofiev, 2016). Clustering of the factor space makes it possible to form a representative sample containing the most unique training examples for training the neural network.

   The most common types of variables are dependent and independent ones, whose possible values are taken from a set of categories. If they are expressed qualitatively, they are yes or no, red, green, or blue values, and if they are expressed quantitatively, they are numeric values.
2. Training (the neural network is trained to understand the logic of the output values of available historical data). When the data was set, the network is ready for training. There are two approaches to learning: supervised learning and adaptive learning.

   In supervised learning, both input and output data are provided, and the network compares the results with the provided output data. This allows to control how well the ANN is configured to predict the correct answer.

   For adaptive learning, only input data is provided. Using self-organizing mechanisms, neural networks benefit from continuous learning to face new situations and environments.

   One of the biggest challenges of the training method is to decide which network to use and how to configure the program execution process. Some networks can be trained in seconds, but in some complex cases with multiple variables and cases, it may take hours just for the training process.

   The results of the learning process are complex formulas that relate input or independent variables to output (dependent variables).
3. Testing (the trained neural network is tested to check the quality of prediction) and sensitivity analysis (to determine the stability of results with different random selections of test cases). After training the network, an image that was not previously presented is submitted to the input, which belongs to the same set of

categories as the set of images used in training (Strokova, 2014). Thanks to the information extracted from the training data, the network can assign the presented image to a specific category (Haikin, 2006).

After training and testing, the neural model is ready to predict future results. The most important information that should be the subject of our research is the contribution of each individual variable to the predicted results and its impact on the reliability of the model.

Let's take a closer look at some of the available neural network configuration models: probabilistic neural networks (PNN), multilayer feed-forward neural network (MLFN), and Best Net Search (Mossalam & Arafa, 2018).

1. PNN are statistical algorithms in which operations are organized into multilayer direct networks with four layers (input, pattern, summation, and output). Training is fast, but it requires a large amount of memory. These networks always have two hidden layers of neurons, with one neuron in the case of training located in the first hidden layer, and the size of the second layer is determined by some facts about the training data. This allows to make the prediction process fast. When an object/phenomenon is presented to the network, each neuron in the template layer of the model calculates the distance between the training object / phenomenon represented by the neuron and the input object/phenomenon. The value passed to the summation layer neurons is a function of the distance and smoothing factors (Mossalam & Arafa, 2018). Each input has its own smoothing factor; these factors determine how quickly the significance of training cases decreases with distance. The summation layer has one neuron per dependent category; each neuron summarizes the output values for the neurons corresponding to the training objects/phenomena in this category. The output values of the summation layer neurons can be interpreted as probability density function estimates for each class (Mossalam & Arafa, 2018). The output neuron selects the category with the highest value of the probability density function as the predicted category.
2. MLFN are the most popular neural networks that are trained using a "backward propagation of errors" learning algorithm, they have one or two hidden node layers. This method is more reliable than PNN, and is usually used if there is not enough time to apply the Best Network Search.

   When training a neural network using the "backward propagation of errors" algorithm, the responsible stage is the formation of the factor space, which is subject to the following requirements:

   - Consistency of the data involved in training is required.
   - The most unique features of the examples that make up the training set should be present.
   - A sufficient amount of training data for the network of the selected architecture is needed (Pastukhov & Prokofiev, 2016).

   To meet the first requirement, the training set should be analyzed for contradictions, it is necessary to find out the causes of errors (the error appeared when entering data or, more seriously, as a result of using an insufficient number of features of the factor space) and, if possible, eliminate them (Pastukhov & Prokofiev, 2016). The second requirement must be met in order to use the training sample as efficiently as possible. The amount of data used for training a neural network is often small, so it is extremely important to correctly form a training set containing data that is most unique in terms of the set of features. The third requirement is to achieve the specified accuracy of training the neural network in a finite number of steps.

3. Best Net Search is the most reliable method in which various neural network configurations are trained and tested, including PNN and MLFN, to generate the one that gives the best predictions for the data.

   Neural networks allow accurate decision-making without an algorithm or formula-based process. To do this, there are various information systems that allow to work with neural networks, create them and adapt them to specific research goals. Among them are, for example, Palisade Neural Tools, Deep Learning Toolbox, Microsoft Power BI, Oracle Crystal Ball, Neural Network Toolbox (MATLAB). For our research, the choice of software product using the neural network method was made based on the following criteria: the ability to process data online, ease of use, speed of operation, availability of output results, and availability for use (price). And as a result, Palisade Neural Tools was selected for future use.

   With the recent development of software tools, the process of building and training neural networks becomes very simple and clear. However, the biggest problem in getting reliable results is the quality of the input information. The whole process is based on real results, and most of the time, labor, and budget require getting enough reliable data to train and test the neural network.

9.6 Different Models of CE

9.6.1 Simulation Model of CE

The model presented in Fig. 9.7 was built using simulation modeling tools, namely, based on the concept of system dynamics and in the AnyLogic program. The model is based on the 6R concept, which includes such resulting indicators as Recycle, Rethink, Reduce, Reuse, Repair, and Recover. In our model, Reduce also includes Refuse, while Repair includes Refurbish, Remanufacture, and Repurpose.

This flow model uses four types of graphic objects that are basic for system dynamics: drives (levels, state variables, parameters containing something), flows (connections between drives, gates that regulate flows), and functional dependencies (determining the mutual influence of flows). The drives are marked by rectangles, flows as arrows, and the auxiliary variables by circles. All arrows indicate cause-and-effect relationships in the model. Streams (green, red, and gray arrows)

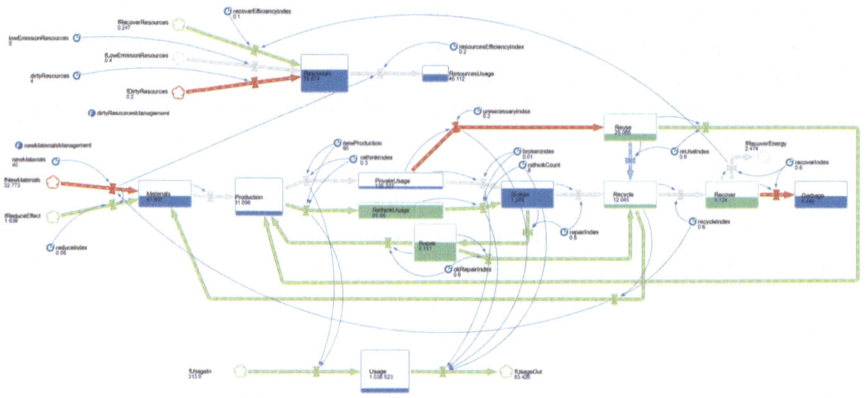

Fig. 9.7 The CE model with recurrent links in the concept of system dynamics. (Compiled by the authors based on the research results)

are the continuous movement of content between drives. Accumulators and flows influence each other through connections that can form chains of positive and negative feedbacks. Auxiliary variables denoted by circles were used to set the effects of parameters. Loops of causal relationships reflect the influence of processes on each other both directly and indirectly (through a long chain of causal relationships).

Thus, the purpose of constructing this model is not to search for a predictive value or identify a single indicator that describes the entire system, but to study the dynamics of the entire CE model with all its connections and parameters, which can lead to a correct understanding of the development processes of the entire system and identify control variables to adjust its operation. This is the main idea of system thinking.

Karpov Y.G. directly linked system dynamics with system thinking in the economy, social sphere, and ecology (Karpov, 2006). Complex models of system dynamics clearly show one of the main reasons for modeling: the direct impact of each specific process on other processes can be understood, analyzed, described, and graphically depicted. However, a person cannot understand the mutual influence of many processes. Only simulation and computer experiments make it possible to understand and evaluate the mutual influence of each process on any other, even if they are not directly related, as well as the influence of process parameters on important system characteristics.

The model consists of a large number of system parameters, drives or processes, flows, and control functions. Let's look at each of the blocks in more detail.

9.6.1.1 System Parameters

A system parameter is a metric or index that is set at the system input. The behaviour of the system can be analyzed with different options for the initial parameters set.

- brokenIndex (0–1) – the percentage of products that fail during a single model cycle.
- dirtyResources (0–100) – the number of newly produced resources that release harmful substances into nature. For example, energy extraction by burning coal or without using sufficient filtration of exhaust or wastewater; open-pit mining in polar regions; mining by explosions or other methods that destroy the structure of the soil.
- lowEmissionResources (0–100) – the number of newly produced resources that do not release harmful substances into nature in significant amounts. Production of electricity and heat by burning natural gas, use of various types of renewable energy (sun, water flow, tides, thermal springs).
- recoverEfficiencyIndex (0–1) – the conditional ratio of the useful volume of incinerated, processed, recoverable waste as a resource to their total amount.
- resourcesEfficiencyIndex (0–1) – the conditional share of material resources (energy, oil, petroleum products, metals, plastic, wood, water, etc.) in the creation of a new product.
- newMaterials (0–100) – the number of newly created materials using certain material resources.
- reduceIndex (0–1) – a conditional value that sets an indicator of resource efficiency, i.e. if earlier 100% of the resource was spent on creating a product, then in the next period it decreases by a specified share. I.e. the amount of metal, paint, and petroleum products decreases, and the cost of processing materials decreases.
- newProduction (0–100) – the number of products that are available for use.
- rethinkIndex (0–1) – the percentage of products that are actively shared. For example, car sharing, public transport, dormitories, and city communications.
- unnecessaryIndex (0–1) – the percentage of products that are out of use because they are not needed. In other words, they remain fully functional, but are not used.
- reUseIndex (0–1) – the percentage of products that were sent for reuse by sale, donation, or other methods. Goods that are not needed but are not used are considered waste and sent for further disposal. Goods that are not used or sold are also considered waste.
- brokenIndex (0–1) – the percentage of products that lose their functionality for various reasons.
- rethinkCount – (0–100) is an indicator of the intensity of product sharing. In other words, the car is used much less in private use than in car sharing, taxi or public transport. At the same time, the load on the product also increases many times, i.e. wear and early failure or loss of product presentation increases.
- repairIndex (0–1) – the percentage of products that lost their presentation or broke down, aimed at repair or restoration. Accordingly, the remaining part of the broken goods is sent for further disposal.
- okRepairIndex (0–1) – percentage of products that were repaired or restored. They are directed to reuse. Items that cannot be repaired or cannot be used to repair any other item are sent for further disposal.

- recycleIndex (0–1) – the percentage of products that were recycled. Recycled goods are materials and are entered into the system. Part of the goods that are not recyclable and cannot be used for further production is sent for further disposal.
- recoverIndex (0–1) – incineration of material with recovery of electrical and thermal energy. As well as reuse of biomaterials and other elements as a resource. The remainder of these processes is transferred to non-recyclable waste.
- resourcesEfficiencyIndex (0–1) – the index sets the effect of increasing the efficiency of using resources to create new materials. For example, if previously 100% of the resource was needed to create a unit of goods, now 98% of the resources are needed to create a unit of goods. That is, increasing the strength of alloys reduces their amount in the final product.
- recoverEfficiencyIndex (0–1) – transfer index for converting processed, recovered goods into resource units.
- lowEmissionResources (0–100) – this parameter sets the amount of use of resources with low emission of harmful substances.
- dirtyResources (0–100) – this parameter specifies the amount of use of resources with a high emission of harmful substances.

9.6.1.2 Drives or Processes

- Resources – resources necessary for the production of materials for production (ore, oil, water, energy, fertile soil, fertilizers, etc.).
- ResourcesUsage – the process of using resources, i.e. this is the part of the resources that really went to the production of any materials.
- Materials – materials that are used for the production of products. This is a generalized indicator of all the material means of production used to create a product. It consists of newly created materials, materials from various parts of already used products, as well as from recycled materials.
- Production – ready-to-use products. It includes all possible variants of products: new, restored, used, etc.
- PrivateUsage – products used for private purposes. They are characterized by a large number, but a small amount of use. On the one hand, the useful life of such products is much higher, but on the other hand, these products do not bring benefits and often clutter the space.
- RethinkUsage – products that are shared. I.e. the product can be rented, public or open for free use. Such products are characterized by increased wear and tear, resulting in faster failure.
- Broken products that have lost their product functions or appearance, i.e. they can no longer be used for their intended purpose.
- Repair – repair and maintenance of a broken product so that it can be used with its original functionality.
- Recycle-recycle materials to get the same high-quality or lower-quality product.
- Recover – the amount of incinerated, recycled, and recoverable waste as a resource.

- Garbage – non-recyclable garbage, from which it is impossible to extract useful properties and must be buried in a specialized landfill.
- Reuse – reuse by another consumer a product that is not needed by the original consumer, but is still in good condition and performs its original function; Active development of resale processes among consumers, especially using popular Internet sites, online auctions, and stores (eBay, Amazon, etc.).
- Usage-a conditional indicator of the use of the entire mass of products. This metric includes personal use products with a weight of one, and shared use products with a weight equal to rethinkCount. In case of failure, the indicator also decreases in accordance with rethinkCount. It shows that when the same product is used together, much more consumers use it. Also, more expensive products can be used by consumers who previously could not afford it, for example, using a car in car sharing and renting complex construction tools or equipment.

9.6.1.3 Flows

Flows show the movement of a resource or product over the network. Each stream has a formula or rule, a direction, and a beginning and end. If there is no initial process in the stream, it is a generator, if there is no final process-a terminator, i.e. the element as a result of the process completely loses its previous functionality and is removed from the network.

$$\text{fNewMaterials} = (\text{newMaterials} - \text{fRecycleMaterials}) * \text{newMaterialsManagement}(\ )$$

Flow of production of new materials. The growth of recycled materials reduces the production of new ones. The quantity of materials is also affected by the newMaterialsManagement control function, which suspends the production of new materials when products that are available for consumption but not used exceed a certain limit.

$$\text{fReduceEffect} = \text{fNewMaterials}^{*}\text{reduceIndex}$$

The flow reflects increased efficiency in the use of materials in the production of new products. The flow depends on the actual release of new fNewMaterials materials and the reduceIndex performance index.

$$\text{Materials}-> \text{Production} = \text{Materials}$$

Flow of materials into the production process. The model assumes that all manufactured and recycled materials are transferred to the production process.

$$\text{Production}->\text{PrivateUsage}=\left(\text{newProduction}-\text{uniform\_discr}\left(1,15\right)\right)^{*}\left(1-\text{rethinkInde}\right.$$

The flow shows the movement of goods from production to private consumption. It depends on the total demand for the product with the addition of a random uniform_discr component, and also decreases depending on the rethinkIndex sharing index.

$$\text{Production}->\text{RethinkUsage}=\left(\text{newProduction}-\text{uniform\_discr}\left(1,15\right)\right)*\text{rethinkIndex}$$

The flow shows the movement of goods from production to joint consumption. It depends on the total demand for the product with the addition of a random uniform_discr component, and also decreases depending on the rethinkIndex sharing index.

$$\text{PrivateUsage}->\text{Broken}=\text{PrivateUsage}^{*}\text{brokenIndex}$$

The flow reflects the movement of a product that is out of order. The flow rate depends on the quantity of the product that is in private use, as well as the brokenIndex reliability index.

$$\text{PrivateUsage}->\text{Reuse}=\text{PrivateUsage}^{*}\text{unnecessaryIndex}$$

The flow of movement of a product that has ceased to be used by consumers and is leaving the process of private use. Depends on the total quantity of the product actually used and the unnecessaryIndex.

$$\text{RethinkUsage}->\text{Broken}=\text{RethinkUsage}^{*}\text{brokenIndex}^{*}\text{rethinkCount}$$

The flow reflects the movement of a product that is out of order. The flow rate depends on the quantity of goods that are shared, the brokenIndex reliability index, and the rethinkCount utilization rate.

$$\text{Broken}->\text{Repair}=\text{Broken}^{*}\text{repairIndex}$$

Products aimed at repair, restoration or reconstruction. Depends on the repairIndex.

$$\text{Repair}->\text{Production}=\text{Repair}^{*}\text{okRepairIndex}$$

Products that have actually been repaired or restored to a state where they can be used either for their original purpose or for other functions.

$$\text{Repair}->\text{Recycle} = \text{Repair}^{*}\left(1-\text{okRepairIndex}\right)$$

Items that cannot be repaired, or waste that appeared during the repair process. The flow depends on the number of items being repaired and the okRepairIndex.

$$\text{Broken}->\text{Recycle} = \text{Broken}^{*}\left(1-\text{repairIndex}\right)$$

The flow of products that are not initially repairable. They are sent for recycling. The flow depends on the number of items being repaired and the repairIndex.

$$\text{Reuse}->\text{Recycle} = \text{Reuse}^{*}\left(1-\text{reUseIndex}\right)$$

The flow of products that for some reason cannot be sent for reuse to other owners. Such goods are sent to the processing process. The flow depends on the total number of unused items and the reuseIndex.

$$\text{Recycle}->\text{Recover} = \text{Recycle}^{*}\left(1-\text{recycleIndex}\right)$$

Goods that can no longer be extracted for reuse in production are sent for incineration, recycling, and recovery as a resource. The flow depends on the total number of products processed and the recycleIndex.

$$\text{Recycle}->\text{Materials} = \text{Recycle}^{*}\text{recycleIndex}$$

The flow reflects the movement of materials extracted from non-working and unused goods, which are returned to the production of new goods. The flow depends on the total number of processed goods and the recycleIndex.

$$\text{fRecoverEnergy} = \text{Recover}^{*}\text{recoverIndex}$$

The flow shows the recovery process, as a result of which the volume of recovered goods is withdrawn from the system, but at the same time, the amount of resources for the production of new materials increases. The flow depends on the recovery volume and the recoverIndex.

$$\text{Recover}->\text{Garbage} = \text{Recover}^{*}\left(1-\text{recoverIndex}\right)$$

The amount of waste from which it is no longer possible to extract useful properties is sent for disposal in specialized landfills. The flow is based on the total amount of waste in the recycling process and the recoverIndex.

$$\text{fUsageIn} = \text{fPrivateProduction} + \left(\text{fRethinkProduction}^{*}\text{rethinkCount}\right)$$

The flow of increasing the utilization rate of manufactured goods. Shows how much real consumption increases among users. This flow depends on the volume of private use with a weight of one and on the flow of shared use with a coefficient of intensity of operation rethinkCount

$$\text{fUsageOut} = \text{fPrivateProductionBroken} + \text{fUnnecessary} + \left(\text{rethinkCount}^{*}\text{fRethinkProductionBroken}\right)$$

The flow of reducing the indicator of use of goods as a result of breakage or refusal to use. The flow depends on the failure of products in private and shared use, as well as on the volume of products that retain their functionality, but for some reason go out of use.

$$\text{fRecoverResources} = \text{fRecoverEnergy}^{*}\text{recoverEfficiencyIndex}$$

Flow of recovered resources. Depends on the flow of waste received for recovery and the efficiency index of this process recoverEfficiencyIndex.

$$\text{fLowEmissionResources} = \text{lowEmissionResources}^{*}\text{dirtyResourcesManagement}(\ )$$

The flow of resource development with insignificant content of toxic substances released into the atmosphere. This flow is regulated by the nature clogging control function dirtyResourcesManagement (), which reduces or completely stops the use of such resources in production.

$$\text{fDirtyResources} = \text{dirtyResources}^{*}\text{dirtyResourcesManagement}(\ )$$

The flow of resource development with a significant content of toxic substances released into the atmosphere. This flow is regulated by the nature clogging control function dirtyResourcesManagement (), which reduces or completely stops the use of such resources in production.

$$\text{Resources}- > \text{ResourcesUsage} = \text{fNewMaterials}^{*}\text{resourcesEfficiencyIndex}$$

The flow of resources. Depends on the release flow of new materials and the resourcesEfficiencyIndex resource efficiency index.

9.6.1.4 Some Control Functions

dirtyResourcesManagement – a control function for nature clogging that reduces or completely stops the use of such resources in production, depending on the need for resources at a certain point in time.

```
if ((Resources/ResourcesUsage)<0.25) return 0.75;
else if ((Resources/ResourcesUsage)<0.5) return 0.5;
else if ((Resources/ResourcesUsage)<0.75) return 0.25;
else if ((Resources/ResourcesUsage)>1) return 0.05;
return 1;
```

```
newMaterialsManagement
if (Production/(PrivateUsage+RethinkUsage) <0.25) return 0.75;
else if (Production/(PrivateUsage+RethinkUsage) <0.5) return 0.5;
else if (Production/(PrivateUsage+RethinkUsage) <0.75) return 0.25;
else if (Production/(PrivateUsage+RethinkUsage)>1) return 0.05;
return 1;
```

9.6.2 Deep Recurrent Neural Network Model of CE

After the CE model in AnyLogic was constructed, we set ranges of numbers and the stochastic order in which the model selected variable values from it. Despite the randomness of the variable values, after the model was launched it became clear that the model stabilizes by the 200th time period (Fig. 9.8) and the indicators tend to constants. Accordingly, 200 data on the following 5 "R" indicators were selected for testing and training the neural model: Reuse, Recover, RethinkUsage, Recycle, and Repair. The 6th "R" – Reduce – could not be taken for training and testing the neural network model as it was used as a function fReduceEffect=fNewMaterials*reduceIndex. Another five parameters are presented in Table 9.1.

Based on the goal of creating a neural network not to make a forecast, but to track the behavior of the system and its stability, the MLFN configuration model and the adaptive learning were used to build and train the RNN. It took some hours for the training process. The results were complex formulas that relate the input or independent variables with the output (dependable variable) (Tables 9.2 and 9.3).

Thus, using RethinkUsage as a dependent variable on 9 independent parameters (Table 9.1), the neural model showed a linear relationship (direct and inverse) between the parameters and calculated their coefficients to determine the regression

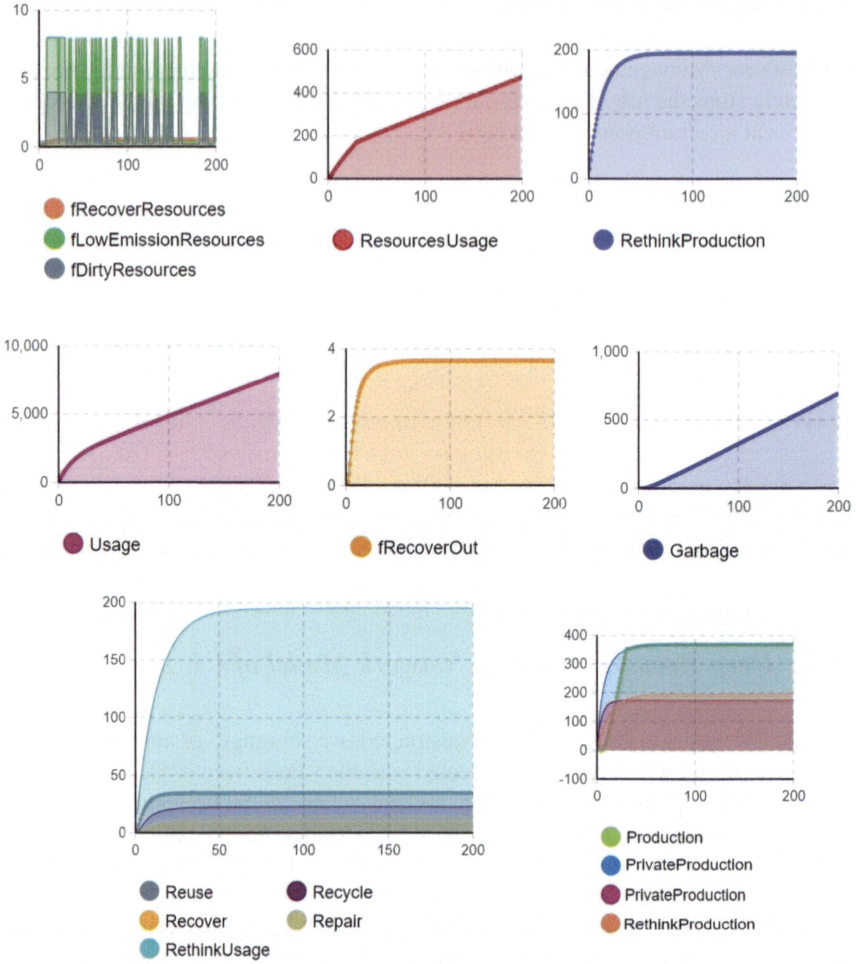

Fig. 9.8 Values of some variables and functions of the model on a time scale (200 time periods). (Compiled by the authors based on the research results)

function (Table 9.2). From the entire set of parameter values over 200 time periods, the neural model itself randomly selected 160 data packets (rows in Table 9.3) for training and 40 data packets – for self-testing. According to the indicators of information suitability of the model based on training and test samples – root mean square error, mean absolute error, and standard deviation of absolute error – it was concluded that the model is suitable for analyzing and explaining economic phenomena and processes (Table 9.3).

Table 9.1 Some of the data used for training and testing the RNN (Palisade Neural Tools software)

Number of cases	Resources Usage	Rethink Production	fRecover Resources	Flow Emission Resources	fDirty Resources	Reuse	Recycle	Recover	Repair	Rethink Usage
1	0.072	15.879	0.002	0.4	0.2	2.61	0.36	0.03	0.15	15.82
2	7.956	29.624	0.017	0.4	0.2	7.23	1.73	0.29	0.69	29.57
3	15.632	42.374	0.05	0.4	0.2	11.89	3.76	0.83	1.50	42.37
4	23.048	54.087	0.095	0.4	0.2	16.01	5.99	1.59	2.41	54.07
5	30.2	64.91	0.145	0.4	0.2	19.48	8.13	2.42	3.32	64.90
6	37.108	74.935	0.194	0.4	0.2	22.35	10.06	3.23	4.19	74.92
7	43.798	84.186	0.239	0.4	0.2	24.69	11.75	3.99	4.99	84.16
8	50.3	92.713	0.279	0.4	0.2	26.60	13.20	4.66	5.73	92.70
9	56.642	100.639	0.314	8	4	28.13	14.43	5.24	6.40	100.60
10	62.848	107.859	0.344	8	4	29.37	15.48	5.73	7.01	107.82
11	68.936	114.628	0.369	8	4	30.37	16.38	6.16	7.57	114.55
12	74.924	120.81	0.391	8	4	31.18	17.14	6.52	8.07	120.74
13	80.827	126.52	0.41	8	4	31.84	17.79	6.83	8.54	126.44
14	86.657	131.809	0.426	8	4	32.39	18.36	7.10	8.97	131.74
…										
198	470.237	194.834	0.549	0.4	0.2	34.68	22.88	9.15	13.87	194.85
199	471.992	194.904	0.549	0.4	0.2	34.68	22.88	9.15	13.86	194.82
200	473.989	194.923	0.549	0.4	0.2	34.68	22.88	9.15	13.86	194.86

Table 9.2 The fragment of a table with the training and testing results (Palisade Neural Tools software)

Tag used	Prediction	Good/Bad	Residual
Train			
Train			
Train			
Train			
Train			
Test	74.90	Good	0.02
Train			
Train			
Train			
Train			
Train			
Train			
Train			
Train			
Train			
Test	141.08	Good	−0.13
Train			
Train			
Train			
Train			
Test	158.81	Good	−0.03
Test	161.64	Good	−0.01
Test	164.21	Good	−0.05
Train			

9.7 Results and Discussion

The results of the analysis of processes in the developed circular economy model can be divided into two groups: positive and negative effects (Table 9.4).

As can be seen from the table, the number of negative factors greatly exceeds the positive ones. Therefore, it is already necessary to develop programs and projects to prevent the occurrence of negative events or minimize the effect of their occurrence. Without a project-based approach at the level of government and at the level of companies, it will be impossible to move painlessly from a linear to a circular economy. The following factors should also be taken into account when planning state transition programs and projects for restructuring production processes:

- The share of "Re"-processes in production is growing, which leads to a shortage of recyclable materials (the volume of flows in reverse loops). If all the return loops are well-defined, the volume of renewable resources and materials is not so large, and competition for the right to use recyclable materials will grow strongly.
- The production of new goods strongly decreases. For the conservation and sustainable development of technologies, the share of scientific and engineering

Table 9.3 The Palisade Neural Tools summary table

Linear function	
	Intercept/coefficient
Intercept	5.809
ResourcesUsage	0.00009995
RethinkProduction	0.1842
fRecoverResources	−23.81
flowEmissionResources	0.2451
fDirtyResources	−0.4914
Reuse	3.064
Recycle	−9.383
Recover	6.072
Repair	15.81
Net information	
Name	Net trained on data set #1
Configuration	Linear predictor
Location	This workbook
Independent category variables	0
Independent numeric variables	9 (Resources usage, RethinkProduction, fRecoverResources, flowEmissionResources, fDirtyResources, reuse, recycle, recover, repair)
Dependent variable	Numeric Var. (RethinkUsage)
Training	
Number of cases	160
Training time	0:05:00
Number of trials	0
Reason stopped	Auto-stopped
% bad prediction (30% tolerance)	0.0000%
Root mean square error	0.05008
Mean absolute error	0.03998
Std. deviation of Abs. Error	0.03015
Testing	
Number of cases	40
% bad prediction (30% tolerance)	0.0000%
Root mean square error	0.04962
Mean absolute error	0.03617
Std. deviation of Abs. Error	0.03397
Data set	
Name	Data set #1
Number of rows	200
Manual case tags	NO

Table 9.4 Positive and negative effects when switching to a CE model

Positive effects	Negative effects
Reduced number of extracted resources and landfills	
Increase the population density without reducing the quality of life	
Decrease the number of credits of the population	Increase the number of regular monthly payments
Increase accessibility to technology	Reduced life cycle of products due to the intensity of their use
	Decrease the number of products
	Increase the intensity of circulation of money in the economy
	Increase competition
	Reduced the number of producers
Increase the services market	Reduce the production market
	Resource-Oriented economies will face inflation due to a decrease in household incomes and a decrease in the number of products sold
	Developed countries with non-resource-oriented economies will face devaluation due to a drop in the competitiveness of manufactured goods in foreign markets

research in production costs grows gradually, but intensively, which leads to a strong increase in the cost of the final product for the consumer. That means that the consumer becomes more profitable to participate in the sharing system and doesn't have a certain product in the property.

- The need for continuous large consumption of new resources and products decreases. There may be an intense drop in demand for primary resources, which may lead to a decrease in their cost on the market. At the same time, resource recycling technologies cannot become cheaper. How can this process be regulated? By setting limits of production? Then countries that do not participate in the recycling process will have a strong competitive advantage, since they will not have restrictions on the use of resources that have fallen in value.
- The actual consumption of various products can afford a much larger number of the population without an increase in production, which means the growth of consumption is not due to an increase in production, but due to more frequent use of the product already produced.
- The reduction of burden on the environment. As a result, the living environment of a large number of people becomes better. Tourist and recreational activity is increasing.
- The emergence of a new technologically intensive and financially active industry for processing waste from industrial and consumer human activity.
- The possibility of closer living of the population without reducing the quality of life. It is important for countries with an existing overpopulation.

References

Aguilar-Hernandez, G. A., Sigüenza-Sanchez, C. P., Donati, F., et al. (2018). Assessing circularity interventions: A review of EEIOA-based studies. *Journal of Economic Structures, 7*, 14. https://doi.org/10.1186/s40008-018-0113-3

Allwood, J. M., Ashby, M. F., Gutowski, T. G., & Worrell, E. (2011). Material efficiency: A white paper. *Resources, Conservation and Recycling, 55*(3), 362–381.

AnyLogic. https://www.anylogic.ru/use-of-simulation/system-dynamics/

Aqib, M., Mehmood, R., Alzahrani, A., Katib, I., Albeshri, A., & Altowaijri, S. (2019). Smarter traffic prediction using big data, in-memory computing, deep learning and GPUs. *Sensors, 19*, 2206. 34 p.

Azadeh, A., Haghighi, S. M., & Panah, M. (2014). A unique intelligent approach for forecasting project completion time in oil refineries. *Journal of Loss Prevention in the Process Industries, 30*, 137–154.

Cheng, M., Tsai, H., & Sudjono, E. (2011). Evaluating subcontractor performance using evolutionary fuzzy hybrid neural network. *International Journal of Project Management, 29*, 349–356.

Circular Economy Concept. https://cloud.anylogic.com/model/f084be8f-9575-4414-bda6-390d402f28e5?mode=SETTINGS

DenHollander, M. C., & Bakker, C. (2012). *A business model framework for product lifeextension*. In Proceedings of Sustainable Innovation 2012, Resource Efficiency, Innovation and Lifestyles, 110–118.

EASAC. (2016). *Indicators for a circular economy*. EASAC policy report 30. https://easac.eu/fileadmin/PDF_s/reports_statements/Circular_Economy/EASAC_Indicators_web_complete.pdf

Elia, V., Gnoni, M. G., & Tornese, F. (2017). Measuring circular economy strategies through index methods: A critical analysis. *Journal of Cleaner Production, 142*, 2741–2751.

EMF. (2013). *Towards the circular economy*. Economic and business rationale for an accelerated transition. Ellen MacArthur foundation. Vol. 1. 96 p.

Filyak, P. Yu., & Zolotarev, V. V. (2015). AnyLogic as an agent-based information security modeling tool. Bulletin of the MFUA. №4. C. 30–34.

Forrester, J. W. (1969). *Urban dynamics*. Productivity Press. 299 p.

Haikin, S. (2006). *Neural networks: A complete course*. - M.: Publishing House "Williams", 1104 p.

Herbrich, R., Keilbach, M., Graepel, T., Bollmann-Sdorra, P., & Obermayer, K. (1999). Neural networks in economics. In T. Brenner (Ed.), *Computational techniques for modelling learning in economics* (Advances in computational economics) (Vol. 11, pp. 169–196). Springer.

Hetemäki, L., Hanewinkel, M., Muys, B., Ollikainen, M., Palahí, M., & Trasobares, A. (2017). Leading the way to a European circular bioeconomy strategy. *From Science to Policy 5*. European Forest Institute. https://doi.org/10.36333/fs05

Karpov, Yu. G. (2005). *Agent modeling - A new paradigm in simulation modeling*. [Electronic resource] http://mas.exponenta.ru/files/npo/texts/karpov.pdf

Karpov, Yu. G. (2006). Simulation of systems. Introduction to modeling with AnyLogic 5. - SPb.: BHV-Petersburg. 400 p.

Kong, H., Fang, Y., Fan, L., Wang, H., Zhang, X., & Hu, J. (2019). A novel torque distribution strategy based on deep recurrent neural network for parallel hybrid electric vehicle. *IEEE Access, 7*, 1–12.

Lorenz, E. N. (1963). Deterministic nonperiodic flow. *Journal of the Atmospheric Sciences, 20*, 130–141.

Masloboev, A. V., Putilov, V. A., & Syutin, A. V. (2014). Multi-level recurrent model of hierarchical management of complex regional security. Scientific and technical Bulletin of information technologies, mechanics and optics. No. 6(94), 163–170.

Moraga, G., Huysveld, S., Mathieux, F., Blengini, G. A., Alaerts, L., Van Acker, K., de Meester, S., & Dewulf, J. (2019). Circular economy indicators: What do they measure? *Resources, Conservation and Recycling, 146*, 452–461.

Mossalam, A., & Arafa, M. (2018). Using artificial neural networks (ANN) in projects monitoring dashboards' formulation. *HBRC Journal, 14*(3), 385–392.

Pastukhov, A. A., & Prokofiev, A. A. (2016). Application of self-organizing Kohonen maps for forming a representative sample when training a multi-layer perceptron. Scientific and technical statements of SPbGPU. *Physical and Mathematical Sciences, 2*(242), 95–107.

Qing, Y., Mingyue, C., & Qiongqiong, G. (2011). Research on the circular economy in West China. *Energy Procedia, 5*, 1425–1432.

Reike, D., Vermeulen, W. J. V., & Witjes, S. (2018). The circular economy: New or refurbished as CE 3.0? - exploring controversies in the conceptualization of the circular economy through a focus on history and resource value retention options. *Resources, Conservation and Recycling, 135*, 246–264.

Sihvonen, S., & Ritola, T. (2015). Conceptualizing ReX for aggregating end-of-life strategies in product development. *Procedia CIRP, 29*, 639–644.

Steffen, W., et al. (2015). Planetary boundaries: Guiding human development on a changing planet. *Science, 347*(6223), 1259855.

Strokova, M. S. (2014). Investigation of the effectiveness of neural networks in the detection of phase-manipulated sequences. *Omsk: Radio Communication Technology, 1*(21), 65–77.

Tarun, K. G., & Khalid, R. (2019). Chapter 7 - Optimization of ANN architecture: A review on nature-inspired techniques. In *Machine learning in bio-signal analysis and diagnostic imaging* (pp. 159–182).

UNEP. (2017). *Resource efficiency: Potential and economic implications*. A report of the International Resource Panel.

Vargas, R. V. (2015). *Applying neural networks and analogous estimating to determine the project budget*. Paper presented at PMI® Global Congress 2015—North America, Orlando, FL. Newtown Square, PA: Project Management Institute. https://www.pmi.org/learning/library/neural-network-estimating-project-budget-9900

Verstyuk, S. (2018). Modeling multivariate time series in economics: From auto-regressions to recurrent neural networks. *Computer Science*. 41 p.

Wanyin, D. (2015). A new project management performance evaluation method based on BP neural network. In *International conference on automation, mechanical control and computational engineering (AMCCE 2015)* (pp. 68–72).

Chapter 10
A Conceptual Framework for Enabling Benefits from Linking Sustainability and Project Management

Danijela Toljaga-Nikolić

Abstract With the aim of identifying and analyzing the impact of sustainable project management on the project's success and the development of values for an organization and society, a conceptual framework for enabling benefits from linking sustainability and project management has been developed and presented in this chapter. The conceptual framework for sustainable project management was developed as a basis for the empirical research in the doctoral thesis of the author. The factors that could contribute to the successful integration of sustainability principles into project management were identified and analyzed in this chapter. Based on the literature review, the following factors were presented: sustainability principles, project success, values for the organization and society, phases of sustainable project management, practices, tools, and techniques for project management, and the competencies of the project manager and the project team members. The starting point of the analysis was the assumption that the integration of the sustainability concept into the concept of project management would encompass all three dimensions of sustainability: social, environmental, and economic. This idea is very close and should contribute to the turning of the idea of a circular economy into reality by providing a better design of a project result in order to make better use of resources and to prevent waste production and pollution. When deciding about energy sources, resources, and materials, it can be seen as a contribution of the circular economy to fully recover resources by closing loops of resource flows instead of wasting them. Social, environmental, and economic dimensions should permeate each phase of sustainable project management by implementing practices, tools, techniques, and competencies for sustainable project management, thus enabling benefits in terms of a project result that better meets the expectations of the stakeholders. The satisfied expectations of stakeholders would result in the project's success, which would lead to the creation of value for the organization and society. When efforts are planned and managed in accordance with the concepts of sustainable development

D. Toljaga-Nikolić (✉)
Faculty of Organizational Sciences, University of Belgrade, Belgrade, Serbia
e-mail: danijela.toljaga.nikolic@fon.bg.ac.rs

V. Obradović (ed.), *Sustainable Business Change*,
https://doi.org/10.1007/978-3-031-23543-6_10

and circular economy, long-term benefits and capacity preservation for future generations are possible.

Keywords Project · Project management · Sustainable development · Sustainability principles · Sustainable project management · Project success

10.1 Introduction

Organizations today are facing accelerated technological change, economic instability, and rising competition in the global marketplace. At the same time, there is a trend of accelerated consumption of resources and raw materials from non-renewable sources, endangering the environment. There is a need to consume responsibly and to reduce the total amount of materials within economies, where the concept of a circular economy should be helpful. The efforts of organizations were directed at establishing a balance between, on the one hand, the need to meet the expectations of their clients and other stakeholders and, on the other hand, contributing to the well-being of the community and environmental protection for generations to come. This balance is considered achievable if organizations decide to integrate the concept of sustainable development into their business strategies, which will lay the foundation for the business system and each individual to act in such a way as to minimize the harmful impact of business activities and maximize the positive impact of business activities on society and the environment. Sustainable development is a philosophy and way of thinking that should permeate every activity in the business system in order to obtain a sustainable output, product, or service, created in a process that will minimally endanger the environment and its stakeholders, and maximally contribute to their benefits. Respecting the concept of circular economy and the benefits it can provide, a long-lasting output that can be reused, repaired, and remanufactured should be designed and promoted.

Many organizations perceive a commitment to the concepts of sustainable development and circular economy as critical to their future survival, growth, and development. The imperative today is to look for ways to conduct business activities that would ensure long-term business stability while also taking care of the benefits of a circular economy, the well-being of the community, and environmental protection. Looking for ways to establish a long-term stable business in unstable circumstances, organizations explore the postulates of sustainable development and ways to integrate it at all levels of management. Integrating the concept of sustainable development into the business system requires flexibility for change, awareness of the impact of business activities on the environment, and a commitment to find answers to social, environmental, and economic issues. The result of this integration at the strategic level of the organization is a strategy of sustainable development, which, like other strategies, needs to be operationalized and implemented. That is achieved through project initiatives. The project objectives should be in line with the strategic goals of the sustainable development of the organization. By managing projects in a sustainable manner, both commercial and projects of wider social significance, more local economic activity could be generated, and relations within communities strengthened. A sustainable project result that can be reused, repaired, and remanufactured could contribute to environmental protection and to the well-being of society.

Sustainable project management is a mechanism that should help the organization transform its business by implementing a strategy of sustainable development, which should result in creating value both for the organization and for society. In order to explore the possibility of integrating the sustainability principles into the project management concept to directly contribute to the project's success and indirectly to the development of previously mentioned values, based on literature reviews, significant factors were identified and a conceptual framework for enabling benefits from linking sustainability and project management was developed (Fig. 10.1). It was decided which sustainability principles are comprehensive enough to examine the possibilities of their integration within the phases important for project management in a sustainable way by applying practices, tools, techniques, and competencies for sustainable project management. The conceptual framework that is presented here is the result of the author's doctoral thesis, and the factors integrated into the framework are defined below.

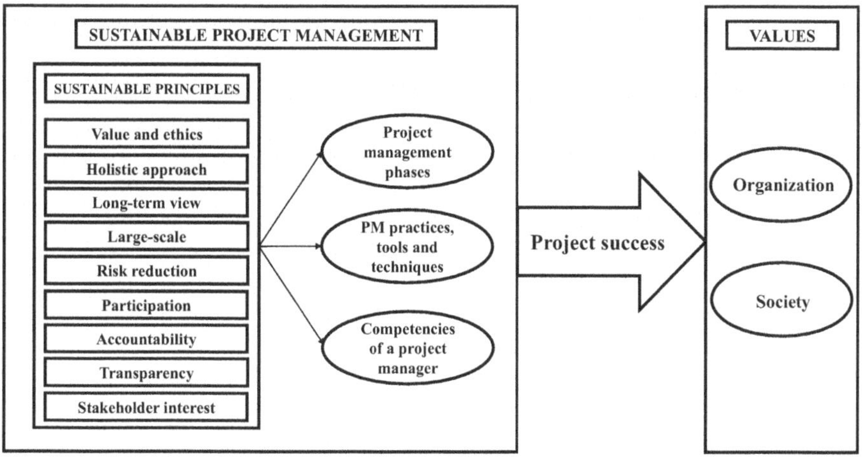

Fig. 10.1 A conceptual framework for enabling benefits from linking sustainability and project management (Toljaga-Nikolić, 2022)

10.2 Sustainable Development and Sustainable Project Management

10.2.1 Sustainable Development

At the end of the twentieth century, the world finally faced the consequences of decades of intensive development, which led to excessive consumption of energy from non-renewable sources and, in return, did not lead to the equal well-being of society at the global level. Additionally, the consequences have become visible in the form of a polluted environment and endangered flora and fauna. At that time, the concept of sustainable development became one of the most important thoughts in society and the business world, because facing an uncertain future, in which the further development of humanity was arguable, led individuals and institutions to address the issue of sustainable development. The highlighted goal was that future generations need to have at least equal, if not greater, opportunities to achieve their development (Toljaga-Nikolić, 2022). The transition from the linear to the circular economy was found to be essential, where the circular economy was seen as a sustainable alternative, compatible with the inherent interests of organizations and capable of coping with many challenges (Sariatli, 2017). According to Sharma et al. (2021), government pressure to implement the concept of circular economy cannot be an effective step towards this transition. The authors conclude that the circular economy is a holistic approach that can contribute to reaching sustainability goals, but needs many prerequisites towards implementation, like strong management will, technology innovation, training, and motivation to support the employees.

Laying the groundwork, the World Commission on Environment and Development published a report in 1987 entitled "Our Common Future", where sustainable development was defined as "development that meets the needs of the present without compromising the ability of future generations to meet their own needs". The report highlighted three perspectives on sustainability: social, ecological, and economic perspectives, since without the simultaneous contribution to progress and development in all three fields, sustainable development cannot be achieved. One of the first authors who highlighted the dimensions of sustainable development and pointed to their role was Elkington (1997), who presented the concept of 3P – "People, Planet, Profit", and explained that the concept of sustainable development is based on a balance of these three dimensions: social equality, environmental protection, and economic prosperity. Dyllick and Hockerts (2002) state that establishing a balance between economic growth and social well-being at the macro and micro levels has been a political and managerial challenge for more than 150 years. Kates et al. (2005) go beyond and analyze the definition of sustainable development when they conclude that it is a system that supports nature, the environment, the individual, society, and the community. According to Oehlmann (2010), organizations that choose sustainable business as their strategic direction and integrate the sustainability concept into their business system will provide numerous benefits in terms of creating sustainable values, improving performance, increasing efficiency and effectiveness, achieving greater flexibility in business and decision making, improving the reputation of the organization, and many other benefits. It is important to emphasize that an organization that integrates the concept of sustainability into its business system and deals with all three dimensions of sustainability equally is ready to take responsibility for the impact of its business activities on customers, employees, management, the community, and the environment.

When it comes to the environment, it is important to address and take the necessary activities to ensure that the project processes and activities, as well as the project result, will not have harmful effects on the environment. These efforts are related to the decisions about the energy sources and resources that will not endanger the sources for the future, the type of transport and transport routes to minimize emissions, as well as dealing with waste disposal, recycling, and other activities that are useful for environmental protection. These have to be seriously considered since Petrović et al. (2017) stated how many natural disasters we are witnessing around us and globally. We are facing many challenges today because nature is returning to the way we treated it and will continue to do so. The concept of circular economy is important because of its ability to connect both the business and policy-making communities to sustainability work (Korhonen et al., 2018) and help in coping with these challenges. Since the general goal of sustainable development is to achieve long-term stability of the economy, which will consequently mean well-being in society as well as preservation and environmental protection, this goal is perceived as achievable through the integration of the sustainability concept in all spheres, while emphasizing the importance of resolving social, environmental, and economic issues during the decision-making process (Toljaga-Nikolić, 2022).

10.2.2 Principles of Sustainable Development and Sustainable Project Management

In order to develop a conceptual framework for sustainable project management (Fig. 10.1), which should contribute to the project's success and indirectly create value for the organization and society, it is important to integrate the concepts of sustainable development and project management. Labuschagne and Brent (2006) were among the first authors who wrote about the integration of sustainability principles into project lifecycle management in the manufacturing industry. Having in mind that sustainability is long-term and future-oriented, they recommended that it is necessary to consider not only the complete project life cycle, from initiation to closure, but also the project result and its subsequent impact. Related to that, they made a distinction between the life cycle of the product or service and the life cycle of the resources used in the processes in which the project result was created.

In the literature, one can find papers that differ in the type and number of sustainability principles that need to be integrated into project management. Agarval and Kalmar (2015) analyzed the available papers from which they pointed out eight sustainability principles that were most often cited as important for integration into project management: establishing balance or harmonization of social, environmental, and economic interests; short-term and long-term orientation; local, regional, and global orientation; values and ethics; transparency and accountability; stakeholder participation; risk reduction and consuming income and not capital. In order to manage the project in a sustainable manner, which means taking into account all short-term and long-term impacts of project results and actively involving all stakeholders, it is necessary to expand the focus from the existing, short-term orientation (traditional project management) to the long-term orientation, which starts with the programming of project goals so that they are in line with the strategies and sustainable development goals. The impact on society and the environment does not end with the delivery of project results but continues into the future, often years after the project closure (Toljaga-Nikolić, 2022).

Sustainable business requires that the organization is well acquainted with its local, regional, and global business environment, and takes all advantages. Often, the great potential for sustainable business is in the local environment, whether that be renewable energy sources or local producers, whose products and services can be included in the organization's processes at significantly lower cost and satisfactory quality. Using the potentials of the local market, the organization at the same time encourages the prosperity of local entities, achieves savings due to lower transport costs, and protects the environment due to shorter transport routes and lower concentrations of harmful gasses (Toljaga-Nikolić, 2022).

In addition to the above-presented classification of sustainability principles (Agarval & Kalmar, 2015), special attention during the development of the conceptual framework (Fig. 10.1) is given to the classification presented by Goedknegt & Silvius (2012), due to its comprehensiveness. The authors conducted the research and listed the following nine principles of sustainability, which is also the largest number of principles that have been pointed out, so this classification was integrated into the conceptual framework (Fig. 10.1) (Goedknegt & Silvius, 2012):

1. *Value and ethics*: An organization should behave ethically in its business, base its business on values such as honesty, equity, and integrity, and actively promote ethical behavior.
2. *Holistic approach*: In order to contribute to sustainable development, an organization should meet all three dimensions of sustainability: social, environmental, and economic. The dimensions of sustainability are interrelated, that is, they affect each other in different ways.
3. *Long-term view*: An organization that operates sustainably should consider both the short-term and long-term effects arising from the implementation of its business activities. This principle focuses on the entire lifespan of the project results.
4. *Large-scale*: Ecological, social, and economic processes that affect our well-being take place simultaneously at various spatial and temporal scales. In order to efficiently address these interlinked processes, sustainable development requires a coordinated effort that takes place at multiple levels, from the global to the regional and the local.

5. *Risk reduction*: The understanding that in the interactions in the environment-society system, characterized by complexity, uncertainty, irreversibility, and nonlinearity, it is more efficient to prevent damage than to mitigate it, has led to the formulation of the so-called precautionary principle. It is becoming increasingly evident that decision-making about complex systems in the conditions of uncertainty, ambiguity, or ignoring them is a significant challenge for how we produce, distribute, and use knowledge.
6. *Participation*: Involving individuals from the social environment in projects that can potentially affect their lives is based on one of the key principles of sustainable development, and that is stakeholder participation. Sustainable development requires a process of dialogue and ultimately reaching a consensus of all stakeholders as partners who together define the problem, design possible solutions, cooperate in their implementation, and monitor and evaluate the outcome. It encourages social and individual learning, which enriches both society and the individual, reduces the uncertainty caused by insufficient scientific knowledge, supports implementation, and mitigates conflicts.
7. *Accountability*: Accountability is logically related to the proactive stakeholder engagement described above. This principle indicates that the organization accepts responsibility for the implementation of its policies, decisions, and actions and the resulting impact on society and the environment. The principle also implies that an organization accepts this responsibility and is willing to be held accountable for these policies, decisions, and actions.
8. *Transparency*: An organization operates openly when it comes to its policies, decisions, and actions, including the effects of those policies and actions on society and the environment.
9. *Stakeholder interest*: An organization should respect international standards of behavior while adhering to the principle of respect for the rule of law. An organization should respect human rights and recognize their importance and universality.

These principles of sustainability were used in developing the conceptual framework (Fig. 10.1), where they are integrated throughout the project management phases in order to examine the impact of sustainable project management on project success and the development of values for organizations and society. The integration should be carried out using project management practices, tools, and techniques, as well as the competencies for sustainable product management.

Integrating the concept of sustainable development into the business system means integrating this concept with other management concepts in the organization so that sustainability can be operationalized and can permeate the entire business system and be integrated into strategies, goals, and projects (Toljaga-Nikolić, 2022). Strategies and strategic changes in organizations are most often implemented through project implementation, and the organization's decision to operate sustainably must be incorporated at all management levels. This is supported by the conclusion of several authors that, if we understand the project as a mechanism for implementing the organization's strategy, then the introduction of the concept of sustainable development at the operational level of project management is crucial for successful strategy implementation (Aarseth et al., 2017; Gareis et al., 2009).

Considering projects as an instrument for achieving sustainable development of the organization and society (Magano et al., 2021), it is necessary to ensure compliance and sustainability of business goals of the organization with global goals of sustainable development, including all sustainability dimensions: social, environmental, and economic. Armenia et al. (2019) considered sustainable project management as a management practice where the project objectives can be achieved by maximizing social, environmental, and economic benefits. According to Schipper and Silvius (2017), linking the concepts of sustainable development and project management leads to fundamental changes in thinking, operations, cooperation, and partnerships at different levels of business and organization. It is important to emphasize here that the implementation of sustainable project management requires flexibility and readiness to introduce and accept continuous changes at the project level, precisely because of the long-term orientation.

Deland (2009) stated that sustainable project management is about minimizing the resources used in every project phase, while Hope (2012) discussed in more detail and emphasized the importance of protecting, sustaining, and enhancing the human and natural resources necessary for achieving a project result and for future generations as well. Different dimensions of sustainability were analyzed and related to sustainable project management. Tam (2010) concluded that having positive economic, environmental, and social impacts within the project management processes contributes to a sustainable society. According to Silvius and Schipper (2014), sustainable project management considers all three dimensions of sustainability within the project life cycle in order to provide benefits for stakeholders that have to be proactively involved. Taking all these into consideration, it can be concluded that the authors emphasized the importance of integrating social, environmental, and economic dimensions of sustainability into the project management concept, as well as the role of stakeholders and their active participation in the project and creating benefits for both organizations and other stakeholders.

When managing a project in a sustainable way, the number of stakeholders increases because the time horizon for which the impacts of project results are identified is extended, as well as the number of dimensions important for management. So in addition to the current traditional economic dimension, equal attention is given to the social and environmental sustainability dimensions. Collaboration with the stakeholders should be transparent, fair, and ethical because of the large number of stakeholders in the project, whose agreement and acceptance of goals, plans, decisions, and influence is crucial for the successful outcome of sustainable project management (Silvius & Schipper, 2014). Importance is given to the project life cycle, which is more long-term in sustainable project management and therefore more uncertain and more susceptible to changes in its specific segments. According to Toljaga-Nikolić (2022), methods and techniques that can be used to integrate sustainability principles into the project management processes were previously rarely considered and therefore included within the conceptual framework for sustainable project management (Fig. 10.1).

10.3 Project Success and Values for the Organization and Society

10.3.1 Critical Success Factors, Success Criteria, and Key Performance Indicators

Identifying and managing critical project success factors is important, whether the project is managed in a sustainable manner or not. This contributes to the project's success, which is, as well as the failure of the project, assessed using the success criteria. Project success criteria have their own indicators called the key performance indicators, which are a set of measurable data for measuring project performance in the implementation phase. Milošević and Patanakul (2005) stated that critical success factors can have a significant impact on the project's success if they are managed adequately. There is a difference in the literature between project management success and project success (Baccarini, 1999; Shenhar et al., 1997; Munns & Bjeirmi, 1996; De Witt, 1988). Munns and Bjeirmi (1996) pointed out that there is a difference between how projects and project management are observed, where the project is aimed to achieve a specific goal by conducting specific activities that require resources, while project management is a process that requires a set of tools and techniques for managing the achievement of a specific project goal. It is concluded that the success of project management is measurable during the project and at the end, where it is important whether the goals are met in terms of time, budget, and quality, which is a short-term orientation in the project and is considered immediately during and after the project's completion.

Project success is oriented on the long-term effects and results related to meeting the expectations of clients and other stakeholders, as well as the impact of project results on society and the environment. This is a direct link to the principle of sustainability in project management, which refers to a long-term orientation. In addition, Pinkerton (2003) believes that if the result of the project is not successful, i.e., it did not meet the expectations of the client, the project was not successful. Since sustainable project management is characterized by the active involvement of a higher number of stakeholders due to the integration of sustainability principles related to establishing a balance of social, environmental, and economic interests, long-term orientation, and local, regional, and global orientation (Eskerod & Huemann, 2013), it is necessary when considering the satisfaction of stakeholders in the project to take into account the success criteria that include all sustainability dimensions.

Related to the social dimension of sustainability of a project managed in a sustainable manner, from the perspective of the client, it is important that the project result, above all, meets the needs and expectations of the client and provides added value and a certain level of innovation. The sustainable project outcome is always observed from the aspect of environmental impact, which can be manifested during the creation of results and also later during their use. As a result, it is critical for the client's satisfaction and perception of success that the project result did not have a negative impact on the environment. This fact significantly positions the result of the project in the minds of stakeholders as successful. Of course, the economic dimension of sustainability must not be neglected, within which the relationship between the market price of project results and the value received by the client is evaluated, and if this relationship is satisfactory for the client, the project result and the project are considered successful (Szabo, 2016). Also, in order to monitor the project's success and direct its progress towards meeting the project goals and indirectly defined strategic goals of the organization, metrics known as key performance indicators (KPI) are used. Kerzner (2017) points out that, in the context of the project, the key performance indicators that are selected and managed are directly related to the success or failure of the project, while Mir and Pinnington (2014) believe that key performance indicators are the most important variables that determine the project's success.

When selecting the KPIs, it is important to ensure that they are (Hristov & Chirico, 2019): in correlation with the strategic goals of the organization; significant, in terms of contributing to the presentation and explanation of the value creation process; and reliable, comprehensive, consistent, and comparable. The literature review highlighted key performance indicators by sustainability dimensions, so that all sustainability dimensions are included. The list clearly points to the conclusion about the complexity of sustainable project management and the responsibility that the project manager has when analyzing, selecting, and managing indicators in order to increase the chance of project success (Toljaga-Nikolić, 2022):

1. Social dimension:
   - Health and safety at work
   - Equality and human rights
   - Ethical behavior
   - Stakeholders satisfaction
   - The well-being of society and the local community
2. Ecological dimension:
   - Gas emissions
   - Transportation
   - Use of energy from renewable sources
   - Use of renewable resources
   - Environmental certification, environmental standards
   - Amount of waste per unit of project results and waste management
3. Economic dimension:
   - Income, profit, return on investment
   - Productivity
   - Costs related to energy consumption
   - Costs of using resources from renewable sources
   - Investment in environmental technology
   - Waste management costs
   - Additional revenue from recycling

When considering the transition to the circular economy and its implementation, Schöggl et al. (2020) concluded that it encompasses a limited number of environmental aspects (waste, resource use, and CO2 emissions), while other environmental and social aspects are not well examined. The conceptual framework for sustainable project management (Fig. 10.1) was created to investigate the impact of sustainable project management on project success and, indirectly, value creation for organizations and societies. Because of that, it is important to consider how the success of a sustainable project is observed and measured, as well as what values are important for the organization and society.

10.3.2 Values for Organization and Society

In order to contribute to the fulfillment of the sustainable development goals, an organization needs to develop strategies, set goals, and prepare plans and projects. Developed strategies are implemented through the projects, and the integration of sustainable principles into project management creates the basis for meeting the goals and creating values for the organization and society. It contributes to business effectiveness because the establishment of a link between strategy, goals, and projects ensures the coherence and purposefulness of all efforts undertaken with the intention of implementing a sustainable development strategy. The concept of circular economy contributes to the sustainability of the project results because the organization has the ability to create the results using energy from renewable sources and materials that bring savings. This concept emphasizes the limitation of the consumption and waste of resources (energy, raw materials, water), as well as the waste production that will increase efficiency. By an effective and efficient implementation of its business goals and plans, the organization provides a result that meets the stakeholders' needs and, as such, will be accepted to a greater extent and provide higher profits, which is one of the organizational values, because profit creates the basis for the survival, growth, and development of the organization (Haffar & Searcy, 2017; Guenster et al., 2011; Jacobs et al., 2010; Gray, 2006; Salzmann et al., 2005). Integration of the social dimension of sustainability into business results in a pleasant and safe work environment where employees and their needs are cared for and continuously improved. Satisfied employees contribute to process efficiency and productivity growth (Lewin & Minton, 1986). Organizations that operate sustainably and based on the pillars of the circular economy are perceived by the environment as organizations that care about their stakeholders, community, and the environment. The achievement of concrete results that confirm it creates the conditions for the growth of the organization's competitive advantage and an improved image (Haffar & Searcy, 2017; Bansal & DesJardine, 2014; Jacobs et al., 2010; Chen, 2001; Menon & Menon, 1997). Due to their comprehensiveness and long-term orientation, sustainability dimensions that are aligned with business strategies, goals, projects, and plans require detailed analysis, testing, assessment, and monitoring. Through the acquisition of knowledge about the sustainability dimensions, their scope, and impacts, individuals and organizations have the opportunity to develop and learn. Organizational learning is one of the values created for the organization through sustainable business and sustainable project management (Sydow et al., 2004; Hobday, 2000). In addition to the assumption that sustainable business and sustainable project management contribute to the achievement of organizational values, certain values for society can be pointed out.

By integrating the sustainability concept into strategies, goals, projects, and plans, the organization also strives to provide results that will create value for the wider environment. In addition to the previously described values for the organization and the economic dimension of sustainability, where dealing with its elements contributes to business profitability, savings, and economies of scale, the elements of social and environmental dimensions are increasingly the focus of interest. A significant contribution to community well-being is achieved through projects that support local initiatives and involve groups and individuals from the local communities, especially in situations where the project result has a direct or indirect impact (Camilleri, 2017; Wong, 2012; Mansuri & Rao, 2004), but these projects require careful planning and implementation of monitoring and evaluation. If there is a possibility of employment within the planned business initiatives (Gallie et al., 2003), a significant contribution can be made to reduce the unemployment rate and to improve the living standards in the local community. Organizations committed to sustainable business care about the well-being of society and analyze and react preventively so that their business processes do not lead to harmful impacts on public health and safety and undermine the trust and intention of the organization to operate transparently (Helne & Hirvilammi, 2015; Rogers et al., 2012; Yamaoka, 2008). According to Schöggl et al. (2020), the social dimension of sustainability is still underrepresented in the concept of circular economy. This leads to the conclusion that this concept can be more widely included to obtain the environmental and economic benefits. In a situation where the processes in the organization can be adapted to the use of resources from renewable sources, the decision to contribute to sustainable development goals and preserve the energy sources for future generations is of great importance (Zaharia et al., 2017; De Marchi, 2012). In addition, by continuously monitoring the consumption of resources, it is necessary to minimize losses and waste. In order to protect and maintain the integrity of ecosystems, plant and animal species, and human species, resources must be used sparingly through the efficient management of natural resources (Moldan et al., 2012).

10.4 Sustainable Project Management Phases

In order to develop a conceptual framework (Fig. 10.1), it was important to examine and define the project management phases in order to integrate the sustainability principles by applying practices, tools, and techniques, as well as the competencies for sustainable project management. The project manager and the project team develop and apply certain competencies and use practices, tools, and techniques for project management in order to realize the sustainable project goals and obtain a sustainable result. In order to define the phases of sustainable project management, it was considered which phases and processes of project management were presented by other authors in their papers and which were recognized in certain methodologies. The analysis was conducted in the context of their support for the integration of the sustainability concept, in order to highlight the phases of project management that are considered to be adequate support for the integration of the sustainable principles. Table 10.1 shows the phases, processes, and groups of project management processes, identified within the existing methodologies and in the papers of internationally recognized authors in the field, and then their analysis and

Table 10.1 Phases, processes, and groups of project management processes

Sources	Phases, processes, and groups of project management processes
PRINCE2 (2017)	Starting up a project Initiating a project Directing a project Controlling a stage Managing product delivery Managing stage boundaries Closing a project
PMBOK (2017)	Initiating Planning Executing Monitoring and controlling Closing
PCM – Project cycle management	Programming Identification Formulation Financing Implementation and monitoring Evaluation
European Commission PM^2 (2018)	Initiating Planning Executing Closing Monitor and control
Fangel (2018)	Project preparation Project start-up Manage project execution Project close-out
Kerzner (2009)	Planning Scheduling Organizing and Staffing Monitoring Controlling Directing

finally the identification of the phases for sustainable project management, which are integrated into the sustainable project management conceptual framework (Fig. 10.1).

For the purpose of developing the conceptual framework (Fig. 10.1), it was analyzed to what extent the existing classifications of project management processes and phases show the possibility for the integration of the sustainable principles. It was important to decide on the project management phases to be integrated into the conceptual framework. The PRINCE2 methodology was concluded to be flexible enough to allow the process to be adapted to the project's needs and to decide to what extent each process will be applied (PRINCE2, 2017). This is important for the integration of the concept of sustainability, which requires flexibility in processes. Project initiation includes harmonization of project goals with the strategic goals of the organization, which is one of the key moments in sustainable project management when the project is planned, initiated as a means to implement the strategy of sustainable development in the organization. It has a programming role, which provides answers to strategic questions about the project, its priority, and importance, and indicates the basic motivation for its implementation. The scope of work that needs to be done in order to implement the project is considered, and since all dimensions of sustainability are integrated through sustainable project management, project planning and directing have a more complex scope of work, which is why monitoring and control activities are especially important (Toljaga-Nikolić, 2022). Project management aims to establish and maintain progress in the project from the beginning of the project to the end, which is helped by the fact that the methodology supports regular progress monitoring of the project plan and continuous monitoring of the quality of project results. In sustainable project management, changes during the implementation of the plan are expected, so the monitoring and control activities that permeate all processes are necessary, and this was taken into account when deciding on project management phases in a sustainable way in developing a conceptual framework. The result of the project is developed in stages, with the control being carried out in the project stages in order to monitor progress and respond in a timely manner (PRINCE2, 2017). Even if the project is not completed but interrupted, project closure activities are carried out, and in the case of completion according to plan, project closure includes delivery of project results and acceptance by stakeholders, preparation of final project documentation, evaluation of the results, and release of the resources.

A process-oriented project management approach, according to the PMBOK Standard (2017), defines five groups of processes that are the result of logical grouping to achieve specific project objectives. Initiating processes and activities in the project, as in the PRINCE2 methodology, relates to establishing a link between the project and the organization's strategy, which aims to position the project as a means of implementing the strategy. This step is extremely important in sustainable project management because the project must be programmed to contribute to sustainable development, analyzing the long-term impacts and including in the preliminary scope of the project all sustainability dimensions. This preliminary scope of the project is elaborated in detail within the group of Planning process. Addressing the sustainability dimensions within the planning activities provides excellent opportunities for selecting resources for processes, preserving resources for future generations, and processing selected resources to minimize or eliminate the possibility of environmental hazards. According to Toljaga-Nikolić (2022), the selection of "green" materials and "green" technologies protects the environment, preserves resources, contributes to the well-being of society, and at the same time positions the organization as responsible to society and the environment. The implementation of the prepared project plans and the development of the project results takes place through a group of Execution processes, with continuous Monitoring and control, in order to ensure that the project result is in line with the expectations of stakeholders (PMBOK, 2017). Through the Closing Process group, the results of the project are handed over and accepted by the client, with the formal closing of the project and the archiving of the lessons learned. Both standards (PRINCE2 and PMBOK) include the closure phase, which was taken into account when deciding on the phases of sustainable project management in the preparation of the conceptual framework, because the closure phase is important for sustainable project management due to the acceptance of project results by stakeholders (Toljaga-Nikolic, 2022).

The PM2 methodology has been developed by integrating the elements of globally accepted standards and methodologies (PMBOK and PRINCE2), which makes it a process-oriented methodology. It builds on best practices in project management and provides significant flexibility in implementation and a willingness to support changes. It inherited the PCM methodology, within which the project management phases are important for the integration of sustainability principles. According to Toljaga-Nikolić (2022), the analysis of the PM2 methodology concluded that stakeholder management, the practice of collecting and archiving lessons learned, as well as the business implementation segment, are potential segments to support the integration of sustainability principles during project management phases. The methodology shows the four phases that each project goes through the initial, planning, executing, and closing phases, with monitoring and control being carried out continuously throughout all four phases. In the initial phase, there is a very significant scope of work in the context of sustainability, and it relates to the alignment of the project and strategic objectives, which is a good basis to integrate sustainability principles into project management. Potential for the integration of sustainability also exists in the segment of business implementation, which is planned, implemented, monitored, and controlled. In the initial phase, framework planning is carried out, which corresponds to the programming and initiation, and later, through the planning phase, detailed project plans are developed. During the implementation phase, the project results are generated, followed by administrative closure, lessons learned, and business implementation, which should ensure that the project results are effectively integrated into the organizational environment after the project is completed, i.e., to implement project results into the system (European Commission, 2018). At the same time, monitoring and control are carried out in all project management phases, but mostly in the implementation phase, which is important because the project result is developed there.

Fangel (2018) defines four phases of project management, which partially coincide with the classifications of processes and phases previously presented. Project preparation is a phase that, in terms of its scope, corresponds to the previously presented initiation processes. Before the implementation starts, the project is analyzed in terms of priorities and alignment with the strategy. This requires verification of the project idea, defining the framework scope of work, and the general agreement of stakeholders to move further with the project idea. This classification (Fangel, 2018) does not explicitly point out the planning phase, but preparation activities also include planning activities. This was taken into account when deciding on the phases of sustainable project management because the previous two classifications (PMBOK, PM2) have a separate planning process, which is important in project management. The next phase in the classification is the project start, which means starting, leading the project at the beginning, and then moving on to the next execution phase, which includes the project evaluation activity. Monitoring and control during the implementation were not explicitly stated. The last phase in the classification is the closure of the project, so it can be concluded that all classifications presented so far include an explicitly separated phase of closure, where in this case, the focus is on accepting project results and lessons learned (Toljaga-Nikolić, 2022).

According to Kerzner (2009), project management begins with the development of project plans based on defined client requirements. It is important to note here that the author does not state the initial alignments of the project with the strategy, which was taken into account when concluding on the phases of sustainable project management in the conceptual framework (Toljaga-Nikolić, 2022). The outcome of the planning process before the start of the project is detailed time and resource plans, organized and with allocated resources. The processes of the project plan execution are necessary in order to provide the project result, which is represented in all the classifications presented so far, as well as the processes of monitoring and controlling the progress of the project implementation. The monitoring aims to compare the achieved with the planned to identify possible deviations and the impact that may arise, and the control takes the necessary measurements and proposes corrective actions. Kerzner (2009) does not explicitly state the project closure and evaluation activities, which were presented previously (PRINCE2, PMBOK, PM2, Fangel), and were taken into account when deciding on the sustainable project management phases.

The compared analysis of previously presented phases, processes, and groups of project management processes, defined by internationally recognized authors and institutions in the field of project management, provided four project management phases which are included in the conceptual framework (Fig. 10.1) (Toljaga-Nikolić, 2022):

1. Programming
2. Planning
3. Execution and monitoring
4. Closing and evaluation

According to the classifications that have been shown in Table 10.1, programming activities are not significantly presented as particularly important in sustainable project management due to long-term orientation. Through the analysis of the current situation at the national and sectoral level, it is necessary to determine and establish long-term priorities, explore available energy sources, establish long-term cooperation with stakeholders, consult the competent institutions, etc. It is necessary to program project goals and align them with strategic goals, give a framework of project work, and define and harmonize the expected project results and their impact on the social, environmental, and economic dimensions of sustainability, all of which should be accepted by the project stakeholders. The *Programming phase* provides a framework for the next *Planning phase*, in which concrete and feasible project plans will be defined. In the planning phase, the scope of the project is defined in detail, the tasks and activities that need to be realized in the project are determined, and precise and clear project plans are defined. These plans are approved after finalization. It is important that the plans can be changed later during the project execution, if necessary, which contributes to the flexibility in sustainable project management. Project risks have to be identified and analyzed in order to develop risk response plans. The extent to which flexibility and implementation of risk analysis are important in sustainable project management is indicated by the fact that the specific competencies related to flexibility and risk analysis are considered important for the project manager and project team members. In the *Execution and monitoring* phase, project plans are implemented, with project performance reports for stakeholders. If necessary, changes are made and project plans are harmonized so that the project result can be accepted and approved by the client, which is also a measure of the project's success. The *Closure and evaluation phase*, in addition to the administrative project closure based on the final approval of the project owner and the client's acceptance of the final results, includes lessons learned and evaluation, in order to examine and assess the impact of project results on sustainability dimensions as well as the future impacts that the project result can have on society and the environment.

10.5 Practices, Tools, and Techniques for Sustainable Project Management

In order to implement the sustainable development strategy, organizations integrate the concept of sustainability into other management concepts, including the project management concept, thus developing sustainable project management. The strategic goals are operationalized through the project goals, and their implementation requires planning and monitoring the implementation of project plans. Integrating sustainability into project management requires the use of practices, tools, and techniques to ensure that sustainability permeates all phases of project management, from programming, through planning, execution, and monitoring, to closure and evaluation. Practices, tools, and techniques should help the project manager, who is responsible for integrating the concept at the project initiative level and the results obtained at the end of the management process, to more easily manage the project in a sustainable manner. The principles of sustainability are integrated into the project management phases by applying certain practices, tools, and techniques that project managers have already used, but now find application in sustainable project management (Toljaga-Nikolić, 2022).

While developing a conceptual framework for sustainable project management, practices, tools, and techniques have been analyzed and integrated as one of the important factors that could contribute to the project's success when managing a project in a sustainable manner. In organizations that operate sustainably and implement a sustainable development strategy, the concept of sustainability permeates all levels of management, and integration with the concept of project management is an essential step in these efforts of the organization, motivated by internal and external drivers (Toljaga-Nikolić, 2022). As previously described, the concept of sustainability permeates all phases of project management through the integration of sustainability principles, for which it is necessary to select and apply appropriate practices, tools, and techniques in project management. Based on the literature review, it is concluded that the application of the following practices, tools, and techniques creates the possibility of integrating the principles of sustainability through project management phases (Table 10.2). It requires further analysis and confirmation through empirical research (Toljaga-Nikolić, 2022).

Practices, tools, and techniques were chosen as a means for further operationalization of the concept of sustainable development to the operational level, which provides the basis for the implementation of planned activities and the realization of project goals that could contribute to realizing the strategic sustainable development goals. These practices, tools, and techniques have been integrated into the

Table 10.2 List of practices, tools, and techniques for sustainable project management

Practices, tools, and techniques for sustainable project management	Sources
Project charter/Canvas	Proença (2019), Kohl (2016)
Balanced scorecard	Rabbani et al. (2014), Dias-Sardinha and Reijnders (2005), Möller and Schaltegger (2005), Bieker (2003), Bieker and Waxenberger (2002), Figge et al. (2002), Kaplan and Norton (1992)
Project risks analysis	Wang et al. (2014), Fernández-Sánchez and Rodríguez-López (2010)
Impact analysis	GPM (2019), Chawla et al. (2018), Wang et al. (2014)
Earned value method	Koke and Moehler (2019)
Evaluation of project results	Økland (2015), Sánchez (2015), Wang et al. (2014), OECD (2012)
Lessons learned	Eken et al. (2020), Hansmann (2010), Sydow et al. (2004), Hobday (2000)

conceptual framework of sustainable project management as a significant factor in order to further investigate whether they enable the integration of sustainability principles into the project management phases. The various standards that provide guidance to organizations, regardless of their size and industry, are described in a practical guide to integrate sustainability principles and practices into business processes, strategies, procedures, systems, and organizational structures. An example is ISO 26000:2010 – Guidelines on Social Responsibility (Silvius et al., 2012), as a standard adapted for application in all organizations, in which the principles of social responsibility are in line with the ideas of sustainability. The standard covers the elements that consider the social dimension of sustainability: accountability, transparency, ethical behavior, respect for stakeholders, respect for laws, respect for international standards, and respect for human rights. The standard helps the organization direct its business through adherence to the guidelines covered by the standard and to establish processes in accordance with the given recommendations. Achieving the results should be followed by respect for human rights, good and fair business practices, the environment, consumer and social protection, and others that can help in developing the community. Also, the P5 standard (GPM, 2019) is recognized in practice as a standard that brings a practical tool for integrating sustainability principles into project management, defining what a sustainability management plan should look like and how the results should be analyzed to improve the project outcome from a sustainability perspective.

10.6 Competencies of a Project Manager for Sustainable Project Management

The integration of sustainability principles into project management phases requires competent project managers and members of project teams who, using their knowledge, experience, and skills, select and apply practices, tools, and techniques, thus enabling the integration of sustainability principles. Sustainable project management has its own specifics that distinguish it from previous project management practices, which stem from the integrated principles of sustainability. The development of project plans is preceded by an analysis of all sustainability dimensions and the identification of possible impacts on society and the environment through project processes and project results. Sustainability is associated with a long-term orientation, which increases uncertainty in decision-making. Also, there is an increasing number of stakeholders who need active communication, so it is obvious that the project manager who sustainably manages the project faces many challenges (Toljaga-Nikolić, 2022).

Over the years, the authors have been researching the roles, responsibilities, and competencies of professionals in the field of sustainable project management (Toljaga-Nikolić et al., 2016, 2020; Todorović & Obradović, 2018; Taherdoost, 2018; Banihashemi et al., 2017; Silvius, 2016; Tabassi et al., 2016), which indicates the importance of this factor and the need to be integrated into the conceptual framework (Fig. 10.1). In addition to working on the development of personal competencies for sustainable project management, the project manager and project team members should motivate other participants and provide an example of their own responsible and honest behavior. The integration of sustainability into project management is challenging for project managers, who need to take on new responsibilities and develop new competencies. They are now responsible for the organization's long-term development in addition to project results.

In the literature, a number of competencies, roles, and responsibilities of the project manager and members of the project team for sustainable project management are represented. In order to examine and conclude whether sustainable project management requires the development and application of specific competencies of project managers and project team members, competencies were selected and will be further examined through empirical research in the field of sustainable project management. The selection was made in order to provide a visible connection with the principles of sustainability, which can be integrated through the phases of project management when these competencies are developed and applied (Szabo, 2016; IPMA ICB4, 2015; Hwang & Tan, 2012; Nixon et al., 2012; Maltzman & Shirley, 2010; Lechler, 2000):

- Adaptability to change
- Ethics at work
- Fair and responsible behavior
- Responsibility for sustainable development in the organization
- Communication about sustainable results that need to be achieved through project processes
- Communication on the short-term and long-term social and environmental effects of the project
- Communication with an expanded list of stakeholders due to the long-term effects of sustainable project management
- Awareness of available and sustainable solutions that can be included in the project
- Analysis of risks related to social, environmental, and economic aspects of project sustainability

Sustainable project management changes the degree of responsibility and the set of required competencies of the project manager and other participants in project management. They become responsible for the sustainability of project results, whose short-term and/or long-term impacts on society and the environment should be identified, considered, assessed, and evaluated. By integrating the principles of sustainability into project management, the list of stakeholders in the project becomes more extensive, which makes communication with stakeholders more demanding and intensive. In addition to being competent in terms of possessing knowledge, skills, and abilities for project management, the integration of sustainability principles requires a project manager and project team members to have ethics in work, honesty, and equality (Toljaga-Nikolić, 2022). The abovementioned competencies are integrated into the conceptual framework (Fig. 10.1) in order to examine whether these competencies of the project manager and project team members could help in enabling benefits when they integrate sustainability principles into the phases of sustainable project management.

10.7 A Conceptual Framework for Enabling Benefits from Linking Sustainability and Project Management

A detailed literature review was conducted in order to examine the benefits that can be obtained from linking the concepts of sustainability and project management. Based on that, several important factors have been examined and integrated into the conceptual framework (Fig. 10.1), where the relations among the given factors are clearly presented. The main goal was to examine the impact of sustainable project management on enabling benefits from this linkage. The assumption was that the integration of sustainability principles into the project management phases could have a positive impact on project success and indirectly contribute to the creation of value for the organization and society. The integration would be possible with the application of practices, tools, and techniques for project management and the development and application of specific competencies of a project manager for sustainable project management.

This research is part of the doctoral thesis of the author, where the conceptual framework will be validated through empirical research. The focus is to examine whether the integration of sustainability principles into project management contributes to improving the project success that will lead to the creation of value for the organization and society. Integration should be supported. Therefore, an important part of research is to examine whether sustainable project management requires the application of defined practices, tools, and techniques that enable the integration. According to the literature review, sustainable project management requires the development and application of specific competencies of project managers and project team members, which are also integrated as a factor in the conceptual framework and will be examined.

10.8 Conclusion

Organizations may decide to incorporate the concepts of sustainable development and circular economy when considering the possibility of sustaining economic progress, achieving social and environmental benefits, and promoting added values. This requires more examination in theory and practice in order to conclude how this integration may contribute to sustainable development. When they decide to contribute to the achievement of sustainable development goals with their business activities, organizations agree that these concepts permeate the entire business system and all management levels, from strategic to operational. This change requires the integration of the concept of sustainable development with other management concepts, and one of them is the concept of project management. Within the chapter, a conceptual framework for enabling benefits from linking sustainability and project management is presented, where the influential factors were analyzed and relations among them were established.

Since the project initiatives operationally implement business strategies and translate the goals from the strategic to the operational ones, in the best interest of implementing a strategy of sustainable development and the realization of sustainable business goals, sustainable project management is developed and implemented. Therefore, it was important to analyze how to integrate sustainability principles into project management, what benefits could be achieved, whether it contributes to the project's success, what practices, tools, and techniques enable the integration, and what project manager competencies are necessary for sustainable management of projects. The project result, whose development process was managed in a sustainable manner, is expected to meet the requirements and expectations of the client and other stakeholders, i.e., to meet the profit interests of the organization and to contribute to community welfare and environmental protection. From the perspective of a circular economy, it is important to understand how the resources flow and optimize them through innovations and new solutions. An organization that operates sustainably takes responsibility for the impact of its processes, activities, and results on its stakeholders, society, and the environment. The conceptual framework that is presented within this chapter, with integrated factors and established relations between them, indicates that sustainable project management contributes to the project's success and the development of values for the organization and society. The conceptual framework will be validated through empirical research. A successful project is considered to be one whose results have been accepted by the client and other stakeholders because they have met their expectations and have had a satisfactory impact on society and the environment.

References

Aarseth, W., Ahola, T., Aaltonen, K., Økland, A., & Andersen, B. (2017). Project sustainability strategies: A systematic literature review. *International Journal of Project Management, 35*(6), 1071–1083. https://doi.org/10.1016/j.ijproman.2016.11.006

Agarval, R. S., & Kalmar, T. (2015). *Sustainability in project management – Eight principles in practice*. Umeå School of Business and Economics.

Armenia, S., Dangelico, R. M., Nonino, F., & Pompei, A. (2019). Sustainable Project Management: A conceptualization-oriented review and a framework proposal for future studies. *Sustainability, 11*, 2664. https://doi.org/10.3390/su11092664

Baccarini, D. (1999). The logical framework methods for defining project success. *Project Management Journal, 30*(4), 25–32.

Banihashemi, S., Hosseini, M. R., Golizadeh, H., & Sankaran, S. (2017). Critical success factors (CSFs) for integration of sustainability into construction project management practices in developing countries. *International Journal of Project Management, 35*, 1103. https://doi.org/10.1016/j.ijproman.2017.01.014

Bansal, P., & DesJardine, M. R. (2014). Business sustainability: It is about time. *Strategic Organization, 12*(1), 70–78. https://doi.org/10.1177/1476127013520265

Bieker, T. (2003). Sustainability management with the balanced scorecard. In *Proceedings of 5th international summer academy on technology studies* (pp. 17–34).

Bieker, T., & Waxenberger, B. (2002). Sustainability balanced scorecard and business ethics. In *10th international conference of the greening of industry network*, Göteborg.

Camilleri, M. A. (2017). Corporate sustainability and responsibility: Creating value for business, society and the environment. *Asian Journal of Sustainability and Social Responsibility, 2*(1), 59–74.

Chawla, V., Chanda, A., Angra, S., & Chawla, G. (2018). The sustainable project management: A review and future possibilities. *Journal of Project Management, 3*(3), 157–170. https://doi.org/10.5267/j.jpm.2018.2.001

Chen, C. (2001). Design for the environment: A quality based model for green product development. *Management Science, 47*(2), 250–263.

De Marchi, V. (2012). Environmental innovation and R&D cooperation: Empirical evidence from Spanish manufacturing firms. *Research Policy, 41*(3), 614–623. https://doi.org/10.1016/j.respol.2011.10.002

De Witt, A. (1988). Measurement of project success. *International Journal of Project Management, 6*(3), 164–170. https://doi.org/10.1016/0263-7863/88/030164-07

Deland, D. (2009). *Sustainability through Project Management and net impact. PMI global congress North America*. Project Management Institute.

Dias-Sardinha, I., & Reijnders, L. (2005). Evaluating environmental and social performance of large Portuguese companies: A balanced scorecard approach. *Business Strategy and the Environment, 14*(2), 73–91.

Dyllick, T., & Hockerts, K. (2002). Beyond the business case for corporate sustainability. *Business Strategy and the Environment, 11*(2), 130–141. https://doi.org/10.1002/bse.323

Eken, G., Bilgin, G., Dikmen, I., & Birgonul, M. T. (2020). A lessons-learned tool for organizational learning in construction. *Automation in Construction, 110*, 102977. https://doi.org/10.1016/j.autcon.2019.102977

Elkington, J. (1997). *Cannibals with forks: The triple bottom line of 21st century business*. Capstone Publishing Ltc.

Eskerod, P., & Huemann, M. (2013). Sustainable development and project stakeholder management: What standards say. *International Journal of Managing Projects in Business, 6*(1), 36–50. https://doi.org/10.1108/17538371311291017

European Commission. (2018). *The open PM² Project Management methodology guide of the European Commission – Guide 3.0*. Publications Office of the European Union.

Fangel, M. (2018). *Proactive project management*. Van Haren.

Fernández-Sánchez, G., & Rodríguez-López, F. (2010). A methodology to identify sustainability indicators in construction project management-application to infrastructure projects in Spain. *Ecological Indicators, 10*(6), 1193–1201. https://doi.org/10.1016/j.ecolind.2010.04.009

Figge, F., Hahn, T., Schaltegger, S., & Wagner, M. (2002). The sustainability balanced scorecard–linking sustainability management to business strategy. *Business Strategy and the Environment, 11*(5), 269–284.

Gallie, D., Paugam, S., & Jacobs, S. (2003). Unemployment, poverty and social isolation: Is there a vicious circle of social exclusion? *European Societies, 5*(1), 1–32. https://doi.org/10.1080/1461669032000057668

Gareis, R., Huemann, M., & Martinuzzi, A. (2009). *Relating sustainable development and project management*. IRNOP IX.

Goedknegt, D., & Silvius, A. J. G. (2012). The implementation of sustainability principles in project management. *Proceedings of the 26th IPMA World Congress, Crete*, 875–882.

GPM Green Project Management. (2019). *P5 Standard for sustainability in project management*. USA. Available at: https://greenprojectmanagement.org/

Gray, R. (2006). Social, environmental and sustainability reporting and organisational value creation? Whose value? Whose creation? *Accounting, Auditing & Accountability Journal, 19*(6), 793–819. https://doi.org/10.1108/09513570610709872

Guenster, N., Bauer, R., Derwall, J., & Koedijk, K. (2011). The economic value of corporate ecoefficiency. *European Financial Management, 17*(4), 679–704.

Haffar, M., & Searcy, C. (2017). Classification of trade-offs encountered in the practice of corporate sustainability. *Journal of Business Ethics, 140*, 495–522.

Hansmann, R. (2010). Sustainability learning: An introduction to the concept and its motivational aspects. *Sustainability, 2*(9), 2873–2897. https://doi.org/10.3390/su2092873

Helne, T., & Hirvilammi, T. (2015). Wellbeing and sustainability: A relational approach. *Sustainable Development, 23*(3), 167–175. https://doi.org/10.1002/sd.1581

Hobday, M. (2000). The project-based organisation: An ideal form for managing complex products and systems? *Research Policy, 29*, 871–893.

Hope, A. (2012). *Project management as if the world matters: At the intersection of sustainable development and project management*. Presented at the 12th Northumbria Research Conference. 16–17 May 2012. Northumbria University, Newcastle upon Tyne.

Hristov, I., & Chirico, A. (2019). The role of sustainability key performance indicators (KPIs) in implementing sustainable strategies. *Sustainability, 11*(20), 1–19.

Hwang, B.-G., & Tan, J. S. (2012). Sustainable project management for green construction: Challenges, impact, and solutions. *CIOB Construction Conference, 2012*.

IPMA ICB4. (2015). *Individual competence baseline for project management*. International Project Management Association.

Jacobs, B. W., Singhal, V. R., & Subramanian, R. (2010). An empirical investigation of environmental performance and the market value of the firm. *Journal of Operations Management, 28*(5), 430–441.

Kaplan, R. S., & Norton, D. P. (1992). The balanced scorecard: Measures that drive performance. *Harvard Business Review, 70*(1), 71–79.

Kates, B. R. W., Parris, T. M., & Leiserowitz, A. (2005). What is sustainable development? Goals, indicators, and practice. *Environment, 47*(3), 8–21.

Kerzner, H. (2009). *Project management: A systems approach to planning, scheduling, and controlling* (10th ed.). John Wiley & Sons.

Kerzner, H. (2017). *Project Management metrics, KPIs, and dashboards: A guide to measuring and monitoring project performance*. JohnWiley & Sons.

Kohl, K. (2016). *Becoming a sustainable organization A project and portfolio management approach*. CRC Press.

Koke, B., & Moehler, R. C. (2019). Earned green value management for project management: A systematic review. *Journal of Cleaner Production, 230*, 180–197. https://doi.org/10.1016/j.jclepro.2019.05.079

Korhonen, J., Honkasalo, A., & Seppälä, J. (2018). Circular economy: The concept and its limitations. *Ecological Economics, 143*, 37–46. https://doi.org/10.1016/j.ecolecon.2017.06.041

Labuschagne, C., & Brent, A. C. (2006). Social indicators for sustainable project and technology life cycle management in the process industry. *International Journal of Life Cycle Assessment, 11*(1), 3–15.

Lechler, T. G. (2000). Empirical evidence of people as determinants of project success projects as business constituents and guiding motives. In *Projects as business constituents and guiding motives* (pp. 217–227), Springer. https://doi.org/10.1007/978-1-4615-4505-7_17

Lewin, A. Y., & Minton, J. W. (1986). Determining organizational effectiveness: Another look, and an agenda for research. *Management Science, 32*(5), 514–538.

Magano, J., Silvius, A. G. J., Sousa e Silva, C., & Leite, A. (2021). The contribution of project management to a more sustainable society: Exploring the perception of project managers. *Project Leadership and Society, 2*, 100020. https://doi.org/10.1016/j.plas.2021.100020

Maltzman, R., & Shirley, D. (2010). *Green project management*. CRC Press.

Mansuri, G., & Rao, V. (2004). *Community-based and driven development: A critical review* (Policy research working paper no. 3209). World Bank.

Menon, A., & Menon, A. (1997). Enviropreneurial marketing strategy: The emergence of corporate environmentalism as market strategy. *Journal of Marketing, 61*(1), 51–67.

Milošević, D., & Patanakul, P. (2005). Standardized project management may increase development projects success. *International Journal of Project Management, 23*(3), 181–192. https://doi.org/10.1016/j.ijproman.2004.11.002

Mir, F. A., & Pinnington, A. (2014). Exploring the value of project management: Linking project management performance and project success. *International Journal of Project Management, 32*, 202–217. https://doi.org/10.1016/j.ijproman.2013.05.012

Moldan, B., Janousková, S., & Hák, T. (2012). How to understand and measure environmental sustainability: Indicators and targets. *Ecological Indicators, 17*, 4–13.

Möller, A., & Schaltegger, S. (2005). The sustainability balanced scorecard as a framework for ecoefficiency analysis. *Journal of Industrial Ecology, 9*(4), 73–83.

Munns, A. K., & Bjeirmi, B. F. (1996). The role of project management in achieving project success. *International Journal of Project Management, 14*(2), 81–87. https://doi.org/10.1016/0263-7863(95)00057-7

Nixon, P., Harrington, M., & Parker, D. (2012). Leadership performance is significant to project success or failure: A critical analysis. *International Journal of Productivity and Performance Management, 61*(2), 204–216.

OECD. (2012). *Glossary of key terms in evaluation and results based management*, dostupno na: http://www.oecd.org (datum pristupa: 15.4.2020).

Oehlmann, I. (2010). *The sustainable footprint methodology*. Delft University of Technology.

Økland, A. (2015). Gap analysis for incorporating sustainability in project management. *Procedia Computer Science, 64*, 103–109. https://doi.org/10.1016/j.procs.2015.08.469

Petrović, N., Bošnjak, I., & Nedeljković, S. (2017). Disaster risk reduction for sustainable development goals. *European Project Management Journal, 7*(2), 27–35.

Pinkerton, W. J. (2003). *Project management: Achieving project bottom-line success*. McGraw-Hill.

PMBOK. (2017). *A guide to the project management body of knowledge* (6th ed.). Project Management Institute.

PRINCE2. (2017). *Managing successful projects with PRINCE2* (6th ed.). Office of Government Commerce.

Proença, J. J. C. (2019). Living lab innovation methodologies. *CAD, 13*(2), 1–38.

Rabbani, A., Zamani, M., Yazdani-Chamzini, A., & Zavadskas, E. K. (2014). Proposing a new integrated model based on sustainability balanced scorecard (SBSC) and MCDM approaches by using linguistic variables for the performance evaluation of oil producing companies. *Expert Systems with Applications, 41*, 7316–7327.

Rogers, D. S., Duraiappah, A. K., Antons, D. C., Munoz, P., Bai, X., Fragkias, M., & Gutscher, H. (2012). A vision for human Well-being: Transition to social sustainability. *Current Opinion in Environmental Sustainability, 4*(1), 61–73. https://doi.org/10.1016/j.cosust.2012.01.013

Salzmann, O., Ionescu-Somers, A., & Steger, U. (2005). The business case for corporate sustainability: Literature review and research options. *European Management Journal, 23*(1), 27–36.

Sánchez, M. A. (2015). Integrating sustainability issues into project management. *Journal of Cleaner Production, 96*, 319–330.

Sariatli, F. (2017). Linear economy versus circular economy: A comparative and analyzer study for optimization of economy for sustainability. *Visegrad Journal on Bioeconomy and Sustainable Development, 6*(1), 31–34. https://doi.org/10.1515/vjbsd-2017-0005

Schipper, R. P., & Silvius, A. J. G. (2017). The sustainable project management canvas. *The Journal of Modern Project Management*. https://doi.org/10.19225/JMPM01206

Schöggl, J.-P., Stumpf, L., & Baumgartner, R. J. (2020). The narrative of sustainability and circular economy – A longitudinal review of two decades of research. *Resources, Conservation and Recycling, 163*, 105073. https://doi.org/10.1016/j.resconrec.2020.105073

Sharma, N. K., Govindan, K., Lai, K. K., Chen, W. K., & Kumar, V. (2021). The transition from linear economy to circular economy for sustainability among SMEs: A study on prospects, impediments, and prerequisites. *Business Strategy and the Environment, 30*(4), 1803–1822. https://doi.org/10.1002/bse.2717

Shenhar, A., Levy, O., & Dvir, D. (1997). Mapping the dimensions of project success. *Project Management Journal, 28*(2), 5–13.

Silvius, A. J. G. (2016). Sustainability as a competence of project managers. *PM World Journal, 5*(9).

Silvius, A. J. G., & Schipper, R. (2014). Sustainability in project management: A literature review and impact analysis. *Social Business, 4*(1), 63–96. https://doi.org/10.1362/204440814X13948909253866

Silvius, A. J. G., Schipper, R., Planko, J., van der Brink, J., & Köhler, A. (2012). *Sustainability in project management*. Gower Publishing.

Sydow, J., Lindkvist, L., & DeFillippi, R. (2004). Project-based organizations, embeddedness and repositories of knowledge: Editorial. *Organization Studies, 25*(9), 1475–1489.

Szabo, L. (2016). Sustainability, creativity and innovation in project management – model development for assessing organizational performance through projects. *Vezetéstudomány-Budapest Management Review, 10*, 3.

Tabassi, A., Argyropoulou, M., Roufechaei, M., & Argyropoulou, R. (2016). Leadership behavior of project managers in sustainable construction projects. *Procedia Computer Science, 100*, 724–730. https://doi.org/10.1016/j.procs.2016.09.217

Taherdoost, H. (2018). *How to lead to sustainable and successful IT project management?* Propose 5Ps Guideline.

Tam, G. (2010). The program management process with sustainability considerations. *Journal of Project, Program & Portfolio Management, 1*(1), 17–27.

Todorović, M., & Obradović, V. (2018). Sustainability in project management: A project manager perspective. In I. Ljumović & A. Éltető (Eds.), *Sustainable growth and development in small open economies* (pp. 88–106). Institute of World Economics, Centre for Economic and Regional Studies of the Hungarian Academy of Sciences. ISBN 978-963-301-664-0 (e-book), ISBN 978-963-301-663-3 (paperback).

Toljaga-Nikolić, D. (2022). *Integrated model for project success improvement based on sustainability principles. Doctoral thesis*. University of Belgrade, Serbia.

Toljaga-Nikolić, D., Todorović, M., & Bjelica, D. (2016). Sustainability and project management – Where is the linkage?. In *XV international symposium – Symorg 2016, symposium proceedings "Reshaping the future through sustainable business development and entrepreneurship"* (pp. 1088–1093).

Toljaga-Nikolić, D., Todorović, M., Dobrota, M., Obradović, T., & Obradović, V. (2020). Project management and sustainability: Playing trick or treat with the planet. *Sustainability, 12*(20), 8619. https://doi.org/10.3390/su12208619

Wang, N., Wei, K., & Sun, H. (2014). Whole life project management approach to sustainability. *Journal of Management in Engineering, 30*(2), 246–255.

Wong, S. (2012). *What have been the impacts of World Bank Community-driven development programs? CDD impact evaluation review and operational & research implications*. World Bank.

World Commission on Environment and Development. (1987). *Our common future*. Oxford University Press.

Yamaoka, K. (2008). Social capital and health and well-being in East Asia: A population-based study. *Social Science & Medicine, 66*(4), 885–899.

Zaharia, M., Pătrascu, A., Gogonea, M. R., Tănăsescu, A., & Popescu, C. (2017). A cluster design on the influence of energy taxation in shaping the new EU-28 economic paradigm. *Energies, 10*, 257.

Chapter 11
The Impact of the Circular Economy Approach on the Project Portfolio and Selection Process

Ivana Berić

Abstract The circular economy (CE) concept has reached increasing attention among practitioners and academia as a way to promote sustainability. The circular economy incorporates different features and contributions from various approaches that share the idea of closed loops. It focuses on improving effectiveness and efficiency and closing the energy and resource flows by changing the organization's interpretation of products and economic value. The companies should redesign products to improve multiple life cycles, reuse, recycle, reduce, and recover materials in production or distribution and consumption processes. Since the project portfolio selection is a tool to reach the organizational strategy, there is a growing need to determine which criteria and characteristics of the circular economy approach to consider ensuring that suitable projects are selected. In selecting and managing the portfolio, the economic, social, and environmental sustainability requirements are suggested. The link between the circular economy's characteristics and the selection criteria and the process is missing in the currently available sources. The author intended to determine whether a specific framework will bridge the knowledge gap between circularity conceptualizations and their application in the project portfolio management field and what such a framework should include. The arising question is as follows: Do the CE approach's characteristics extend the criteria and factors considered in the project portfolio selection?

Keywords Circular economy · Project portfolio · Portfolio selection · Sustainability · Selection criteria

---

I. Berić (✉)
Faculty of Project and Innovation Management, University Educons, Belgrade, Republic of Serbia
e-mail: ivana.beric@pmc.edu.rs

V. Obradović (ed.), *Sustainable Business Change*,
https://doi.org/10.1007/978-3-031-23543-6_11

11.1 Introduction

There is no doubt that organizations can succeed by implementing projects in today's business environment. Considering that one-third of the world's gross domestic product is generated by projects (Turner & Müller, 2003; Økland, 2015), it is evident that the integration of sustainability into projects is significant. Although the notion of sustainability in project management has gained considerable ground over the last few years, many project management frameworks do not effectively address sustainability's three pillars.

Organizations willing to adopt new approaches and innovations in business can improve sustainability performance and achieve their ambitions in the markets they operate in (Geissdoerfer et al., 2020).

According to the Green Project Management Global (2019), the demand for sustainable business practices has grown over the past decade since the global focus redirects sustainable development, social responsibility, climate changes, and ethical behavior. GPM's recent study, *Insights on Sustainable Project Management*, found that 96% believe that projects and project management are integral to sustainable development among the over one thousand executives surveyed. 100% of these same executives acknowledge that project managers should understand sustainability's importance to their projects.

The circular economy is often cited as one of the best solutions to sustainable development (Ngan et al., 2019). The implementation of circularity should be taken into consideration as a way of reducing resource usage and waste. A circular economy-oriented business model encompasses principles or practices as guidelines for business model design. It aims to boost the effectiveness and efficiency of resources and close the flow of energy and resources by changing how we interpret products and economic value (Pieroni et al., 2019, p. 201). So far, we lack a more robust integration of value network and business model portfolio considerations. Despite numerous circular methods, principles, and strategies in the literature, implementing these approaches into practice is challenging.

CE transition might require radical changes, such as developing circular products and finding a new revenue stream. Managing a portfolio from a circularity perspective and integrating circularity indicators and portfolio management are necessary for supporting the CE transition. Despite a significant number of papers related to the circular economy topic, no research work has examined whether there is some CE inevitable element that needs to be included in the selection process.

Introducing a framework that would help project portfolio managers, organizational decision-makers, and policymakers select and prioritize CE actions with respect to the relevant stakeholders within and outside their organizations could enable such a transition (Coenen et al., 2020).

Practitioners and researchers should direct their efforts to create a framework used as a guideline for undertaking activities that contribute to circularity. It should be suitable for use as an independent tool by individuals who are not experts in CE. It should also help portfolio managers present, select, and prioritize strategies, actions, and directions for the practical implementation of circular principles for organizations.

Possibilities for the application of circularity actions differ for each lifecycle phase. CE embodies a multi-lifecycle approach – the end of one lifecycle is the beginning of a new one. CE actions can range from affecting large-scale systems to only specific system elements (Coenen et al., 2020). The concept of life cycle orientation extends the time dimension by considering this horizon within the scope of the strategic planning process (Schipper & Silvius, 2018).

The following sections present the literature review on project portfolio management, sustainability, and circular economy approaches, followed by a detailed description of the selection process, selection techniques and criteria, different project types, prioritization, and difficulties in portfolio selection. A special part is dedicated to the characteristics of a potential framework that could help select CE initiatives.

11.1.1 Project Portfolio Management and Sustainability

Numerous complex problems that businesses and other organizations face daily require implementing modern management methods and disciplines for the more efficient functioning of the organizations. Implementation of project management is necessary for the efficient execution of various projects and enterprises (Jovanović & Berić, 2018).

Obradović et al. (2012) stated that project management is no longer just a subdiscipline of engineering; the management of projects – including program management and portfolio management – is now the dominant model in many organizations for strategy implementation, business transformation, continuous improvement, and new product development.

Achieving organizational goals is not related to only one project but rather to a set of related or independent projects that, each in its way, contribute to achieving the goals and the overall organizational results. It is not possible to observe and measure individual projects' outcomes, but rather their overall impact upon the corporate results should be taken into consideration (Berić et al., 2012).

Project management, in general, could respond to numerous global environmental issues, but the current standards fail to include sustainability seriously. Sustainability is accepted as a success source; therefore, the project portfolio must demonstrate how this sustainability is effectively addressed. As sustainability compliance is of interest to many stakeholders, the project manager must look at the project results beyond its completion date. Sustainability has to be an integrated part of the portfolio process to achieve organizational goals (Hope & Moehler, 2014).

Which projects will be implemented has become a critical issue that significantly affects the success. As project portfolio management represents a dominant model for strategy implementation, business transformation, and continuous improvement, the portfolio offers an ideal opportunity to introduce sustainable development principles. PPM can provide opportunities to enhance the integration of sustainability within organizations that run multiple projects. Also, considerations at a portfolio level can help achieve more holistic tools for corporate, sustainable, and circular economy settings.

It is important to understand sustainability issues, and project managers should take a long-term view of product development, from its initial creation and deployment to its final retirement, known as whole lifecycle thinking. The standardized routines and processes for project and portfolio delivery are independently associated with better information production and higher portfolio success levels. This, in turn, improves transparency and comparability in the portfolio environment, and with a defined process comes defined information requirements, leading to improved availability and comprehensiveness of data.

The sustainable project portfolio management ensures that the business and commercial benefits balance society and the environment.

The objectives of project portfolios suggested by the work of Cooper et al. (2001) are as follows:

- Value maximization according to business objectives
- Strategic direction
- Portfolio balancing in alignment with strategy

Twenty years later, this definition has not yet been primarily surpassed or modified.

Many project management approaches do not consider sustainability factors in available papers and the current practice. From a P5 perspective, they would be viewed as ineffective even if they effectively achieve the traditional objectives of cost, time, and scope. Increased project processes' effectiveness would help achieve sustainable project outcomes, such as minimal losses from rework and other wasted resources, improved project governance, and increased project execution capacity leading to increased profitability (GPM, 2019).

Recognizing that a project is a mechanism for the implementation of an organization's strategy and that the establishment of the concept of sustainable development on the project management level is of crucial importance for the successful strategy implementation have already been a part of many papers and research initiatives (Toljaga-Nikolić et al., 2020). Consequently, we should assume that the impact of circularity characteristics must be considered when selecting the beforementioned projects.

Effective project selection has been a topic that has attracted academics and practitioners' attention over the past few decades, but a universal model that would help all companies find answers to this question has not yet been discovered. Project portfolio selection presupposes using several multi-criteria decision-making methods to analyze and compare different projects' potential success in the organization's dimensions.

Aarseth et al. (2017) carried out a systematic literature review covering all research published in five leading journals in project management and sustainable production before 2016 and identified eight distinct sustainability strategies used by either the project or its host or by both of them to support sustainability goals. According to their findings, a sustainability emphasis in project portfolio management from a project perspective relies on either using a framework for project selection or actively including sustainability as a dimension in early-phase appraisals. It includes emphasizing sustainability issues when deciding which projects to fund and approve from a host perspective.

Due to its increasing importance, sustainability is considered a factor that needs to be included in all selection processes regardless of the type of project or industry in which it operates. Recently, certain studies have emerged that adopt the three-pillar concept: economic, environmental, and social sustainability and the selection of targeted projects from the perspective of sustainability in an uncertain decision-making environment (Ma et al., 2020; Dobrovolskienė et al., 2017). There are also studies dealing with the sustainable selection of project portfolios (Khalili-Damghani et al., 2013; Khalili-Damghani & Tavana, 2014). Despite the apparent advantages, the portfolio elements are not sufficiently represented in the most reviewed literature. Portfolio managers take only a limited number of sustainability criteria into account in their decisions (Dobrovolskienė et al., 2017); despite its proven contribution to business success in general, the current methods and practices for project management and PPM hardly address sustainability (Schipper & Silvius, 2018).

The most comprehensive approaches to the relationship between PPM and sustainability were presented in the S-PPM framework developed by Schipper and Silvius (2018); the four-dimensional conceptual framework for managing sustainable projects proposed by Marcelino-Sádaba et al. (2015), and Sanchez's approach described as integrating sustainability issues into project management (Sanchez, 2015). They could develop a similar framework related to the relationship between PPM and circular economy.

11.1.2 Project Portfolio Management and Circular Economy

According to the research of numerous authors, the circular economy (CE) is a crucial industrial economics model to pursue sustainable development. (Bakker et al., 2014; Bocken et al., 2014; Mac Arthur, 2013; Rashid et al., 2013; Webster, 2015). Furthermore, it is necessary for sustaining economic output.

The recent studies advocate that the CE principles' application may increase the European GDP by about 11%, bringing to net benefits of about € 1.8 trillion by 2030 without compromising the environment (Ellen MacArthur Foundation, 2016).

A detailed definition of CE is still missing in the literature. Kirchherr et al. (2017) define CE as 'an economic system based on business models which replace the end-of-life concept with reducing, alternatively reusing, recycling and recovering materials in production/distribution and consumption processes [...]to accomplish sustainable development, which implies creating environmental quality, economic prosperity and social equity, to the benefit of current and future generations'.

Products should be redesigned to improve multiple lifecycles, improving product reuse, refurbishment, remanufacturing, and recycling.

The CE's adoption represents a fundamental systemic change, regardless of industry, location, scale, nature of the business, etc. (Kirchherr et al., 2017). CE has been usually indicated as a promising approach to reduce global sustainability pressures (Ellen MacArthur Foundation, 2016) and to promote economic development and sustainability.

The successful development of CE requires a system of indicators for its assessment. However, different implementation levels of the CE and various characteristics of enterprises, industries, or regions require other assessment indicators. Each enterprise needs to tailor the firm-specific indicators according to its features, conditions, and existing problems (Su et al., 2013).

Although a wide range of possible measures is part of the transition toward a CE, no circularity criteria are considered in portfolio reviews. One of the issues that deserve special attention is scoring circularity and balancing (i.e., earning profit in the short term and becoming more sustainable in a long time).

To achieve reasonable consideration of CE within the selection, the organization should discuss the following:

1. The capability of scoring circularity – CE assessment
2. The ability to evaluate circular products and circular business models
3. The amount and nature of additional criteria – resource circularity, assessing circularity from both sustainability and profitability perspectives, considering environmental impact drawbacks to ensure that circular benefits take over the ecological impact, evaluate value proposition novelty to judge if projects can be called innovation or not (Hamano et al., 2020).

The CE framework's existence can support a company in evaluating and managing CE projects, facilitating CE portfolio management, and considering environmental sustainability and circularity in a single tool to help portfolio management.

From a companies' portfolio circular innovation perspective (both for entirely new products and incremental improvements of existing products), the circular KPIs can support not only the decision-making process along with the design of new products but also the comparison of different versions of the same product based on their degree of circularity and the benefits they can bring. Companies would be able to compare different products based on their circularity and their benefits (Sassanelli et al., 2019).

The CE approach in relation to sustainable development has increasingly gained practitioners' and academics' attention during the last decade. Nevertheless, CE's scientific research content remains mostly unexplored (Korhonen et al., 2018).

According to the findings available in the existing literature, some challenges and shortcomings have been identified, such as follows:

1. Regardless of the growing need for guidelines in selecting CE projects, both in practice and the scientific research domain, the concrete, and comprehensive methods and models are still undeveloped.

2. The literature lacks insights on integrating CE characteristics while evaluating projects in the selection process.
3. There are no conclusions on the impact of CE characteristics on product design and the possibility to evaluate project outputs that could be returned after project closing as a part of new projects in the selection process.
4. There is a lack of appropriate selection criteria for projects concerning the circular economy.
5. Guidelines to select an appropriate combination of methods for each industry have not been developed.
6. Limited attention has been paid to integrating CE into the PPM decision-making and selection process.
7. The existing studies fell short in explaining which method is suitable for various projects, and there is no consensus on methods appropriate for each category of projects.
8. Lack of industry-specific evaluation criteria to assist in the business case of CE-driven projects that should be considered as a part of the future portfolio. Consequently, a theoretical framework for evaluating projects through the PPM process is lacking.

Recent studies indirectly indicate that managing a portfolio from a circularity perspective is necessary. Future research should develop a mapping tool for integrating circularity indicators into portfolio management and finding the link between circularity indicators, portfolio management, and business strategy. Although the PPM is divided into numerous sub-processes, the emphasis will be on the selection as a critical phase since it is generally accepted that inappropriate project evaluation usually leads to ineffective PPM.

11.2 Project Portfolio Selection

11.2.1 General and CE-Driven Project Evaluation and Selection

More than 30 years ago, the project selection's central issue – choosing the best project or portfolio – was considered. It is a complex decision-making process influenced by many critical factors, such as market conditions and technical success probability (Bard et al., 1988). It can also involve many different groups – from the top management level to the project management level, with different experiences, cultures, and other preferences (Jiang & Klein, 1999; Machacha & Bhattacharya, 2000), as well as a high level of risk due to uncertainty or incompleteness of information. They all affect the selection process.

Existing project selection research results concentrate on strategic mechanisms and methods for project selection.

The selection should be analyzed according to the collected project information, comparing the advantages/disadvantages of the projects and the final determination of the projects (King & Mercer, 1987; Skitmore & Pemberton, 1994; Slade, 1991). The decision on project selection is based on the company's long-term development goals and in compliance with the organizational strategy, so it does not focus only on the project's economic benefits (Jiang & Klein, 1999; Rothkopf & Engelbrecht-Wiggans, 1993). Some researchers suggest three principles for project selection (Wang et al., 2009):

1. The project must remain consistent with the goals of the company and the current development strategy of the company (Alvey et al., 1998).
2. Profit maximization (Tam, 1999).
3. Proper allocation can improve projects' quality and build better society projects (Liu, 1999).

Effective and efficient project selection in strategically – focused environments is essential for the organization's sustainability. Project portfolio selection and activities related to managing selected projects through their life cycles are important activities in many organizations (Martino, 1995; Cooper, 2001; Meredith & Mantel, 2012). Project management approaches are widespread and applied in many areas, such as R&D of a new product, implementing new systems and processes in production and IS, engineering, construction, and other industries. Since there are usually more projects available for selection than can be realized within the firm's physical and financial constraints, the organization must create an appropriate portfolio of projects.

Why is project selection important? The most challenging phase in strategic management is not forming a strategy but in its implementation (Hitt & Ireland, 2001; Kaplan & Norton, 2001), and one of the most critical steps in implementing a strategy is selecting projects (or programs or activities) aligned with the organizational goals.

Companies that want to be competitive by selecting the most appropriate projects must use portfolio selection techniques and procedures based on the most critical project measures. However, these techniques cannot be used if decision-makers do not understand them.

Project portfolio selection is the routine activity of selecting a portfolio from available project proposals and projects currently being implemented that respond to the organization's goals appropriately, without exceeding available resources or violating other constraints (Archer & Ghasemzadeh, 1999). Some of the issues relevant to this process are organizational goals, priorities, financial benefits, intangible benefits, availability of resources, and portfolio risk level (Archer & Ghasemzadeh, 2000).

Is improper management or a complete lack of project portfolio management a problem in today's corporate world? According to numerous analyses, it is.

Today, the public and private sectors spend approximately ¼ of the GDP on projects in most countries. Nevertheless, at the same time, roughly in around 80% of cases, companies either do not collect business cases for their projects or run them on selected key projects, companies do not have measures except for financial data, and in many instances, they are unable to adjust the budget to their business needs.

Such information imposes that modern business is not methodical and systematic enough to assess and select project mixes properly.

Regardless of the business area, there are often situations where projects are initiated only based on a superior decision made without prior detailed analysis.

The three essential requirements that portfolio professionals should expect from each project candidate are:

- Every project and a portfolio of projects should maximize the company's value.
- Candidate projects should maintain the desired balance in the portfolio mix (Moustafaev, 2010).
- The final portfolio of projects should be strategically (Moustafaev, 2010) linked and genuinely reflect the business strategy.

Generating a company project portfolio is often seen as a random process rather than a systematic and value-oriented process. Many variables should be considered and evaluated (Lereim, 2008). In many companies, the primary value drivers are economic and social/organizational. However, when optimizing, it is necessary to consider the maintenance of competitive advantage (i.e., special attention should be paid to the dimension of competitiveness).

Portfolio management refers to the processes and strategies for project prioritization, effective use of resources, and contribution to achieving strategic goals (Stawicki & Müller, 2007). Selection and prioritization criteria ensure respect for organizational strategy. When selecting projects, it is necessary to observe the whole set of potential projects and their interactions through resource constraints.

The rule for selecting appropriate product development projects is to choose projects that enable the organization to achieve its goals in the most effective way (Schelle et al., 2006), which is very difficult to accomplish for methodological and other reasons.

Selection is crucial to an organization's success and a problem that requires much time. The reason for this lies in conclusion, with which almost all authors in this field agree that the four most significant universal issues in the project portfolio are precisely those that are the result of poor project selection.

These problems are (Kendall & Rollins, 2003) as follows:

1. Too many active projects (often twice as many as the organization needs)
2. Projects that will not provide value for the organization
3. Projects are not related to strategic goals
4. Unbalanced portfolio

The selection process includes identifying opportunities, assessing organizational alignment, cost-benefit-risk analysis, and portfolio development and selection. In this process, we should ask: Which projects should we choose? How do projects relate to the portfolio, and how can the project mix be optimized?

Apart from these universal questions, the possible additional question related to the CE approach could be: How to achieve an effective balance of the portfolio of projects concerning sustainability and circular economy?

The portfolio selection process should be organized logically so that each step takes place from top to bottom (strategic considerations) or from bottom to top (individual project considerations) to select the most appropriate projects. Each step must have a theoretical basis and generate relevant data to be the "input" for the next step. Data must be available to users to make decisions, but at the same time, they must not be overwhelmed with unnecessary details. An overall balance must be struck between simplification and generating well-founded and logical solutions. We should also consider CE's principles – less product complexity and more manageable life cycles when it comes to CE.

To simplify the portfolio selection process, the organization should organize it in several phases, allowing decision-makers to move logically in considering projects, which they will choose based on theoretical models.

To allow comparisons of projects during the portfolio selection process, the organization will choose standard measures and calculate them separately.

Portfolio selection and adjustment are repetitive processes. Existing projects require resources from the available database, and therefore their schedules and resource requirements must have interacted with new projects.

Current projects that have reached a particular key event must be reviewed simultaneously as new projects are being considered for selection, which will enable the creation of the portfolio generated within available resources, and at certain time intervals, following (a) completion or abandonment of the project, (b) new project proposals, (c) changes in strategic focus, (d) revision of available resources, (e) changes in the environment (Archer & Ghasemzadeh, 1999).

The number of projects that the organization can propose for a portfolio can be vast (Cooper et al., 1997). The complexity of the decision-making process and the time required to select a portfolio grows exponentially with the number of projects that need consideration. Therefore, it is necessary to eliminate insufficient projects through review processes before the phase of the portfolio selection process begins. For instance, projects that do not correspond to the company's strategic focus, for which there is not enough information to make logical decisions, and do not meet the minimum requirements, such as the minimum return rate, are eliminated.

Besides numerous variables included in the analysis, project interactions through direct dependencies or resource bidding are considered in portfolio selection.

In the published methodologies for portfolio selection, little progress has been made in reaching an integrated framework that decomposes the process into a flexible and logical set of activities involving selection boards' full participation. Such an approach could have an advantage in combining the best characteristics of existing methods theoretically well developed.

It is necessary to connect the strategic and project selection processes to reach the best possible portfolio under different conditions. Pre-screening ensures that the organization evaluates individual projects under the intended requirements. Once the appropriate selection criteria have been established, the evaluation of individual projects begins. The assessment uses selection criteria with certain minimum acceptable limits.

After the first check, the organization obtains a list of the best-ranked individual projects according to the set criteria.

When the pre-audit is completed, individual projects are reviewed from the portfolio point of view. For each project already evaluated, there are two additional criteria/dimensions included in the portfolio selection optimization:

1. Is the project autonomous, independent, or part of a program?
2. How will the project contribute to increasing the competitive strength of the business unit and corporation? (economic indicators and functionality) (Lereim, 2008).

The choice of the best project depends on the following:

1. Quality of defined requirements
2. Quality of evaluation
3. Quality of understanding of what alternative projects can "produce" (Tiong & Alum, 1997).

11.2.2 Selection Techniques

Many published articles and books on project evaluation and selection describe more than 100 different techniques. Common to all the methods used in project selection is their common goal (i.e., a ranking of options).

One of the approaches proposes project selection and evaluation using techniques that go through 3 phases: strategic considerations, individual project evaluation, and portfolio selection (Archer & Ghasemzadeh, 1999).

In the first phase, organizations use the techniques to determine the portfolio's strategic focus and overall budget allocation. The second use techniques to evaluate individual projects independently of other projects. The third phase refers to portfolio selection based on project candidate parameters, including interactions with other projects, resource constraints, or interdependencies (Archer & Ghasemzadeh, 1999).

Since there is no consensus on the most effective methodology, organizations choose methodologies that suit their culture and consider project attributes that they believe are most important (Hall & Nauda, 1990; Krumm & Rolle, 1992; Mukherjee, 1994). The methodologies most useful in portfolio development for one class of projects may not be suitable for another.

The project selection framework must be flexible enough to enable stakeholders to choose predetermined techniques or methodologies familiar to them, which helps them analyze relevant data and make appropriate choices.

Decision-makers must be provided with interactive mechanisms for controlling portfolio selection generated by algorithms or models and receive feedback on changes' consequences.

Portfolio selection is a process in which objective criteria, such as expected project costs, are mixed with subjective ones related to the needs of different organizations presented on the project selection committee.

The selection of project portfolios must be adapted to group decision-making environments. The result is a portfolio that responds to the organization's goals optimally or close to optimal. Portfolio selection is a strategic decision, and relevant information must be presented in such a way as to enable decision-makers to evaluate the portfolio without the danger of being "overwhelmed" by unnecessary details.

The metrics used by organizations significantly impact the selected and proposed projects.

Using financial metrics to evaluate project applicants requires forecasting how the project may affect cash flows, which is very difficult to do. The most significant limitation of these methods is that they can provide only a partial presentation of what is essential. Financial measures do not consider all the organization's goals, especially for public companies with non-financial goals (e.g., environmental protection). Financial measures cannot measure the total value of projects that achieve non-financial goals.

It is essential to include other indicators; otherwise, the project's value may be underestimated. It is critical to define the time in which the project's benefits and risks will appear (expressed in monetary values if possible).

Most currently available prioritization and portfolio management software allows the financial and non-financial parameters to be defined simultaneously.

Different organizations implement different types of projects and create value in different ways. Therefore, organizations' measures for project evaluation are also different. Consequently, each organization should develop a model that explains how their projects create value.

Almost all organizations take care of liquidity and security, but many do not treat risk adequately for project selection. Carelessness is one of the reasons for choosing the wrong projects. The growing competitive economic environment puts a lot of pressure on managers to get results quickly. Organizations must maintain a high standard towards shareholders, clients, and the public. There is less and less tolerance for exceeding the budget, and unsuccessful projects can jeopardize the organization's reputation and profitability. Realization of projects within the budget and on time is no longer enough.

Project selection aims to select the portfolio that creates the most significant possible value, taking the risk and available resources into account. While studies have suggested that some selection methods might be difficult to use due to the complexity of mathematical programming work, there is a lack of a comparative overview of approaches using a common set of evaluation parameters (Brook & Pagnanelli, 2014).

11.3 Selection Criteria

11.3.1 Defining Project Categories

Practical experience during decades of managing many types (or categories) of projects has led to:

1. Defining, recognizing, and understanding the principles of project management and practices that are "common to all (or a large number of) projects" in all types of human endeavors
2. Recognition of differences within many existing and potential projects that require projects to be separated in several ways and for several purposes, to continue to improve how both project owners and buyers (beneficiaries):
   (a) Strategically select and prioritize projects
   (b) Operationally plan and implement their projects individually, within the program, or the project portfolio
   (c) Educate and train managers and specialists involved in projects
   (d) Develop and manage the criteria of managers and specialists involved in projects (Archibald, 2008)

Regardless of the type and structure, projects can be grouped based on different criteria.

One of the benefits of project classification is a better understanding of projects, an important input for portfolio selection. The weakness is that there is no tool to help practitioners classify projects.

The project sponsor determines the differentiation between external and internal projects. External project sponsors are persons or institutions outside the organization implementing the projects. These projects are also known as sponsored (construction, mechanical engineering, heavy industry, software, consulting) (Schelle et al., 2006). these projects are undertaken for an external client and bring profit but do not change the organization. For internal projects, sponsors are persons or entities within the organization implementing the projects. These projects were created to change the organization (reengineering, new information system, reorganization, etc.). The economic effect on the company is part of business success because they contribute to improving and advancing business processes and activities. Projects can exist in any sector, including retail, insurance, banking, and transportation.

Based on project results, according to (Schelle et al., 2006), projects can be:

1. Investment
2. Organizational
3. Research and development

Classification can be done concerning the project's size (the number of staff engaged or the implementation time).

Also, there are different projects where the staff has a common or different culture. Many organizations implement several projects with varying interdependencies at the same time.

According to Sommer (1999), the highest level of portfolio analysis is the classification of projects. He believes that there are two primary categories in which all projects can be classified: survival and growth. The organization must review the investment justification if the project does not fall into these categories. The explanation of these categories is as follows:

- "Must-do" projects are survival projects that must be realized. Projects fall into this category only if they can contribute to the organization's health.
- Growth – all other projects belong to this category. These projects can be exclusively internal and represent support or external and financially oriented.

Types of projects, according to (Ivan & Sandu, 2008), differ concerning the area in which they are implemented:

1. Projects financed by the government for rural or urban communities: hospitals, bridges, schools
2. Industry-specific projects
3. Building construction projects
4. Research projects
5. Infrastructure projects: highways, transnational oil pipelines

The project management principles can be applied to any project and any branch. Different projects have different characteristics. The main variables/parameters/attributes by which projects are distinguished (Youker, 1999) should include: Stability encompasses, Degree of uncertainty or risk, Type of worker, Significance of time (tempo), Significance of costs, Degree of new technology, Series of projects or one in type, Form of commitment (External contract or internal work), Level of detail of plans.

The list of presented criteria for project categorization does not end here. The Project Management Body of Knowledge of the German Association, a member of the IPMA (International Project Management Association), distinguishes between investment projects (construction and systems engineering), research and development/innovation projects, and organizational projects.

The method used to categorize projects usually depends on specific attributes. Projects have many characteristics that can define categories and classify projects within a particular category.

Numerous factors are significant for projects but are not specific to any particular type of project on the list. They can refer to any type and can be used in other types of project classifications:

1. Size
2. Duration
3. Industrial sector
4. Geographical location

5. Number of employees hired
6. Costs (large, medium, small)
7. Complexity
8. Urgency
9. Organizational design

The challenge is to select the most appropriate characteristics to define the best category for a specific purpose. The project categorization method is defined as the procedure to be applied in identifying a set of features (or attributes), which will be used for:

- Placing a particular project in a specific category
- Classification of projects within a category (or subcategory)

At the PMI Research Conference in Montreal, it was proposed that the ideal portfolio contain 75% of projects focused on mission, vision, and strategy, about 20% of projects related to improvements in business units (still aligned with mission and vision), and about 5% of problem-solving projects.

Apart from presenting different approaches and categorizations, if we refer to the European Commission's conclusion, each industry sector is different regarding resource use, waste generation, and management (EUR – Lex, European Commission, 2015). Consequently, we should consider industries' specificities that will impact each industry's possible and available CE project types.

11.3.2 Criteria for Project Selection in Different Projects Types

There are many studies on the project selection process, and each of them explores the selection from different perspectives (Baker, 1974; Jiang & Klein, 1999). Only a few investigated the factors influencing project selection (Mohanty, 1992; Jiang & Klein, 1999). Mohanty (1992) developed a project selection process through multi-criteria decision-making methods. He defined an attractive project with the following characteristics:

1. Sustainable proposal
2. Competent team and well-structured organization
3. A reliable source of capital and other resources
4. A potentially high return on investment

Organizations prefer to choose projects that require a minimum investment, a low level of competencies, a minimum time, and the potential for the highest return. In reality, ideal projects are challenging to find, so the most acceptable project is selected based on a comparison with existing proposed projects.

Mohanty (1992) classifies the criteria influencing selection into internal and external categories. Okpala (1991) defined ten factors for construction projects; Rengarajan and Jagannathan (1997) described 13 factors for research and development projects; Jiang and Klein (1999) – 6 groups of factors, the World Bank (2003) evaluates infrastructure projects based on nine factors. The existing examples are numerous.

All factors can be classified depending on the type of project or phase in the selection process. It is also possible to adjust these factors to the organization and its approach.

Some authors (Puthamont & Charoenngam, 2007) also explained certain significant factors for the selection:

1. Factors related to project explanation (Objectives, Justification, Urgency, Concept, Previous performances related to the project)
2. Factors related to project feasibility (Budget, Duration, Staff and equipment, Plan and timeline)
3. Factors related to investment analysis (Investment analysis, Project alternatives, Overlap with other projects, Relationship between procedures and budget, Future operations and maintenance, Suitability of project size, method, and technology)
4. Factors related to readiness for implementation (Readiness for implementation, Compliance with regulations and laws)
5. Factors related to project benefits and evaluation (Benefits for the target population, Performance measurement, Evaluation plan)
6. Factors associated with the impact of the project (Impact on society and the local population, Impact on the environment, Impact on human resources in the organization)
7. Factors related to the socio-economic and political environment (Economic and political climate, Public opinion on the project)

The last two categories include the sustainability issues in consideration.

When selecting a portfolio, the following project/portfolio characteristics should also be considered (Archer & Ghasemzadeh, 1996): multiple goals, links between projects, mutually exclusive projects, resource constraints, qualitative attributes, number of projects, project phases, portfolio risk, and uncertainty.

Once the selection criteria are determined, the weights of the criteria are determined. It is crucial to enable mutual comparison from a portfolio perspective. Once the organization defines a priority list of projects, the measurement can begin and rebalancing if the organization decides to introduce new priorities. By rebalancing, new projects are evaluated based on the current portfolio. The organizations should change the portfolio by increasing the investment or replacing the project based on the same weighting criteria.

Appropriate projects are organizational goals, mission and vision, culture, and priorities. Based on the strategic plan, the first step in the project portfolio management process involves generating a standardized understanding or minimum eligibility criteria, to which projects must respond and developing a list of existing and proposed projects.

The information collected may vary but usually includes estimated financial impact, risk factors, resource requirements, key participants, dependencies, project type, timelines, key events, user benefits, and basic project identification.

Also, it would be good to have organizational data: resources (labor, material, equipment), availability of skills, and standard financial information.

The organization must also establish weighting criteria for evaluating and ranking projects from its strategic plan. The model must be easy to understand. Another vital thing is optimization, which focuses on making the optimal mix (refers to optimizing the entire portfolio, not individual projects). Since portfolio management is an iterative process, once the portfolio is optimized, it must be monitored to be consistent with the objectives, even if they change.

Although each organization can create its specialized criteria, according to Frame (2002), experience has shown that the four most important selection criteria (consideration on a cost-benefit basis) are as follows:

1. Financial
2. Technical
3. Developmental
4. Organizational

Based on the literature research, each author who dealt with the selection criteria made a personal contribution to the problem, and there are a large number of classifications of criteria that are applied.

The question that arises in this area is: What are the criteria for portfolio selection to move toward the CE approach?

What should be included in the evaluation is a consideration of the extent to which CE practices are relevant for implementing sustainable development goals since there are numerous findings reported on the potential direct contribution of the CE practices to achieving a significant number of SDGs targets. These considerations impose the question of factors related to CE that should be included to meet sustainability criteria.

Silvius (2017) stated that sustainability is irrelevant to all project types. Consequently, different aspects and sustainability criteria have different relevance for different projects. Further, he explained that water usage criteria might be relevant to a construction project but more minor to a software development project. This argument also applies to considering sustainability on an organizational or corporate level. Organizations should therefore assess which aspects and indicators are relevant to their industry. The same applies to integrating sustainability into project management. Based on the project's characteristics and consultation or participation of stakeholders, the project should transparently assess which sustainability indicators or aspects are most relevant.

After all general and available consideration, if we would try to identify the scrutinizing items that should be used in the evaluation of projects and are related to sustainability and CE, the list of criteria should include as a minimum:

1. The lifespan of project design.

2. Environmental sustainability requirements may include the alignment with the "green" strategies, reuse of existing resources, consumption of electricity, paper, cooling, and space, consolidations of infrastructure and related technologies, end-user endorsements, agreement on the environmental aspects of the technology, to name a few.
3. Social sustainability requirements include equal community access to the project facilities, consistency with provincial, sectoral, and national environmental priorities, health and safety (employees, contractors, customers, citizens), and adherence to human rights standards (e.g., child labor).
4. Economic sustainability requirements – consistency with the national priorities, equal opportunities, discrimination (in sharing expected benefits), community benefits, taking benefits from previous projects, possible negative effect on GDP per capita or local economic opportunities.

The possibility for the application of circularity actions differs for each lifecycle phase. CE embodies a multi-lifecycle approach: the end of one cycle is the beginning of a new one. The most challenging task for organizations is to evaluate how their assets can be made more circular. Every domain or field has characteristics that offer particular circularity opportunities regarding the production process, product lifecycles, and markets.

Although there is a definite need to develop indicators that can be used in decision-making processes to ensure that projects are managed according to practices that will contribute to sustainable development, a prerequisite for Sustainable Project Life Cycle Management is a clear understanding of the various life cycles involved in a project and their interactions (Labuschagne & Brent, 2005).

Future research should concentrate on providing guidelines for the following issues to help organizations deal with all challenges mentioned above:

1. How to use circularity indicators to determine the extent of circularity?
2. How to design for multiple-use cycles?
3. How to anticipate how the circular offering will evolve over multiple life cycles?
4. How to set circular criteria (design competencies for a circular economy)?
5. Which of the multi-criteria decision-making methods to use in the selection?

The additional consideration should take into account well-known CE strategies for the lifespan extension of entities:

1. Refurbish – restore end-of-life product to become functional again
2. Remanufacture – use elements of the discarded entity with the same function
3. Repurpose – use a discarded entity in a new (lower) function (Coenen et al., 2020)

According to Bocken et al. (2016), all big companies will have some part of the portfolio where circular business practices could be profitable regardless of their products' nature.

Despite all the above mentioned, several sectors face specific challenges in the context of the circular economy. Since the innovative forms of consumption can also support the development of the circular economy, e.g., sharing products or infrastructure (collaborative economy), consuming services rather than products, or using IT or digital platforms (EUR-Lex, European Commission, 2015), all of these should be taken into consideration before defining the relevant criteria.

11.3.3 A Theoretical Framework

According to Brenner (1994), the project selection process consists of:

1. Identification and selection of criteria and their aggregation into a logical framework
2. Determining the weights of criteria and reaching consensus on their relative importance
3. Evaluation of project proposals using weighting criteria
4. Resource allocation to maximize project progress

Although Brenner reported on this issue more than two decades ago, his statement could be accepted nowadays.

The first sub-process is critical for efficient selection.

As mentioned earlier, the conceptual framework's existence will positively contribute to the success of this phase. The framework would guide decision-makers in the portfolio selection and open up a future research agenda for the CE approach.

The question that arises in this stage would be: Is it possible to create a reliable, coherent model for the CE project selection, and what are the characteristics of such a framework?

The framework that will help decision-makers select and prioritize CE actions concerning relevant stakeholders within and outside their organizations should:

1. Provide guidelines for activities that contribute to circularity
2. Refer to specific stakeholders in the field
3. Illustrate clear links between circular activities and stakeholders
4. Be suitable for use as an independent tool that is user-friendly to individuals who are not experts in the CE field

Despite an abundance of circular methods, principles, and strategies provided in the literature, professionals without a CE background often struggle with the implementation of circularity due to the high level of abstraction and ambiguity of the concept (Coenen et al., 2020).

The previous research on selection frameworks concludes that there is a need for a framework grounded in empirical data to encourage fact-based decisions.

An interesting view on a possible framework that would help integrate sustainability into innovation project portfolio management from a strategic perspective was proposed by Brook and Pagnanelli (2014).

A similar model could be developed to integrate CE characteristics into PPM. As a minimum, such a framework should consist of:

1. *Strategic analysis*

   As stated in (Brook & Pagnanelli, 2014), sustainability should not be managed as a separate strategic initiative. What is needed is balancing strategic priorities related to 3 pillars with possibilities for each industry and project type. In this first phase, when speaking about the CE approach, the organization should consider how the circularity could be included depending on the industry and project types in the respective organization.
2. *Evaluation of projects*

   The purpose of CE-driven projects is to develop a business case and derive meaningful judgment on the value of each project proposal that will lead to effective selection decisions within the resource boundaries.

   Projects may differ in their nature and expected contribution to the firm performance based on sustainability objectives and CE characteristics. Also, project outputs should be assessed during the evaluation of projects so that the projects with the possibility of reuse in future projects receive specific weights.
3. *The split of resources among initiatives*

   How to balance the investment on CE initiatives that will strengthen the competitiveness in the short term while creating opportunities to sustain competitiveness in the future? The CE characteristics should be integrated into the project evaluation scope and considered along with other initiatives within the organizational capabilities.
4. *Prioritization and selection of CE-driven projects*

   Define what criteria should be used in selection to reflect the attempt to balance the selection criteria based on CE and sustainability. Each criterion's relative importance varies and depends on the weights (AHP or other multi-criteria decision methods can determine that).

   The balance of the complete portfolio should be judged concerning the corporation's strategic objectives.
5. *Performance management of the portfolio over time*

   The critical element is benefits management. There has to be a defined way in which the progress of ongoing projects in the portfolio and new project proposals will be conducted.

All items mentioned above should be further explored. The concrete proposal of techniques and criteria in each step should contribute to developing the model to help in a more structured selection process.

The evaluation of project proposals' value is performed based on multi-criteria weighting models (Ghorbani & Rabbani, 2009). Given the complexity of contemporary project selection problems, both analytical and statistical modeling tools can be applied to these problems. Researchers who dealt with the project selection process tried to make mathematical models for this process but mainly dealt with only one project type. The literature in the field has presented many attempts to create models and define their characteristics. The most frequently considered type of project is research and development projects. Project selection is performed by considering different objectives that can be opposed. Various methods are used to make project selection decisions, and mathematical programming models can be used.

In one of the previous research works, the author of this paper created a model based on structural equations and the ANFIS method. This model was designed to examine the hypotheses after a conducted survey related to the examination of the existing application of project selection methods and their focus; consideration of interdependencies that exist between strategic and operational portfolio components; and the possibility to combine criteria elements of portfolio selection, presented in several quantitative and qualitative methods, to form mathematical model applicable in project portfolio management (Berić, 2013).

In the period ahead, similarly to this research, the author will conduct an empirical study related to examining the impact of CE characteristics on PPM processes, emphasizing portfolio selection.

Structural Equation Modeling (a multivariate statistical analysis technique) will be used in the first research step. SEM is a coherent statistical approach for testing hypotheses about the relationships between observed (measured) and latent variables. It is also used to test hypotheses about direct and indirect relationships between a set of measured and latent variables. LISREL software can be used for link analysis in the model. SEM relies on several statistical tests to determine the model's adequacy and how well the model agrees with the data.

The proposed SEM analysis approach consists of the following processes: a review of relevant theory and research of literature supporting model creation, model specification, model identification, selection of measures for variables presented in the model, data collection, preparation of a preliminary descriptive statistical analysis, estimation of parameters in the model, assessment of model fitting, re-specifying the model, if necessary, interpretation and presentation of the model.

After a literature review, forming a questionnaire, conducting a survey, the descriptive statistical analysis, and selecting the model's variables will be implemented. The variables within each group represent questions that need to be asked and answered in the decision-making process in project selection. The proposed variables can be presented as independent or latent depending on the research goal. This way, whether they have a set of linear interdependencies can be tested by examining the variables' variances and covariances.

Based on the obtained level of linearity, modeling is further selected based on linear statistical methods (MLRA type) or nonlinear statistical methods (ANN or FIS/ANFIS type). In the second phase, Adaptive Neuro-Fuzzy Inference System (ANFIS) method will be included. The basic idea of this neuro-adaptive learning technique is very simple. These techniques are based on network modeling and learning phase methods based on a given data set. Calculating the membership functions' parameters occurs so that the appropriate fuzzy inference system (FIS) with the smallest possible error corresponds to the given pairs of input-output data. This method of learning is similar to the method of learning neural networks. The final model will consider the connection between the methods used to evaluate CE projects within the portfolio and determine their compliance with the organization's strategic goals and the success of a created portfolio.

11.3.4 Project Prioritization and Portfolio Optimization

Regardless of the industry and project types, the goal of prioritization is to select the best projects. These projects are expected to lead to significant benefits and a high probability of success.

Compliance with strategic goals ensures that the company invests cash and other resources on strategic interest.

After the final selection, the projects are appropriately linked to the company's resources in a simple portfolio management model, and the process moves to the second stage. Portfolio considerations imply a periodic review of selected mix and questions to be asked after each examination, especially at the end of the initiation and project planning phase, including the following:

1. Is the original business case for the project (Moustafaev, 2010) supported in terms of value, balance, and strategic alignment?
2. Are there drastic changes in budget, duration, return projections, or any other factor considered during the selection?
3. Which projects should be discontinued because they no longer fit the baseline criteria?
4. What projects should be added to the mix due to changes in conditions, new ideas, and market demands? (Moustafaev, 2010).

Symptoms characteristic of organizations that do not analyze, select, and manage projects appropriately are as follows:

1. Project and functional managers often clash over resources.
2. Project priorities change frequently and resources are continually shifting.
3. Senior managers have the authority to approve and release projects independently.
4. Projects start as soon as they are approved by senior managers, regardless of the availability of resources.
5. Senior managers regularly complain about how long projects last or how much they cost.
6. Even if a strategic proposal is implemented, the organization often does not achieve the desired improvements.
7. There is no comprehensive document linking an organization's endeavors to a strategic plan.
8. The strategic plan is written as a list of projects. There is a lack of cause-and-effect links for the logical connection of these projects with strategic organizational goals.
9. The list of projects is not prioritized appropriately, so it seems that all ideas should be implemented simultaneously (Moustafaev, 2010).

If the criterion of strategic compliance is not on the list of criteria when selecting projects, many projects that the company implements end up as useless expenses that do not benefit the organization.

The organization can use financial measures, decision analysis models, AHP, or scoring models. The best results are obtained by classifying proposals into strategic groups, followed by prioritization and selection. Optimization models can also be used.

Two aspects of project portfolio management affect the quality of project selection. The first is the quality of prioritization, which is influenced by uncertainty, the complexity of proposals, and evaluation techniques. Uncertainty and complexity cause evaluation errors, affecting the quality of prioritization. As uncertainty and complexity grow, the quality of prioritization declines.

The quality of prioritization is caused by another aspect, which can be controlled entirely: the number of proposals. The rate of project selection could be reduced if the organization deals with selecting a larger number of projects.

When it comes to prioritizing projects selected for the portfolio, four main areas are:

1. Economic justification of the project. ROI, NSV. They play a role in setting relative priorities so that projects with the most attractive returns have a higher priority.
2. Compliance between the project and the strategic goals of the organization. The strategic plan's corporate level articulates the goals and actions necessary to achieve strategic objectives.
3. Risks associated with the successful implementation of projects. Project managers must make assumptions about the money, dates, results, and resources related to project development. Project risks should be a project selection factor because if a portfolio is made of high-risk projects, it can be too difficult a challenge for management.
4. Risks related to the realization of project benefits. The project's benefits are also based on assumptions (Wessels, 2008), such as the market response to the new product, cost savings, etc.

What can be added to this list is the extent to which the selected portfolio supports CE initiatives and can contribute to the organization's sustainability.

11.3.5 Difficulties and Mistakes in Portfolio Selection

Difficulties in portfolio selection are the result of several factors:

- There are many often conflicting goals or criteria related to portfolio selection
- Even when all the targets are identified, there are still problems associated with determining different criteria
- The evaluation of the proposed individual measures is complicated due to two additional factors. First, some criteria are qualitative. Their comparison is mainly based on assessing one or more stakeholders and differs from quantitative aspects, for which analytical models can be made and data collected. Second,

each project has a risk (probability of failure) associated with its implementation, and there can be significant uncertainty in the degree of risk and determining the value of individual projects for each criterion.

- Projects can be significantly related to other projects
- In addition to the difficulties associated with the project objectives, some limitations must be considered. The main constraints include budget, timing, and program review. Additional ones are the market, labor constraints, and technological capabilities.
- The number of feasible projects, especially in large organizations, is generally large, and there can be a considerable number of possible combinations of projects being considered for a portfolio (if there are 100 individual projects, then there are about 2100 possible portfolios)
- Portfolio selection or adjustment is a process that should be repeated at more or less regular intervals. Projects in the portfolio must be reviewed when they reach a specific critical event to decide whether to proceed compared to projects not yet included in the portfolio. Project cancellation is probably the most difficult to implement, as it usually involves severe organizational and behavioral consequences.
- Portfolio selection is usually not the responsibility of one person. It is often a process in which the entire board participates (Archer & Ghasemzadeh, 2000)

The main reason why some projects were selected but not completed is that resource constraints are not always included (formally) in the selection process. Portfolio selection becomes even more complicated if the availability and use of resources change over time.

Methodologies useful for portfolio development for one class of projects may not be the best for another project type. Another important assumption is that there is no best single way for portfolio selection.

Each organization must choose the methodologies that best suit its culture within the class of projects it considers and the project attributes it believes that are most important for selecting.

Since numerous difficulties exist in the selection process, specific decision support can help if possessing the following characteristics:

1. Sensitivity analysis – the value of the portfolio objective function estimates the sum of the contributions of all projects in the portfolio, which will depend on the values used for each variable or attribute of each project. Sensitivity analysis provides ways to measure how sensitive the target function is to changes in these parameters.
2. Portfolio balancing – necessary for certain portfolio dimensions such as risk, project size, short-term versus long-term projects, etc.
3. User-friendly interface – the interface must be easy to use, understandably present information, and be based on a graphical interface for ease of use. A decision support system that does not possess these characteristics is unlikely to be used by managers if optional.

4. Overall perspective – the choice of project portfolio is a strategic decision, and relevant information must be presented in such a way as to enable decision-makers to develop insight into the portfolio without cluttering it with unnecessary details, which is possible by the use of simple matrices with dimensions relevant to decision-makers.
5. Group support – all managers involved in the selection process must access relevant information and contribute their knowledge to the decision-making process.
6. Strategic considerations – for instance, the Q-sort method enables strategic considerations of all proposed projects. It allows a high level of classification of projects into strategic categories. Strategic planning approaches such as cognitive modeling and cluster analysis also have a broad strategic perspective. (Souder, 1978)

The reasons for bad choices are lack of the right metrics, not paying attention to risk, and inability to set boundaries.

Many organizations make an effort to make an individual project successful but do not make enough effort to make the entire project portfolio as successful as it could be. Just because an organization may have projects completed on time and within a budget does not mean it has the best possible portfolio or efficiently allocated resources between the projects it implements. Senior managers need to think about the total cost, risk, and value of the overall portfolio.

Project selection decisions should be made at the portfolio level; if not, the project portfolio will result from individual project choices made for each other without considering the impact that one project has on another. Regardless of which tool is used for the formal evaluation of project proposals, the basis for project approval, when decisions are made at the project-by-project level, is whether the proposed project meets some eligibility threshold.

Projects can be classified in different ways, according to size, duration, degree of risk, geographical location, necessary skills or technologies, sponsors, clients or market, phase of the project life cycle, type of product produced, etc. Different schemes can be used to classify each project in several different ways. No approach is best for every organization. The key is to choose the classification that brings the most information to decision-makers.

Each type of project has its characteristics, which cause different project management approaches to each type.

The project management profession needs a classification system for different projects (Youker, 2017), for effective communication, across the entire spectrum of projects and the globe. There are many potential purposes for a classification system. One of the worthwhile goals of listing different types of projects is market segmentation. The second is to define the different management approaches required for different projects.

There is usually a large volume of projects within each category or subcategory in large organizations. Each category's project management processes must provide flexibility in choosing the appropriate level of planning and control for large, complex, high-risk projects compared to small projects. Systematic categorization and classification of projects should be researched by all people involved in project management.

11.4 Discussion and Conclusions

The challenges and potentials of relating the circular economy approach and project portfolio selection have not yet been researched in-depth (Gareis et al., 2010), pointing out the need for additional research. Although increasing interest in the circular economy is observable in the project management practice and research community, direct integration of the CE characteristics in project management and operational methods is missing. Despite numerous circular methods, principles, and strategies in the literature, implementing these approaches into practice is still challenging.

Based on the analysis of all available sources related to the portfolio selection and issues considered within this process, there is no doubt that the CE approach's characteristics will extend the list of criteria and factors that should be considered in the project portfolio selection. The concrete proposals and guidelines are still missing, and their absence opens up space for further examinations by practitioners and academics.

Integrating the CE concept into the processes, standards, and practices of project portfolio management should be researched and discussed since the CE concept has different relevance to different organizations and project types.

Recent research has already considered the sustainability elements and key variables, but the CE topic's attractiveness imposes the need to include some reusability criteria within this process.

This paper summarizes the relevant characteristics of the selection process and all its related elements, which is necessary awareness through the decision-making process and additions that need to be incorporated if we deal with the CE approach. Further research should aim to find the models that support the practical selection of portfolios that meet decision-maker preferences and are simple to interpret by individuals who are not CE experts.

The CE framework's existence could support a company in evaluating and managing CE projects, facilitating CE portfolio management, and considering environmental sustainability and circularity in a single tool to help portfolio management. Although the CE transition might lead to sweeping changes, creating and managing a portfolio from a circularity perspective is necessary. Also, considerations at a portfolio level can help achieve more holistic tools for corporate, sustainable, and circular economy settings.

Future research into developing guidelines that would enable organizations to deal with challenges would be helpful. Some of the identified challenges are how to use circularity indicators to determine the extent of circularity; how to design for multiple-use cycles; how to anticipate how the circular offering will evolve over multiple life cycles; how to set circular criteria (design competencies for a circular economy); which of the multi-criteria decision-making methods to use in the selection.

The framework would guide decision-makers in the portfolio selection, open up a future research agenda for the CE approach, and help organizations carry out sustainable projects and promote a higher level of environmental responsibility.

A proposal of a possible framework that could integrate CE characteristics into PPM was made, but the concrete proposal of techniques and criteria in each step could contribute to developing the model that helps in a more structured selection process.

Also, when it comes to prioritization, besides the main areas already recognized in the literature, for the organizations willing to make the CE transition, the list of criteria should include the extent to which the selected portfolio supports CE initiatives and can contribute to the organization's sustainability.

Apart from the practical aspects and implementation attempts, circular economy thinking should be embedded as an indispensable part of companies' strategic visions and missions.

This research aimed to explore whether there is a need to include additional criteria within selecting projects that could indicate the connections between outputs of realized projects and future project proposals. The empirical part of the study is under development. The author believes that further research should expand to provide empirical and more profound statistical analysis with the correlation between the industry, project types, CE characteristics, selection criteria, and the creation of a reliable, coherent model for the CE project selection.

References

Aarseth, W., Ahola, T., Aaltonen, K., Økland, A., & Andersen, B. (2017). Project sustainability strategies: A systematic literature review. *International Journal of Project Management, 35*(6), 1071–1083.

Alvey, T., Goodwin, D., Ma, X., Streiffert, D., & Sun, D. (1998). A security-constrained bid-clearing system for the New Zealand wholesale electricity market. *IEEE Transaction on Power Systems, 13*(2), 340–346.

Archer, N., & Ghasemzadeh, F. (1996). *Project portfolio selection techniques: A review and a suggested integrated approach*. Innovation Research Centre.

Archer, N., & Ghasemzadeh, F. (1999). An integrated framework for project portfolio selection. *International Journal of Project Management, 17*, 207–216.

Archer, N., & Ghasemzadeh, F. (2000). Project portfolio selection through decision support. *Decision Support Systems, 29*, 73–88.

Archibald, R. D. (2008). The interfaces between strategic management of an enterprise and project portfolio management within the enterprise. In *22nd IPMA world congress project management to run, Roma*.

Baker, N. R. (1974). R&D project selection models: An assessment. *IEEE Transactions on Engineering Management, EM-21*(4), 165–171.

Bakker, C. A., den Hollander, M. C., van Hinte, E., & Zijlstra, Y. (2014). *Products that last: Product design for circular business models*. TU Delft Library Delft, Marcel den Hollander IDRC.

Bard, J. F., Balachandra, R., & Kaufmann, P. E. (1988). An interactive approach to R&D project selection and termination. *IEEE Transaction on Engineering Management, 35*(3), 139–146.

Berić, I. (2013). *Analiza primene kvantitativnih i kvalitativnih metoda u selekciji i optimizaciji portfolija projekata. PhD Dissertation*. University of Belgrade, Belgrade, Serbia.

Berić, I., Jovanović, F., & Jovanović, P. (2012). Implementation of quantitative methods in project selection. *Metalurgia International, 17*(5), 190–194.

Bocken, N., Short, S., Rana, P., & Evans, S. (2014). A literature and practice review to develop sustainable business model archetypes. *Journal of Cleaner Production, 65*, 42–56.

Bocken, N. M. P., de Pauw, I., Bakker, C., & van der Grinten, B. (2016). Product design and business model strategies for a circular economy. *Journal of Industrial and Production Engineering, 33*(5), 308–320. https://doi.org/10.1080/21681015.2016.1172124

Brenner, M. S. (1994). Practical R&D project prioritization. *Research Technology Management, 37*(5), 38–42.

Brook, J. W., & Pagnanelli, F. (2014). Integrating sustainability into innovation project portfolio management – A strategic perspective. *Journal of Engineering and Technology Management, 34*, 46–62.

Coenen, T. B. J., Haanstra, W., Jan Braaksma, A. J. J., & Santos, J. (2020). CEIMA: A framework for identifying critical interfaces between the circular economy and stakeholders in the lifecycle of infrastructure assets. *Resources, Conservation and Recycling, 155*, 104552.

Cooper, R. G. (2001). *Winning at new products: Accelerating the process from idea to launch* (3rd ed.). Perseus Publishing.

Cooper, R. G., Edgett, S. J., & Kleinschmidt, E. J. (1997). *Portfolio management for new products*. Innovation Research Centre, McMaster School of Business.

Cooper, R. G., Edgett, S., & Kleinschmidt, E. (2001). Portfolio management for new product development: Results of an industry practices study. *R&D Management, 31*(4), 361–380.

Dobrovolskienė, N., Tvaronaviciene, M., & Tamošiūnienė, R. (2017). Tackling projects on sustainability: A Lithuanian case study. *Journal of Entrepreneurship and Sustainability, 4*(4), 477–488.

Ellen MacArthur Foundation. (2016). *Intelligent assets: Unlocking the circular economy potential*. Ellen MacArthur Found.

European Commission. (2015). Communication from the Commission to the European Parliament, The Council, The European Economic and Social Committee and The Committee of the Regions Closing the Loop – An EU Action Plan for the Circular Economy.

Frame, J. D. (2002). *The new project management; tools for an age of rapid change, complexity and other business realities*. John Wiley & Sons.

Gareis, R., Huemann, M., & Martinuzzi, A. (2010). Relating sustainable development and project management: A conceptual model. In *Paper presented at PMI® research conference: Defining the future of Project Management, Washington, DC*. Project Management Institute.

Geissdoerfer, M., Pieroni, M. P. P., Pigosso, D. C. A., & Soufani, K. (2020). Circular business models: A review. *Journal of Cleaner Production, 277*, 123741.

Ghorbani, S., & Rabbani, M. (2009). A new multi-objective algorithm for a project selection problem. *Advances in Engineering Software, 20*, 9–14.

GPM Global. (2019). *The GPM P5™ standard for sustainability in project management, USA*.

Hall, D. L., & Nauda, A. (1990). An interactive approach for selecting R&D projects. *IEEE Transaction on Engineering Management, 37*(2), 126–133.

Hamano, M., Arekrans, J., & Ölundh Sandström, G. (2020). Toward circular economy implementation: A tool for integrating circularity indicators into portfolio management. (English) In *Electronics goes Green 2020+: The Story of Daisy*. Alexa and Greta, 2020.

Hitt, M. A., & Ireland, R. E. (2001). *Strategic management: Competitiveness and globalization*. South-Western College Publishing.

Hope, A. J., & Moehler, R. (2014). Balancing projects with society and the environment: A project, programme and portfolio approach. In *27th IPMA world congress. Procedia – social and behavioral sciences* (vol. 119, pp. 358–367).

Ivan, I., & Sandu, A. (2008). Project hierarchy based on duration and complexity. In *22nd IPMA world congress project management to run, Roma.*

Jiang, J. J., & Klein, G. (1999). Information system project-selection criteria variations within strategic classes. *IEEE Transaction on Engineering Management, 46*(2), 171–176.

Jovanović, P., & Berić, I. (2018). Analysis of the available project management methodologies. *Management: Journal of Sustainable Business and Management Solutions in Emerging Economies, 23*(3), 1–13.

Kaplan, R. S., & Norton, D. P. (2001). *The strategy-focused organization: How balanced scorecard companies thrive in the new business environment* (Vol. 29). Harvard Business School Press.

Kendall, G. I., & Rollins, S. C. (2003). *Advanced project portfolio management and the PMO: Multiplying ROI at warp speed.* J. Ross Publishing.

Khalili-Damghani, K., & Tavana, M. (2014). A comprehensive framework for sustainable project portfolio selection based on structural equation modeling. *Project Management Journal, 45*(2), 83–97.

Khalili-Damghani, K., Sadi-Nezhad, S., Lotfi, F. H., & Tavana, M. (2013). A hybrid fuzzy rule-based multi-criteria framework for sustainable project portfolio selection. *Information Sciences, 220*, 442–462. ISSN 0020-0255.

King, M., & Mercer, A. (1987). Note on a conflict of assumptions in bidding models. *European Journal for Operational Research, 32*(3), 462–466.

Kirchherr, J., Reike, D., & Hekkert, M. (2017). Conceptualizing the circular economy: An analysis of 114 definitions. *Resources, Conservation and Recycling, 127*, 221–232.

Korhonen, J., Honkasalo, A., & Seppälä, J. (2018). Circular economy: The concept and its limitations. *Ecological Economics, 143*, 37–46.

Krumm, F., & Rolle, C. F. (1992). Management and application of decision and risk analysis in Du Pont. *Interfaces, 22*(6), 84–93.

Labuschagne, C., & Brent, A. C. (2005). Sustainable project life cycle management: The need to integrate life cycles in the manufacturing sector. *International Journal of Project Management, 23*(2), 159–168.

Lereim, J. (2008). Optimization of project portfolios from a competition strategy point of view. In *22nd IPMA world congress "project management to run", Roma.*

Liu, X. S. (1999). Research on project selection of enterprise product development. *Industrial Technology and Economy, 18*(5), 63–64.

Ma, J., Harstvedt, J. D., Jaradat, R., & Smith, B. (2020). Sustainability driven multi-criteria project portfolio selection under uncertain decision-making environment. *Computers & Industrial Engineering, 140*, 106236. ISSN 0360-8352.

MacArthur, E. (2013). Towards the circular economy. *Journal of Industrial Ecology, 23–44.*

Machacha, L. L., & Bhattacharya, P. (2000). A fuzzy-logic-based approach to project selection. *IEEE Transaction on Engineering Management, 47*(1), 65–73.

Marcelino-Sádaba, S., Pérez-Ezcurdia, A., & González-Jaen, L. F. (2015). Using project management as a way to sustainability. From a comprehensive review to a framework definition. *Journal of Cleaner Production, 99*, 1–16.

Martino, J. (1995). *Research and development in project selection.* Wiley.

Meredith, J. R., & Mantel, S. J. (2012). *Project Management: A managerial approach* (8th ed.). John Wiley& Sons.

Mohanty, R. P. (1992). Project selection by a multiple-criteria decision-making method: An example from a developing country. *International Journal of Project Management, 10*(1), 31–38.

Moustafaev J. (2010). Chapter 15: Project portfolio management part 1: PPM illustrated. In J. Moustafaev (Ed.), *Delivering exceptional project results: A practical guide to project selection, scoping, estimation and management.* Available at www.maxwideman.com/guests

Mukherjee, K. (1994). Application of an interactive method for MOLIP in project selection decision: Case from Indian coal mining industry. *International Journal of Production Economics, 36*, 203–211.

Ngan, S. L., How, B. S., Teng, S. Y., Promentilla, M. A. B., Yatim, P., Er, A. C., & Lam, H. L. (2019). Prioritization of sustainability indicators for promoting the circular economy: The case of developing countries. *Renewable and Sustainable Energy Reviews, 111*, 314–331.
Obradović, V., Jovanović, P., Đorđević, N., Berić, I., & Jovanović, F. (2012). Using project management as a way to excellence in healthcare. *HealthMED, 6*(6), 2100–2107.
Økland, A. (2015). Gap analysis for incorporating sustainability in project management. *Procedia Computer Science, 64*, 103–109.
Okpala, D. C. (1991). Evaluation and selection of construction project in Nigeria. *Construction Management and Economics, 9*(1), 51–61.
Pieroni, M., McAloone, T., & Pigosso, D. (2019). Configuring new business models for circular economy through product–service systems. *Sustainability, 11*(13), 3727.
Puthamont, S., & Charoenngam, C. (2007). Strategic project selection in public sector: Construction projects of the ministry of defence in Thailand. *International Journal of Project Management, 25*(2007), 178–188.
Rashid, A., Asif, F. M., Krajnik, P., & Nicolescu, C. M. (2013). Resource conservative manufacturing: An essential change in business and technology paradigm for sustainable manufacturing. *Journal of Cleaner Production, 57*, 166–177.
Rengarajan, S., & Jagannathan, P. (1997). Project selection by scoring for a large R&D organization in a developing country. *R&D Management, 27*(2), 155–164.
Rothkopf, M. H., & Engelbrecht-Wiggans, R. (1993). Misapplications reviews getting model right – The case of competitive bidding. *Interfaces, 23*(3), 99–106.
Sánchez, M. A. (2015). Integrating sustainability issues into project management. *Journal of Cleaner Production, 96*, 319–330.
Sassanelli, C., Rosa, P., Rocca, R., & Terzi, S. (2019). Circular economy performance assessment methods: A systematic literature review. *Journal of Cleaner Production, 229*, 440–453.
Schelle, H., Ottmann, R., & Pfeiffer, A. (2006). *Project manager*. German Association for Project Management.
Schipper, R. P. J. R., & Silvius, A. J. G. (2018). Towards a conceptual framework for sustainable project portfolio management. *International Journal of Project Organisation and Management, 10*(3), 191.
Silvius, G. (2017). Sustainability as a new school of thought in project management. *Journal of Cleaner Production, 166*, 1479–1493.
Skitmore, M., & Pemberton, J. A. (1994). A multivariate approach to construction contract bidding make-up strategies. *Journal of the Operational Research Society, 45*(11), 1263–1272.
Slade, S. (1991). Case-based reasoning. *AI (Artificial Intelligence) Magazine, 12*(1), 42–45.
Sommer, R. J. (1999). Portfolio management for projects: A new paradigm. In L. D. Dye & J. S. Pennypacker (Eds.), *Project portfolio management – Selecting and prioritizing projects for competitive advantage*. Center for Business Practices.
Souder, W. E. (1978). A system for using R&D project evaluation methods. *Research Management, 21*(5), 33.
Stawicki, J., & Müller, R. (2007). From standards to execution: Implementing program and portfolio management. In *21st IPMA world congress, Cracow, Poland.*
Su, B., Heshmati, A., Geng, Y., & Yu, X. (2013). A review of the circular economy in China: Moving from rhetoric to implementation. *Journal of Cleaner Production, 42*, 215–227.
Tam, C. M. (1999). Build-operate-transfer model for infrastructure developments in Asia: Reasons for successes and failures. *International Journal of Project Management, 17*(6), 377–382.
Tiong, R. L. K., & Alum, J. (1997). Evaluation of proposals for BOT projects. *International Journal of Project Management, 15*(2), 67–72.
Toljaga-Nikolić, D., Todorović, M., Dobrota, M., Obradović, T., & Obradović, V. (2020). Project management and sustainability: Playing trick or treat with the planet. *Sustainability, 12*(20), 8619. https://doi.org/10.3390/su12208619
Turner, J. R., & Müller, R. (2003). On the nature of the project as a temporary organization. *International Journal of Project Management, 21*, 1–8.

Wang, J., Xu, Y., & Li, Z. (2009). Research on project selection system of pre-evaluation of engineering design project bidding. *International Journal of Project Management, 27*, 584–599.

Webster, K. (2015). *The circular economy: A wealth of flows*. Cowes Isle of Wight.

Wessels D. J. (2008). *Project Prioritizing. Guest article in Portfolio Management: Aligning projects with corporate strategy and values*. Available at http://www.maxwideman.com/guests/portfolio_mgmt/prioritizing.htm

World Bank. (2003, November). *Project appraisal document. Highway management project in Thailand.*

Youker, R. (1999). The difference between different types of projects. In *The Project Management Institute 1999 conference in Philadelphia*.

Youker, R. (2017, April). The difference between different types of projects. *PM World Journal, VI*(IV).

Chapter 12
Circular Economy and Project Management: The Road Ahead

Marija Todorović and Vladimir Obradović

Abstract Sustainability and circularity open many areas for practitioners, academia, and policy decision makers on how to run business and economy. Sustainable management and circular economy are nowadays present in different ways and levels in a society; therefore, it seems necessary to explore how projects and project management can contribute to sustainable business and circular economy and what projects can bring to society if managed based on circular economy principles. The purpose of this chapter is an overview of opportunity for integration of project management and circular economy and the benefits that could arise from this integration. Based on profound literature review, circular economy business models are in expansion and project management as a discipline has a potential to bring value for circular economy. The aim of this chapter is to contribute with an overview of existing forms of circular economy and project management potential to create the road ahead. Key findings are mapped project management areas as project strategy, processes, tools, and competencies, for integration of circularity.

Keywords Circular · Project · Project management · Economy

12.1 Introduction

A distinguished global institution provided data according to which we can conclude that a global circularity is less than 10% at the global level. The measuring started in 2018, when the measure of circularity was 2.5% (Reuters, 2019; PACE, 2021; OECD, 2020). In addition to this data, it is estimated that the use of resources will be tripled until 2050, compared to 1970 (PACE, 2021). The whole society became aware and concerned about the resource consumption for the same or different motives, hence the concern is overall present. Government and civil societies

M. Todorović (✉) · V. Obradović
Faculty of Organizational Sciences, University of Belgrade, Belgrade, Serbia
e-mail: marija.todorovic@fon.bg.ac.rs; vladimir.obradovic@fon.bg.ac.rs

V. Obradović (ed.), *Sustainable Business Change*,
https://doi.org/10.1007/978-3-031-23543-6_12

are engaged and contributing to environment, society, and economy. On the other side, companies are part of a society ecosystem and are very much aware of how important are resources and benefits that circular economy (CE) can provide their business, especially on the profit side. From the aspect of economy, CE opens a completely new field for business. Many multinational companies have already incorporated circular economy business model: IKEA, Renault, P&G, IKEA, H&M, Apple, PH, Bridgestone, Xerox, Rolls Royce, Forlin & Scholz, and many others. In addition, start-ups are considered important pillars of business and could positively contribute to develop the technological innovation for the CE (Pruijsen, 2019).

The key megatrends of a society are demographic growth, climate change, urbanism, shift in economic power, and technological breakthroughs (PwC, 2019; OECD, 2020; Florence, 2019; Roland Berger Institute, 2020; Ezzat, 2016). International institutions provided research, initiatives, and action plans for CE. The concept of circular economy (CE) has been largely promoted by European Commission which has adopted New Circular Economy Action Plan, in 2020 (European Green Deal, 2019). One more European initiative is the European Circular Economy Stakeholder Platform, a joint initiative by the European Commission and the European Economic and Social Committee. Several governments at various levels have been establishing CE strategies/roadmaps/action plans/framework/white papers (Colombia, Denmark, Finland, the Netherlands, Spain, Sweden, France Belgium, Chile, France, Slovenia, Spain, Portugal, Italy, Norway) (OECD, 2020). CE in Europe could generate savings of even €600 billion per year and achieve €1.8 trillion in economic benefits by 2030 (Ellen McArthur Foundation and McKinsey, 2015).

For business and a society, it is very important to:

- Measure the performances and the effect of CE business model (Laubscher & Marinelli, 2014; Beulque & Aggeri, 2016; Scarpellini, 2021; Vegter et al., 2020; Rossi et al., 2020; Sassanelli et al., 2019; Saidani et al., 2019)
- Understand what the impact of the changes on the society is (Bocken et al., 2019; OECD, 2020; Pitkänen et al., 2020; Banaitė, 2016; Kyriakopoulos et al., 2019)
- Learn, develop, and become more circular with sharing knowledge, best practices and know-how (Bolger & Doyon, 2019; Hedlund et al., 2020; Marra et al., 2018; Atiku, 2020; Mullins et al., 2020; Jiménez-Rivero & García-Navarro, 2017; Kirchherr & Piscicelli, 2019; Petit-Boix & Leipold, 2018).

The key challenge is to discover if project management logic could support business and society needs and how we can be more circular in project management. The purpose of this chapter is an overview of an opportunity for integration of project management and circular economy and the benefits that could arise of this integration.

12.2 Circular Economy Development and Perspective

CE is perceived as a systematic change, not just the nature of a business but the change of the entire value chain. It is oriented on replacement of source materials with alternative resources (more durable, bio-degradable, recycled), design out waste and pollution, keeping products and materials in use, and regeneration of natural systems. It is a model that regenerates itself, production, and consumption of goods through closed-loop material flows (Lewandowski, 2016; Linder & Williander, 2017). The circular economy concept significantly contributes to sustainable development (Schroeder et al., 2019; Suárez-Eiroa et al., 2019) that explains the reason for becoming a hot topic in all aspects of a society.

12.2.1 Circular Business Models

As mentioned above, companies have become aware of the pressure associated with the resource availability and consumption. Upon that, companies are oriented to new business models to make a profit and contribute to social well-being as a socially responsible entity. One of the main competitive advantages of a company today is knowledge, resilience, and ability to create value using opportunities for development. The literature provides insights into CE business model drivers that could be summed up in internationalization and strict environmental regulations, institutional factors such as governmental support, laws, tax policy, global standards and goals, technological development, and information transparency (Tura et al., 2019).

There are many definitions of a circular business model. One of them presents it as "a model in which the conceptual logic for value creation is based on utilizing the economic value retained in products after use in the production of new offerings" (Linder & Williander, 2017). Geissdoerfer et al. (2020) has collected many definitions in order to describe the types of circular business models. Based on his analysis circular business models are classified into four categories:

- Cycling – providing product through reuse, remanufacturing, refurbishing of material, and energy
- Extending – extended product usage due to design, marketing, maintenance, and repair
- Intensifying – providing product through sharing economy solution
- Dematerializing – providing product through service and software solution.

Very similar logic could be found in Lewandowski (2016) where circular business models are classified based on the next criteria: regenerate (energy recovery, circular supplies, sustainable product location, etc.); share (product lease, maintaining ad repair, upgrading, etc.); optimize (assets management, waste management; produce on demand, etc.); loop (recycling, resource recovering, etc.); virtualize (dematerialized services); exchange (new technologies). Laubscher and Marinelli (2014) have defined the next areas for integration of the CE principles in business in sales, supply, IT management, partnership, and human resources (HR): (a) Sales in terms of shifting from selling products toward selling services; (b) Product designed and material and components in the production phase; (c) IT/data management to enable resource optimization using technologies: (d) Supply loops as the maximization of the recovery of own assets where profitable and to maximization of the use of recycled materials/used components; (e) Partnerships with suppliers and customers; (f) HR incentives toward culture adaptation and building capacities of human resources.

Most of the authors accepted the approach of 10R framework for companies to support circular economy: reuse, rethink, reduce, refuse, repair, refurbish, remanufacture, repurpose, recycle, and recover (Blomsma et al., 2019; Campbell-Johnston et al., 2020), which are directed at application of materials, manufacturing, and product lifespan.

There are many examples of companies who have successfully applied CE business model strategies, some of them are Renault with implemented solutions of using car sharing platform, integrating plastics from recycling into its vehicles and open first European factory dedicated to the circular economy of mobility; HP and Xerox are (among other initiatives) moving into the product as a service business model: HP is focused on leasing, renting and other service contracts for ink, print and PC services Xerox, use "pay-per-copy" instead selling actual machine): Ikea organizes workshop to educate customer about the product, recycling, and repairing options; has opened the second hand store, where customers can find repaired or refurnished products, previously acquired through "take back" program; H&M from textile industry is reusing and repairing them as much as possible, before finally recycling items.

There are also companies that are providing platforms from intermediation platforms either business-to-business (B2B) or business-to customers (B2C) sharing marketplaces. These companies enable other companies to communicate with each other, for example, Kalundborg Symbiosis allows public and private companies to sell and buy waste of industrial production (Urbinati et al., 2017).

Very important role in CE ecosystem play start-ups (D'Amatao et al., 2020). Start-ups are important pillar of business and could positively contribute to develop the technological innovation for the CE (Pruijsen, 2019). Large companies due to the system complexity are more and more oriented on the open innovation concept and cooperation with start-up companies to provide innovative solutions faster (Chesbrough, 2006; Todorović, 2020). The other market opportunity for CE start-ups is of course new market demands. London, it is estimated that 40,000 new jobs will be created by 2036 in the areas of reuse, remanufacturing, and materials (OECD, 2020).

There are six types of CE start-ups (Henry et al., 2020):

- Design-based – pre-market phase through source material minimization
- Waste-based – seeking additional value from waste

- Platform-based – pursuing sharing/trading business models
- Service-based – to increase usage efficiency
- Nature-based – based on nature-based systemic solutions

However, the implementation of CE business models has many barriers. The key ones related to business environment and industry are the lack of knowledge on how to transform companies' operations into circular business, the lack of knowledge of new technologies and materials, lack of databases for sharing waste information, industrial focus on linear model, and the lack of suitable partners to establish supply chains meeting CE principles (Tura et al., 2019).

As mentioned at the beginning of the paper organizations might become sustainable by implementing circular business models; however, it requires many intraorganizational and interorganizational initiatives. Organizational barriers are related to management support, organizational agility, risk awareness (Liu & Bai, 2014), capabilities and to incorporate CE principles in existing operations, communication, organization culture, strategic thinking, lack of knowledge on CE benefits for business, technology solutions, materials, product design, etc. (Kirchherr et al., 2018).

12.2.2 National and Local CE Initiatives

To introduce the paradigm of CE, to create good business environment and to raise awareness of people related to transition to CE, it is necessary to implement initiatives at the government level. The success of economic activity is very dependent on national governmental actions, goals and its incorporation in the education system (Kirchherr & Piscicelli, 2019). European Commission, OECD, PACE; UNEP are providing research, measuring of CE effect, guides, events, forums, and platform to enable countries and cities to implement CE concept. Nowadays, countries and cities have strategic document up to 2030 and 2050 where the promotion of sustainable development is a common aim including various national circular economy initiatives. Based on OECD (2020) study, countries that have implemented CE are mainly focused on these objectives: protect environment, decrease consumption, and rethink production and business models, favours innovations, boost behavioural change. Actions are focused on climate change, waste reduction and more efficient and optimal use of resources, reduction in the use of primary raw materials, neutrality, stimulating employment through new jobs related to recycling, remanufacturing, etc.

A good example is Germany, who has benefited from the implementation of CE. The number of people who have been employed is 270,000 employees, 11,000 companies started the business, and the turnover in 2018 was €70 billion. In addition, greenhouse gas emissions seriously harmful for human health were significantly reduced (Mohajan, 2021a, b).

CE present a system change which is the reason of involving all stakeholder in the process that cannot be done only by companies. Strategic orientation of many countries confirms this, as well as the leading education institutions on the world, introducing CE as a course and even study programmes.

12.2.3 *Technological Breakthrough*

Digital technologies are one of the megatrends in the society. It is recognized in literature and in practice that information technologies can significantly contribute to CE, especially in regard to life cycle stages. Digital technologies can help close the material loop and further to, monitoring, control, and optimize stocks.

Digital technologies can be divided into (1) data collection (Radio Frequency Identification (RFID) and Internet on Things (IoT), (2) data integration (Relational Database Management Systems (RDBMS) and Product Lifecycle Management (PLM) systems, and (3) data analysis (Product Lifecycle Management (PLM) systems, Artificial intelligence, and Machine learning (Pagoropoulos et al., 2017).

Digital technologies can support smart manufacturing, smart remanufacturing, and smart recycling (Alcayaga et al., 2019). Introducing technologies changes not only business models but also the way of value creation, enabling capturing of data, storage, and analysis of product information (Pagoropoulos et al., 2017).

From the perspective of CE, it is important to track the current state of the product to predict future condition and to plan actions. This further enables on-time remanufacturing planning, reduction of time waste, waste of work, inventory, disposal, etc..

Smart technologies and all the opportunities of Industry 4.0 significantly improve circular business models, especially Big data analytics and Artificial intelligence (Alcayaga et al., 2019; Todorović, 2020). Concepts of CE, Internet on Things, and product-service system can be synthesized on a framework of smart-circular systems. The smart-circular system consists of smart use, smart maintenance, smart reuse, smart remanufacturing, and smart recycling (Alcayaga et al., 2019).

Hence, the implementation of technologies directly affects business processes, procedures, working practices, employees, organizational values, system, and strategies. Therefore, the success of the smart CE implementation depends largely on organizational factors.

12.3 Project Management Development and Perspective

The evolution of project management has shown that numerous methodologies, approaches, concepts, and standards have been defined. The first globally accepted approach, known as the "waterfall" approach, is a linear and incremental approach with defined phases. This approach has been proposed by the leading institutions of the Institute for Project Management (PMI), the International Project Management Association (IPMA), and the European Commission. The advantage of this approach is that it emphasizes the development and requirements of the project, and the disadvantage that it is difficult to predict all the circumstances at the beginning of the project, i.e. the client is not able to always state all the requirements. The International Organization for Standardization presented the international standard ISO 21500:2012 to provide general guidelines and principles for Project Management, that confirm the presence of projects in every aspect of a society.

With various changes in the business world and society, it has been concluded that many projects have never been completed and/or did not deliver all expected results (Zwikael & Globerson, 2006). Various studies conducted in the early 2000s have shown that between 25% and 40% of the time on a project needed rework; that 40% of errors were detected by users, and that 2/3 of IT projects fail (Hass, 2007). In response to this Agile Project Management was presented as a non-linear approach in which processes are carried out iteratively. Agile approach focuses on integrated communication, user involvement, document minimization, incremental, and iterative development for fast delivery of results in a constantly changing environment. It is often emphasized that the agile approach gives good results on small and medium projects in which communication with an experienced team is easily facilitated (Dyba & Dingsoyr, 2008). Still, the application of agile methodologies depends on organizational factors.

Cross-border project, international and public projects mostly have high level of complexity, significant impact, great number of beneficiaries, many stakeholders, and sources of funding. This led to significant changes in the approach to project management. To cope with the trend of projectification of a society and new demands, European Commission presented a new methodology for project management in the public sector: Open Project Management Methodology – Open PM^2 (European Commission, 2018).

Following global trend of sustainable management, project management evolved with a new standard for sustainability in project management: GPM P5 Standard for Sustainability in Project Management, provided by Green Project Management Institute (GPM, 2018).

Projects are a tool for the implementation of organizational strategy, enabling the organization to implement planned activities and achieve change through a special organization, project. For this reason, the project management system should be set up to meet the needs of the organization. The needs of organizations today are to have special business models that are competitive and create positive social change. For almost 10 years, the authors have been exploring the need to combine existing models and find solutions that will ensure business dynamics and business sustainability (Haigh and Hoffman, 2011; Doherty et al., 2014). Based on the previous chapter we can conclude that CE contributes to sustainable development. Further, it is crucial for research and practice (Velenturf et al., 2019).

Projects have been recognized as one of the crucial mechanisms to enable organizational learning, provide innovation, and/or change implementation (Scipioni et al., 2021). Many industries, companies, non-governmental organizations, and public institutions nowadays are project-based (Artto & Wikström, 2005; Lundin et al., 2015; Packendorff & Lindgren, 2014; Jensen et al., 2016). This is especially important for the CE implementation, not only from a company perspective but from the perspective of the economy. CE requires partnership in the supply chain and is not just a question of one entity. The main barriers to CE implementation are industry development, knowledge and business processes of all partners in the value chain, institutional regulation, and global standards. Projects are the mechanism used by international institutions such as European Commission, UNDP, OECD to implement circularity at the local and national level, to create ecosystem for circular business. From business perspective, one of the most popular trends is open innovations where large companies work on projects with small companies to provide innovations increase company's productivity and agility (Leemann, 2002; Bagherzadeh et al., 2019). The next chapter analyses the option to integrate CE principles in project management and the ways project management could contribute to CE business.

12.4 Project Management Meets Circular Economy

From 2010 to 2020, 78 studies were published on the topic of CE (Piscitelli et al., 2020). At the project level, it also recognized how important it is to apply circular economy principles and philosophy on projects. One of the industries where project management research is very present is the construction industry. This industry frequently requires multifunctional teams and collaboration between different partner companies that make this industry complex, yet very suitable to demonstrate the implementation of circularity. In addition, the construction industry is welcome for circular economy principles due to resource usage. Senaratne et al. (2021) have published a study on circular economy principles for sustainable construction, where it is clearly emphasized the importance of stakeholder collaboration to meet project goals for circular buildings. From this study, we can see that attaining circularity along with other project objectives is a necessity. Further, the authors stated the significance of CE mindset of designers, collaboration with suppliers, manufacturers, client, and end users in promoting CE are key factors that enable sustainability and circularity on project.

Modern business development highlights the strategy that aims to provide value for customers. As seen in the previous chapter, most of the companies, non-governmental organizations, and public institutions implement their strategies using projects. But strategies do not fail when they are being analysed or when the objectives are being set. They fail during implementation and, more particularly, due to the lack of proper project management (Van Der Merwe, 2002). To deal with this issue an approach of strategic project management has been developed with the purpose to focus project implementation on achieving business performances (Patanakul & Shenhar, 2012).

One of the options to integrate CE principles at the project level could be using project strategy (top-down approach). Strategic project management includes strategic alignment of the projects including: project selection and portfolio management, leading, monitoring, and controlling projects to achieve strategic goals of the company. Project strategy presents a strategic focus for project teams, that is designed to help achieve business goals (Milosevic & Srivannaboon, 2006), and a project direction that enables project success in its environment (Artto et al., 2008). The main components of project strategy are (1) perspective (business background, business objectives, strategic concept); (2) position (product definition, competitive advantage, success/failure criteria; (3) guidelines (project definition, strategic focus) (Shenhar et al., 2007). It can be concluded that if a company follows CE principles and uses CE strategy and business model, the CE philosophy could be implemented to a project level, therefore project design, resource usage, resource management, and other processes could be managed in a circular way and contribute to the overall circular business performances.

CE principles could be integrated via project management processes: initiation, planning, execution, monitoring, and closing. Through the project initiation, project goals, scope, and results could be stated in terms to integrate circularity. Good planning could improve collaboration and communication on the project, which is one of the recognized organizational barriers to CE implementation. Execution and monitoring could follow CE principles to ensure circular business performances. This integration via project management process depends on project management approach and methodology.

Circularity could be implemented using different project management tools. This is very much related with the previous options since typical project management tools could be used to incorporate CE principles. For example, project charter, different project plans, and especially monitoring system. Project charter can present the intention toward circularity since it presents the identity of a project. Project plans could include a set of activities on collaboration with suppliers, manufacturers, client end users and other partners in promoting CE (Senaratne et al., 2021). Since projects are oriented on achieving business performances, the monitoring and controlling systems should be an important tool to track project results. Monitoring system could include key performance indicators that are consistent with circular performance at the organizational level. Further, Mayer et al. (2019) emphasized the importance of measuring progress toward circular economy meaning that the project evaluation could also include criteria that illustrate desired impact on circular economy. Lessons learned could be a significant tool to acquire knowledge and create new knowledge at the project and organizational levels (Scipioni et al., 2021).

One of the levels where CE can be incorporated in project management is the level of competencies. Leading project management associations (IPMA and PMI) are developing a framework of project managers competencies. Competencies are in every industry crucial for value creation (Todorović & Obradović, 2018). On the other hand, one of the barriers of CE implementation is lack of knowledge on CE business models, CE benefits, technology solutions, materials, product design, etc. It is stated by Scipioni et al. (2021) that the knowledge and the learning process in the organization are crucial for CE implementation. Project manager competencies framework includes competencies in three areas: perspective, practice, and behavioural. Based on previous conclusion, the set of standardized set of project management competencies is in line with CE requirement, and still it requires new knowledge on CE practice, context, and attitude to cope up with all CE requirements regarding new solutions, green technologies, product/service design, options of recycling, reusing, etc.

The opposite point of view is to consider projects as a mechanism to implement CE, to contribute to a company that does not have CE business model or does not implement CE principle in any form. In this case, projects can be a vehicle to test and introduce circularity in the organization. Many companies are using open innovation approach that has been shown as a very effective approach, especially for complex systems. Companies which use open innovation are developing external connections and spread network, get new knowledge, and proven solutions, shared. It can be concluded that strong ties to partners lead to superior firm performance (Caputo et al., 2016). If the market accepts CE concept and there are reliable partners for the cooperating, e.g. CE start-ups that are very agile, a company can consider an option of developing a project in cooperation with partner(s) and create an innovation that follows CE principles. Project could present a roadmap for the company on how to use CE in its business (bottom-up approach). In addition, these projects can be a knowledge provider to a company.

12.5 Discussion and Conclusion

Circular economy business models are in expansion and project management as a discipline has a potential to bring a value for circular economy. Project management is present in all aspects of a society and many business companies as well as non-governmental and public institutions are project-based. The main international institutions are developing circular ecosystem at the national and local levels in many countries based on projects. Therefore, the importance of integrating circularity into project management could bring many benefits for the system.

This chapter analyses the impact that CE has on the companies, the development of CE business models, and their presence at the global level. It also presents the evolution of the project management discipline that has already incorporated market and a society needs in terms of agility, sustainability, and digitization. The concept has proven flexibility which was the starting point of initiation of circularity in project management.

Projectification of a society, on the other hand, and barriers for the CE implementation on the other hand provide an opportunity to define the area of improvement of project management to contribute to the CE trend. The chapter defines the option of implementing CE at the project level as top-down approach. CE principles could be manifested on project via project strategy, project management processes, project management tools, and project management competencies. Considering the fact that project could be used to introduce CE in a company, we presented the concept of the open innovation as a mechanism to introduce CE to the company via collaboration.

Future direction of the research could be towards each of the suggested options for integrating CE in project management. The integration of CE via competencies could be extremely significant for organizations in selecting and developing project managers to manage projects based on circularity. Cascading from organizational strategy to a project strategy could enable the system to implement circularity through all business processes. This could be even more supported if the circularity is implemented via project management tools. Different project management methodologies have different possibilities to incorporate circularity, still a proper way could be found. The main conclusions are that CE integration in project management is a valuable topic for researchers and practitioners equally and that there is a significant need for developing project management framework that includes circularity.

References

Alcayaga, A., Wiener, M., & Hansen, E. G. (2019). Towards a framework of smart-circular systems: An integrative literature review. *Journal of Cleaner Production, 221*, 622–634.

Artto, K. A., & Wikström, K. (2005). What is project business? *International Journal of Project Management, 23*(5), 343–353.

Artto, K., Kujala, J., Dietrich, P., & Martinsuo, M. (2008). What is project strategy? *International Journal of Project Management, 26*(1), 4–12.

Atiku, S. O. (2020). Knowledge management for the circular economy. In *Handbook of research on entrepreneurship development and opportunities in circular economy* (pp. 520–537). IGI Global.

Bagherzadeh, M., Markovic, S., & Bogers, M. (2019). Managing open innovation: A project-level perspective. *IEEE Transactions on Engineering Management, 68*(1), 301–316.

Banaitė, D. (2016). Towards circular economy: analysis of indicators in the context of sustainable development. *Social Transformation in Contemporary Society, 4*(9), 142–150.

Beulque, R., & Aggeri, F. (2016). Circular business model innovation: Key patterns and challenges to unleash recycling value creation potential. In *EGOS*.

Blomsma, F., Pieroni, M., Kravchenko, M., Pigosso, D. C., Hildenbrand, J., Kristinsdottir, A. R., et al. (2019). Developing a circular strategies framework for manufacturing companies to support circular economy-oriented innovation. *Journal of Cleaner Production, 241*, 118271.

Bocken, N., Strupeit, L., Whalen, K., & Nußholz, J. (2019). A review and evaluation of circular business model innovation tools. *Sustainability, 11*(8), 2210.

Bolger, K., & Doyon, A. (2019). Circular cities: exploring local government strategies to facilitate a circular economy. *European Planning Studies, 27*(11), 2184–2205.

Campbell-Johnston, K., Vermeulen, W. J., Reike, D., & Brullot, S. (2020). The circular economy and cascading: towards a framework. *Resources, Conservation & Recycling: X, 7*, 100038.

Caputo, M., Lambert, E., Cammarano, A., & Michelino, F. (2016). Exploring the impact of open innovation on firm performances. *Management Decision, 54*(7), 1788–1812.

Chesbrough, H. (2006). Open innovation: a new paradigm for understanding industrial innovation. *Open Innovation: Researching a New Paradigm, 400*, 0–19.

D'Amatao, D., Veijonaho, S., & Toppinen, A. (2020). Towards sustainability? Forest-based circular bioeconomy business models in Finnish SMEs. *Forest Policy and Economics, 110*, 101848.

Doherty, B., Haugh, H., & Lyon, F. (2014). Social enterprises as hybrid organizations: A review and research agenda. *International Journal of Management Reviews, 16*(4), 417–436.
Dyba, T., & Dingsoyr, T. (2008). Empirical studies of agile software development: A systematic review. *Information and Software Technology, 50*(9–10), 833–859.
EMF (Ellen Macarthur Foundation). (2015). *Growth within: A circular economy vision for a competitive Europe*. Retrieved from: https://emf.thirdlight.com/link/8izw1qhml4ga-404tsz/@/preview/1?o
European Commission. (2018). PM² Project management methodology. Guide 2.0.
European Green Deal. (2019). Available at: https://eur-lex.europa.eu/legal-content/EN/TXT/?qid=1588580774040&uri=CELEX:52019DC0640
Ezzat, A. M. (2016). Sustainable development of seaport cities through circular economy: A comparative study with implications to Suez Canal corridor project. *European Journal of Sustainable Development, 5*(4), 509–509.
Florence, G. (2019). *Global Trends to 2030 Challenges and Chices for Europe*. European Strategy and Policy Analysis System.
Geissdoerfer, M., Pieroni, M. P., Pigosso, D. C., & Soufani, K. (2020). Circular business models: A review. *Journal of Cleaner Production, 277*, 123741.
Green Project Management Institute. (2018). The GPM P5 Standard for Sustainability in Project Management, USA.
Haigh, N., & Hoffman, A. J. (2011). Hybrid organizations: the next chapter in sustainable business. *Organizational Dynamics, 41*(2), 126–134.
Hass, K. B. (2007). The blending of traditional and agile project management. *PM World Today, IX*(V), 1–8.
Hedlund, C., Stenmark, P., Noaksson, E., & Lilja, J. (2020). More value from fewer resources: how to expand value stream mapping with ideas from circular economy. *International Journal of Quality and Service Sciences, 12*, 447–459.
Henry, M., Bauwens, T., Hekkert, M., & Kirchherr, J. (2020). A typology of circular start-ups: An Analysis of 128 circular business models. *Journal of Cleaner Production, 245*, 118528.
Jensen, A., Thuesen, C., & Geraldi, J. (2016). The projectification of everything: Projects as a human condition. *Project Management Journal, 47*(3), 21–34.
Jiménez-Rivero, A., & García-Navarro, J. (2017). Best practices for the management of end-of-life gypsum in a circular economy. *Journal of Cleaner Production, 167*, 1335–1344.
Kirchherr, J., & Piscicelli, L. (2019). Towards an education for the circular economy (ECE): five teaching principles and a case study. *Resources, Conservation and Recycling, 150*, 104406.
Kirchherr, J., Piscicelli, L., Bour, R., Kostense-Smit, E., Muller, J., Huibrechtse-Truijens, A., & Hekkert, M. (2018). Barriers to the circular economy: Evidence from the European Union (EU). *Ecological Economics, 150*, 264–272.
Kyriakopoulos, G. L., Kapsalis, V. C., Aravossis, K. G., Zamparas, M., & Mitsikas, A. (2019). Evaluating circular economy under a multi-parametric approach: A technological review. *Sustainability, 11*(21), 6139.
Laubscher, M., & Marinelli, T. (2014). Integration of circular economy in business. In *Proceedings of the conference: Going green–care innovation* (pp. 1–7).
Leemann, T. (2002). Project management: managing the chaos of change. *Journal of Business Strategy, 23*, 11–15.
Lewandowski, M. (2016). Designing the business models for circular economy—Towards the conceptual framework. *Sustainability, 8*, 1–28.
Linder, M., & Williander, M. (2017). Circular business model innovation: inherent uncertainties. *Business Strategy and the Environment, 26*(2), 182–196.
Liu, Y., & Bai, Y. (2014). An exploration of firms' awareness and behavior of developing circular economy: An empirical research in China. *Resources, Conservation and Recycling, 87*, 145–152.
Lundin, R. A., Arvidsson, N., Brady, T., Ekstedt, E., Midler, C., & Sydow, J. (2015). *Managing and working in project society*. Cambridge University Press.

Marra, A., Mazzocchitti, M., & Sarra, A. (2018). Knowledge sharing and scientific cooperation in the design of research-based policies: The case of the circular economy. *Journal of Cleaner Production, 194*, 800–812.

Mayer, A., Haas, W., Wiedenhofer, D., Krausmann, F., Nuss, P., & Blengini, G. A. (2019). Measuring progress towards a circular economy: a monitoring framework for economy-wide material loop closing in the EU28. *Journal of Industrial Ecology, 23*(1), 62–76.

Milosevic, D. Z., & Srivannaboon, S. (2006). A theoretical framework for aligning project management with business strategy. *Project Management Journal, 37*(3), 98–110.

Mohajan, H. (2021a). *Germany is ahead to implement sustainable circular economy*. Spiru Haret University.

Mohajan, H. (2021b). Germany is ahead to implement sustainable circular economy. *Journal of Economic Development, Environment and People, 10*(2), 46–64.

Mullins, R., Dettmer, S., Eisenbardt, M., & Ziemba, E. (2020). Using knowledge exchange between prosumers and enterprises to implement circular economy activities in businesses. *Online Journal of Applied Knowledge Management (OJAKM), 8*(2), 71–90.

OECD. (2020). The Circular Economy in Cities and Regions, OECD Programme on Cities and Circular Economy, Cities, Urban Policies, and Sustainable Development Division

OECD. (2021). *Centre for entrepreneurship, SMEs, regions and cities*. Retrieved form: https://www.oecd.org/cfe/CFE-Brochure-2021-Final-web.pdf

PACE (The Platform for Accelerating the Circular Economy). (2021). *Circularity GAP report, 2021*. Ruparo.

Packendorff, J., & Lindgren, M. (2014). Projectification and its consequences: Narrow and broad conceptualisations. *South African Journal of Economic and Management Sciences, 17*(1), 7–21.

Pagoropoulos, A., Pigosso, D. C., & McAloone, T. C. (2017). The emergent role of digital technologies in the circular economy: A review. *Procedia CIRP, 24*, 19–24.

Patanakul, P., & Shenhar, A. J. (2012). What project strategy really is: The fundamental building block in strategic project management. *Project Management Journal, 43*(1), 4–20.

Petit-Boix, A., & Leipold, S. (2018). Circular economy in cities: Reviewing how environmental research aligns with local practices. *Journal of Cleaner Production, 195*, 1270–1281.

Piscitelli, G., Ferazzoli, A., Petrillo, A., Cioffi, R., Parmentola, A., & Travaglioni, M. (2020). Circular economy models in the industry 4.0 era: a review of the last decade. *Procedia Manufacturing, 42*, 227–234.

Pitkänen, K., Karppinen, T. K. M., Kautto, P., Turunen, S., Judl, J., & Myllymaa, T. (2020). Sex, drugs and the circular economy: the social impacts of the circular economy and how to measure them. In *Handbook of the circular economy*. Edward Elgar Publishing.

Pruijsen, J. (2019). *The role of startups in the circular economy of 2050*. Utrecht University, Faculty of Humanities Theses.

PwC. (2019). *5 megatrends affecting your business in 2019*. Retrieved from: https://brandminds.ro/pwc-5-megatrends-affecting-your-business-in-2019-1-of-2/

RBI (Roland Berger Institute). (2020). *Trend compendium 2050: Megatrends shaping the coming decades*. Roland Berger Institute.

Reuters Events. (2019). *Our throwaway planet: Can we end the scourge of single use?* Ethical Corporation.

Rossi, E., Bertassini, A. C., dos Santos Ferreira, C., do Amaral, W. A. N., & Ometto, A. R. (2020). Circular economy indicators for organizations considering sustainability and business models: Plastic, textile and electro-electronic cases. *Journal of Cleaner Production, 247*, 119137.

Saidani, M., Yannou, B., Leroy, Y., Cluzel, F., & Kendall, A. (2019). A taxonomy of circular economy indicators. *Journal of Cleaner Production, 207*, 542–559.

Sassanelli, C., Rosa, P., Rocca, R., & Terzi, S. (2019). Circular economy performance assessment methods: A systematic literature review. *Journal of Cleaner Production, 229*, 440–453.

Scarpellini, S. (2021). Social impacts of a circular business model: An approach from a sustainability accounting and reporting perspective. *Corporate Social Responsibility and Environmental Management, 29*, 646–656.

Schroeder, P., Anggraeni, K., & Weber, U. (2019). The relevance of circular economy practices to the sustainable development goals. *Journal of Industrial Ecology, 23*(1), 77–95.

Scipioni, S., Russ, M., & Niccolini, F. (2021). From barriers to enablers: The role of organizational learning in transitioning SMEs into the circular economy. *Sustainability, 13*(3), 1021.

Senaratne, S., Abhishek, K. C., Perera, S., & Almeida, L. (2021). Promoting stakeholder collaboration in adopting circular economy principles for sustainable construction. In: Y. G. Sandanayake, S. Gunatilake, & K. G. A. S. Waidyasekara (Eds.), *Proceedings of the 9th world construction symposium, 9–10 July 2021* (pp. 471–482). Sri Lanka. https://doi.org/10.31705/WCS.2021.41.

Shenhar, A. J., Dvir, D., Guth, W., Lechler, T., Milosevic, D., Patanakul, P., Poli, M., & Stefanovic, J. (2007). Project strategy: The missing link. In: A. J. Shenhar, D. Milosevic, D. Dvir, & H. Thamhain (Eds.), *Linking project management to business strategy* (pp. 57–76). Project Management Institute.

Suárez-Eiroa, B., Fernández, E., Méndez-Martínez, G., & Soto-Oñate, D. (2019). Operational principles of circular economy for sustainable development: Linking theory and practice. *Journal of Cleaner Production, 214*, 952–961.

Todorović, M. (2020). Circular economy: 4.0–An Entrepreneurial approach. *An international serial publication for theory and practice of Management Science*. In *International may conference on strategic management – IMCSM20*. Bor, Serbia.

Todorović, M., & Obradović, V. (2018). Sustainability in project management: a project manager's perspective. *Sustainable Growth and Development in Small Open Economies*, 88.

Tura, N., Hanski, J., Ahola, T., Ståhle, M., Piiparinen, S., & Valkokari, P. (2019). Unlocking circular business: A framework of barriers and drivers. *Journal of Cleaner Production, 212*, 90–98.

Urbinati, A., Chiaroni, D., & Chiesa, V. (2017). Towards a new taxonomy of circular economy business models. *Journal of Cleaner Production, 168*, 487–498.

Van Der Merwe, A. P. (2002). Project management and business development: integrating strategy, structure, processes and projects. *International Journal of Project Management, 20*(5), 401–411.

Vegter, D., van Hillegersberg, J., & Olthaar, M. (2020). Supply chains in circular business models: processes and performance objectives. *Resources, Conservation and Recycling, 162*, 105046.

Velenturf, A. P., Archer, S. A., Gomes, H. I., Christgen, B., Lag-Brotons, A. J., & Purnell, P. (2019). Circular economy and the matter of integrated resources. *Science of the Total Environment, 689*, 963–969.

Zwikael, O., & Globerson, S. (2006). Benchmarking of Project Planning and Success in Selected Industries. *Benchmarking: An International Journal, 13*(6), 688–700.

Index

C

Circular, 9, 10, 17, 20, 21, 28, 30, 31, 42–46, 50, 82, 100, 113, 114, 136, 149, 154–159, 167, 174–177, 179, 180, 183, 185, 186, 189, 197, 198, 200, 202, 207, 212, 214, 217, 270, 271, 274, 286, 287, 294, 295, 302–306, 308–310

Circular economy, 6, 7, 10, 12, 13, 16, 17, 19–21, 23, 27–29, 42–46, 50, 51, 60, 62, 69, 70, 75, 78, 98–100, 102–104, 106, 107, 112–115, 118, 119, 121–125, 128, 135–138, 146–156, 158, 160–169, 173–175, 178, 180, 182–188, 198, 212, 213, 216–219, 223, 238, 244, 246, 247, 253, 254, 263, 264, 270–295, 301–311

Circular economy indicators, 42, 147, 151–158, 160–162, 164, 168, 169

Circular economy projects, 169, 183

Construction waste (CW), 60–63, 68–88

Corporate Social Responsibility (CSR), 97, 98, 103, 112–124, 128–132, 135–138

Crowdfunding, 174, 177–183, 187–189

Curriculum, 7, 8, 11, 13, 15, 17, 27, 28, 31

E

Economy, 4–7, 11, 12, 17, 19, 20, 30, 31, 41, 42, 44, 48, 59, 65, 102, 112–116, 119, 120, 135, 137, 138, 146–151, 153, 154, 157, 159, 161–163, 165, 168, 174, 175, 179, 180, 183, 186, 187, 212–214, 217, 221, 228, 240, 244, 247, 254, 287, 302, 303, 308

Entropy, 125–127, 132, 202–204

European Union indicators, 162–164, 217

G

Green human resource management (GHRM), 43, 46, 47, 49–51

H

Higher education, v, vi, 3–32

Higher education for circular economy, 12, 17, 28, 30, 31

K

Kickstarter campaigns, 181

M

Methodology, 12, 62, 63, 65, 68, 100, 113, 124, 125, 127, 132, 156, 159, 174, 197–207, 255–257, 278, 279, 292, 306, 307, 309, 311

P

Portfolio selection, 271, 272, 276, 278–281, 285, 287, 289, 291, 292, 294, 295

Procurement, 9, 42, 70, 71, 98–107, 157, 165–167

V. Obradović (ed.), *Sustainable Business Change*, https://doi.org/10.1007/978-3-031-23543-6

Program, 13, 25, 60, 79, 83, 106, 115, 119, 122, 135, 152, 153, 174, 198, 199, 202, 205–207, 220, 225, 227, 238, 259, 271, 276, 279, 281, 292, 304
Project, 4, 5, 11, 13, 17, 31, 32, 59–61, 69–73, 75, 77–79, 81, 82, 84–87, 98–107, 113, 115, 119–122, 136–138, 162, 169, 173–189, 197–202, 204–207, 212, 224, 238, 245, 247–264, 270–295, 301–311
Project management, 197, 260, 270–272, 275, 281, 282, 285, 293, 294
Project-oriented organizations, 120–122, 130
Project portfolio, 198, 270–295
Project success, 122, 249, 251–253, 263, 309

R
Recurrent neural network (RNN), 224, 225, 235, 237
Responsible project management, 98, 106, 107

S
Selection criteria, 275, 278, 279, 284, 285, 288, 295
Simulation modeling, 218, 227
Start-ups, 174–177, 179, 187, 302, 304, 310
Sustainability, 5–13, 15–17, 23, 31, 43, 48–51, 61, 64, 69, 77–79, 85, 87–89, 98–107, 112, 114–117, 119–125, 128, 131, 136–138, 146, 148, 149, 151, 158, 160, 166, 168, 178–180, 188, 199, 201, 205, 214, 244–264, 270–274, 276, 278, 284–288, 291, 294, 295, 307, 308, 310
Sustainability principles, 115, 119, 120, 122, 245, 247, 248, 250, 251, 254, 257, 260–264
Sustainable development (SD), 6–10, 12, 13, 15, 16, 20–30, 43, 48, 50, 61, 69, 70, 75, 112, 113, 115, 116, 118–123, 138, 146, 148, 149, 160, 166, 175, 178, 200–202, 238, 244–250, 253, 254, 256, 259, 260, 262–264, 270–274, 285, 286, 303, 305, 307
Sustainable project management (SPM), 79, 119, 122, 245, 247, 249–264, 270
Sustainable strategies, 200, 201, 207
System dynamics, 212–240

T
Tensions, 98, 100–107, 207

W
Waste management, 11, 13, 19, 27–29, 42, 59–89, 102, 152, 156, 158, 161, 162, 164–167, 169, 176, 252, 303

The manufacturer's authorised representative in the EU is Springer Nature Customer Service Centre GmbH, Europaplatz 3, 69115 Heidelberg, Germany. If you have any concerns regarding our products, please contact ProductSafety@springernature.com

Printed and bound by CPI Group (UK) Ltd, Croydon, CR0 4YY
01/10/2026
02225892-0008